HOLMES, Ernest, ed.   Handbook of industrial pipework engineering.
Halsted, a div. of Wiley, 1974.   570p   il   tab   73-21538.   47.50.
ISBN 0-470-40769-7.   C.I.P.
An excellent book on pipework engineering that may be used by engineers and students as a complete handbook. It contains chapters on
plant start-up, general responsibilities, theory of fluid mechanics (including many useful charts and formulas), pipes and pipe fittings,
gaskets and jointing, valves, operating mechanisms, automatic process
control, safety, piping design and drafting by computer, piping models,
and piping fabrication. A discussion of SI units is included, and these
units are used wherever possible. Conversions to Imperial and U.S.
units are provided. There are liberal use of illustrations (including computer isometrics), flow diagrams, organizational charts, and many
sketches. This book is recommended for college and industrial li-
braries as a "how-to-do-it" and reference book. Since it assumes a
basic knowledge of fluid mechanics it would be recommended for students beyond that level.

INTRODUCTION TO
PETROLEUM GEOLOGY

HOBSON, G. D. and E. N. Tiratsoo. Introduction to petroleum geology. Scientific Press (dist. by Geology Press, 420 Riverside Dr., 2B, N.Y. 10025), 1975. 300p il map tab. 28.50. ISBN 0-901360-07-4

CHOICE   *OCT. '75*

Earth Science

TN 870.5
H68

The latest of several books authored by Hobson and Tiratsoo, this is a most valuable addition to the abundant literature on the geology of petroleum. In only 300 pages, the full spectrum of knowledge required by petroleum geologists is skillfully compressed in clear, substantial writing that is reinforced by carefully chosen illustrations and charts. Each chapter ends with a select and extensive list of meaningful references, both old and new. The treatment is up to date, and modern tectonic theory related to petroleum is discussed. This compendium would be useful to undergraduate or graduate petroleum geology students and would provide a condensed refresher course for all professional petroleum geologists. Excellent for the student and professional alike and highly recommended for both, whether specialist or generalist within the earth science field.

Introduction to
PETROLEUM GEOLOGY

by

G. D. Hobson, Ph.D.
Reader in Oil Technology, Imperial College, University of London

and

E. N. Tiratsoo, Ph.D.
Consulting Geologist

First Edition, 1975

Scientific Press Ltd.
Beaconsfield, England

© Copyright 1975 by Scientific Press Ltd.

Printed in Great Britain by The Creative Press (Reading) Ltd.

ISBN 0 901360 07 4

PREFACE

GEOLOGY must inevitably be a relatively inexact science, inasmuch as the dimension of time—in terms of millions of years—is commonly involved, and this consideration, together with size and rock variability, limits the exactness and experimental repeatability which characterize most other scientific disciplines. However, petroleum geology is a branch of applied geology which is susceptible to rigorous checks, because its premises and conclusions are continually being tested by the results of thousands of wells which are drilled every year, with the expenditure of considerable effort and money, on the basis of geological advice.

The success that has been achieved by petroleum geologists in the last quarter-century or so may probably best be measured by the remarkable increase in estimates of world "published proved" oil reserves over this period, to which attention is drawn at the end of Chapter 1. Thus, in spite of the consumption of some 250 billion barrels of oil in the intervening years, world known oil reserves at the end of 1973 were some *ten times* higher than they were at the end of 1945. Furthermore, although a very large financial investment has had to be made to establish and develop these greatly increased reserves, their capital worth has increased enormously as a consequence of the rises in the price of crude oil resulting from recent events—notably the action of the OPEC countries at the end of 1973: thus, for example, the "posted price" of Saudi Arabian "light" crude oil (which sets the general price pattern) rose from $1·8/barrel in January, 1971 to $11·65/barrel in January, 1974.

Petroleum geologists have, therefore, been outstandingly successful in the period that has elapsed since the end of the 1939–45 war, a time during which the demand for crude oil has grown rapidly in all the developed and developing countries of the world, with oil replacing coal in an increasing number of applications. This period of rapidly growing consumption coincided with the inevitable decline of many of the older oilfields, whose place has had to be filled by new discoveries. As in other walks of life, the discovery of new petroleum accumulations is inevitably subject to a law of diminishing returns in particular areas or through the application of customary techniques; further discoveries can then only be achieved by the invention or application of more sophisticated methods, or by use of the existing techniques in previously undeveloped areas. Thus, in the last quarter-century, there have been notable advances in offshore operations and in applied geophysics, as well as the acquisition of more detailed knowledge concerning the origin of petroleum, and, more recently, the development of ideas on plate tectonics, which open up new prospects of making useful syntheses and reconstructions relating to oil occurrence.

The application of petroleum geology does not end with the drilling of a discovery well. Indeed, from that stage onwards, the geological data available about any particular oilfield will grow markedly, and geological studies will become possible which previously could not be undertaken. The careful consideration of such data will, in some cases, lead to new discoveries. It is vital also, from the point of view of the most effective development of an oilfield, that geological knowledge should be adequately applied at this stage, as well as later if secondary recovery operations are undertaken. The maximum return in terms of subsurface information must be gained from the increasingly costly drilling operations.

v

The petroleum geologist must always have in mind quantitative as well as qualitative aspects of the data obtained. Clearly, fluids, and their behaviour over time, are as important as the rocks involved. There are skills which he may not practise, but of which he should be aware with respect to their applications and limitations, not least because he is basically a member of a team. Equally, other members of the team—geophysicists, reservoir engineers, palaeontologists, and other specialists—can benefit by learning something about the special knowledge and interests of the petroleum geologist.

Turning to the future, it is clear that in the years ahead the demand for crude oil will inevitably continue to grow, albeit at perhaps a slower rate than in the immediate past, and at least until the end of the century oil and natural gas will provide the largest proportion of the world's energy and an increasing proportion of the chemical feedstocks required. The task of petroleum geologists everywhere will be not only to discover new accumulations of petroleum to match this need, but also to ensure that the already known oil and natural gas reserves are systematically and economically developed. There is much yet to be done in this important area of scientific endeavour, and it is the authors' hope that this work will be of assistance to those wishing to equip themselves for the tasks ahead, in as many as possible of the sixty-two nations in which "commercial" oilfields are already being developed, and in the others where oil will undoubtedly be discovered in the future.

G.D.H., E.N.T.

NOTES

The references given at the end of every Chapter are designed to provide the student with further relevant information, as well as to indicate the sources of particular data or ideas.

The general plan of this work follows that of an earlier book by one of the present authors ("Principles of Petroleum Geology" by E. N. Tiratsoo, Methuen, London, 1951), from which some illustrative material has been drawn.

The authors wish to acknowledge with thanks permission given by the American Association of Petroleum Geologists, Tulsa, for the reproduction of a number of diagrams, which are individually acknowledged in the relevant captions. The British Petroleum Co. Ltd. and Shell International Petroleum Co. Ltd. kindly gave permission for the reproduction of the material on which some of the figures in Chapters 6, 7 and 10 are based. We are also grateful to Mr. John Tiratsoo for his assistance with many of the diagrams.

CONTENTS

The drill- ship "Offshore Mercury"—276 ft. long and 130 ft. wide—moving through the English Channel en route for the Irish Sea, to drill exploratory wells on behalf of the British Gas Corporation.

CHAPTER 1

Petroleum Accumulations

THE NATURE OF ACCUMULATIONS

PETROLEUM in the solid, liquid or gaseous form is common in small amounts, but relatively rare in substantial quantities. An oil or gas *field* is developed where a hydrocarbon *accumulation* has been discovered which is estimated to be capable of producing a quantity of oil and/or gas sufficient to more than pay for the essential operations of drilling and completing the wells, providing any preliminary treatment needed by the oil or gas, and transporting the hydrocarbons to the point at which a sale can be effected. Ideally, the discovery costs (including acquisition of the concessions) should also be covered, but this is not necessarily true for every oil- and gasfield, although it must be the case for oil- and gasfields as a class. Indeed, the return from the sale of the oil and gas must more than repay the expenditure involved; it must provide an adequate surplus to justify the use of the funds in this way.

The factors which determine whether an oil or gas accumulation is of "commercial" value are technical, economic and political. They comprise the depth and geographical position, including proximity to other accumulations, the amount and properties of oil or gas that can be produced; the costs of the various operations involved in drilling wells, extracting the hydrocarbons and handling them, the price at which the hydrocarbons are sold, the concession terms, taxation and the political stability of the area. Some of these factors are susceptible to rapid changes.

A "commercial" oilfield can therefore really only be defined as one whose development *in the existing circumstances* would be worth the investment of the necessary capital and effort. Obviously, an oil accumulation which is discovered in or near densely populated and industrialised areas could be relatively small both in its reserves and productivity and yet be profitable, whereas another accumulation situated in a remote region where the factors of accessibility, distance, local demand, and perhaps political stability were considerably less favourable would need to be proportionately larger and more prolific to be worth developing. Similarly, offshore fields must be large enough to offset the extra costs and risks that their development entails compared with onshore fields.

1

The existence of an oil or gas accumulation calls for the presence of several types of rock, suitably arranged, each type having certain characteristic properties: they are the "reservoir rocks", the "cap rocks" and the "seat seals". Rocks in the last two of these categories may sometimes also fulfil the vital role of "source rocks" in which the oil and gas are generated.

The essential properties of reservoir rocks are their porosity and permeability. The porosity provides the storage space for the oil and gas: while the permeability is needed to allow the movement of fluids through the rocks and into wells which are drilled into them. Permeability is provided by the interconnection of the pore spaces and other openings in the rocks. The bulk of the storage space consists of the small openings (commonly only a small fraction of a millimetre across) between the constituent grains of the rock; openings inside fossils or created by partial solution or mineralogical changes of the rock, together with fractures which have been formed in various ways, also contribute. Quantitatively, porosity is the ratio of the openings to the bulk volume of the rock in which they occur. Multiplication of this ratio by a factor of one hundred enables the porosity to be expressed as a percentage. Reservoir rocks commonly have porosities in the range from about 5% to 30%.

The permeability, the fluid transmitting capacity of the rock, depends on the numbers, sizes and degree of inter-connection of the openings. It is defined by:

$$K = \eta VL/PA$$

where V is rate of flow, ml/sec
η is viscosity of the liquid, centipoises
L is length of rock prism, cm
A is cross-section of rock prism, cm^2
P is pressure drop in length L, atm.

These units lead to the specific (single-fluid saturation) permeability being expressed in terms of *darcys,* and the permeabilities of reservoir rocks usually lie in the range from a few millidarcys to several darcys.

The hydrocarbon-bearing part of the reservoir rock has two and sometimes three fluids in its openings (natural gas, crude oil and salt water) and the most important consideration is its effective permeability to oil or gas from the point of view of the potential production of these fluids. The effective permeability falls progressively below the specific permeability as the proportion of the particular fluid in the pore space decreases, and it becomes zero well before the proportion of that fluid reaches zero.

A rock must be porous in order to be permeable, but a rock can be porous without being permeable; otherwise there is no quantitative link between porosity and permeability. The interconnected openings in a rock provide the permeability, and the effective porosity, whereas the total porosity is that provided by all of the openings in the rock. When the oil-containing zone of a reservoir rock is thick, even quite small permeabilities can give considerable total flow rates. Hence, the significant quantity is the summation of the products of thickness and permeability, commonly quoted as millidarcy-feet, since a reservoir rock may be non-uniform vertically as well as laterally.

The coarser-grained sedimentary rocks are the most typical reservoir rocks, because they usually have relatively large pores. Sands, sandstones, and various kinds of limestones are therefore the commoner reservoir rocks, with some kinds of limestones tending to have the greater irregularities in the distribution of their porosity and permeability. Fractures have comparatively high fluid-transmitting capacities, even though they do not contribute much to the bulk porosity. Low-porosity and quite fine-pored limestones can be effective reservoir rocks when they have joints or other fractures, because fluids can pass from the pores into the fractures and then move relatively easily along them to wells. Fractures may be developed in rocks in various ways, and they can make rocks which would otherwise not function as satis-factory reservoir rocks fulfil that role, e.g. so-called fractured shales, igneous and metamorphic rocks, which are the rarer kinds of reservoir rocks, possess these secondary openings. (The different kinds of reservoir rocks are described in greater detail in Chapter 3).

The cap rock overlying the reservoir rock acts as a seal that prevents the upward escape of hydrocarbons from the reservoir rock. Cap rocks are fine-pored, and hence normally fine-grained rocks. They are essentially of very low permeability (values of 2×10^{-9} darcys have been reported), but low permeability is not the effective property, for in itself it could still allow the hydrocarbons to escape, albeit at a very slow rate. The important characteristic of a cap rock is the very high "capillary pressure" (in contrast to the low capillary pressure of reservoir rocks), a property which precludes, under certain conditions, the entry of a non-wetting fluid, oil or gas, into the water-filled pores of the cap rock. Capillary pressure is dependent on pore-throat sizes and the interfacial tension between water and oil or gas; it is proportional to the interfacial tension and inversely proportional to the radius of the pore throat. Provided that the height of the oil or gas column in the reservoir rock does not exceed a critical value, the oil or gas will not enter the cap rock. The critical height is dependent on the density difference between the water and hydrocarbons, the interfacial tension, and the pore sizes in the reservoir rock and pore throat sizes in the cap rock.

It is advantageous for the cap rock to be plastic, for then it is less likely to develop fractures when the rocks are folded or are put under tension by other means. Clays, shales, salt, gypsum and anhydrite are the commonest cap rocks. Compact limestones can also act as seals.

The seat seal underlies the reservoir rock, but only has a sealing function where the hydrocarbons extend from the top to the bottom of the reservoir rock. Its properties must be the same as those of the cap rock, i.e. it must not admit oil or gas. Seat seals can therefore be lithologically the same as cap rocks.

When erosion has exposed an oil-bearing reservoir rock, vaporization of the low-boiling components of the crude oil may leave behind asphaltic material which blocks the rock pores, thereby halting further oil escape. This "plugged" rock then functions locally as a cap rock.

Within the reservoir rock itself, the fluids which are present are distributed vertically in the order of their densities in all cases where there are no marked variations in pore sizes which might result in layering. Gas overlies salt water, when there is no crude oil present, or a gas/oil solution, which in turn rests on salt water. The distribution must be considered as being in the form of zones with pores largely filled by gas or gas/oil solution, rather than zones completely

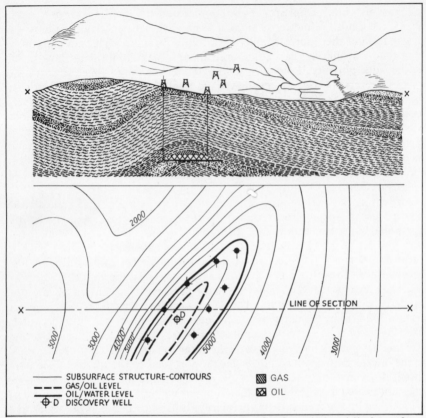

Fig. 1. **Cross-section and subsurface stratum contours of a theoretical anticlinal petroleum accumulation** (after F. R. S. Henson).

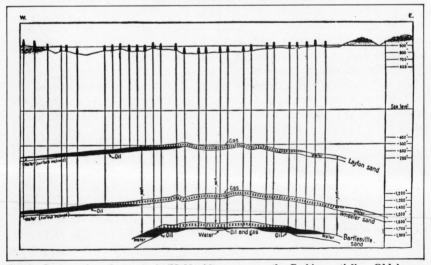

Fig. 2. **Cross-section of an actual oilfield with a gas cap—the Cushing anticline, Oklahoma.** (after Beal).

filled by these substances, for a certain amount of water is always present. This is referred to as the "interstitial water", which can fill 10% or more of the pore space without entering the wells when oil or gas production is under way. Most reservoir rocks are water-wet, but the wetting film is of molecular dimensions and hence occupies only a very small part of the pore space. However, most of the interstitial water is present as collars round grain contacts, or it completely fills pores bounded by unusually small throats, including any micro-pores.

The gas zone is not sharply separated from the oil zone; there is a transition between the oil and water zones, or between the gas and water zones in the absence of oil. The thicknesses of the transition zones depend on the pore sizes, interfacial tensions and density differences between the particular pairs of fluids. For given fluid properties, the finer the pores the thicker will be the transition zone. The transition zones will be horizontal in the absence of water flow or lateral variations in pore size for the reservoir rock. (The causes of tilted fluid contacts are discussed on p. 117.)

The solubility of natural gas in crude oil depends on the temperature and pressure, and on the composition of the gas and crude oil. Increase in pressure raises gas solubility; rise in temperature decreases it. Gas is more soluble in light than in heavy crudes, and methane is less soluble than gases containing considerable amounts of the higher members of the paraffin series.

For an accumulation of hydrocarbons to be formed there must be a *trap*, that is to say a suitable shape of the boundary between the reservoir rock and the cap rock. Except when there is an asphalt seal or wholly hydrodynamic trapping (p. 115) that boundary must be convex upwards. A "dome" is the best known shape which satisfies this condition, but there are a number of other arrangements of the rocks which provide a basically equivalent form. Folding, faulting, depositional features, and various combinations of these factors can all lead to the formation of traps. The maximum storage capacity of a trap is fixed by the vertical "closure", which is defined as the distance between its highest point and the "spill point", i.e. the level at which hydrocarbons could spill under to enter an adjacent structure or escape to the surface of the ground; this may coincide with the level of the highest adjacent saddle for folds, or the highest level at which communication between rocks with reservoir properties exists in the case of fault closure. For some faults it may involve questions of capillary pressure and other factors. Traps need not be filled with hydrocarbons to the limit of their storage capacity.

Reservoir rocks range in age from Pre-Cambrian to Plio-Pleistocene, but these extremes are rare, and most of the world's oil occurs in Tertiary and Mesozoic beds. Different areas tend to have reservoir rocks of fairly limited age ranges. The age of the accumulation can be very different from that of the reservoir rock in which it is now found.

Reservoir rocks may be only a few feet thick, or they may be many hundreds or even thousands of feet thick. Oil and gas columns range in thickness from tens of feet to several thousand feet. A field may have only one reservoir rock, or there may be many reservoir rocks;* and there are cases known of very

---

* *Thus, in the* Oficina *field of East Venezuela, there are as many as 85 productive oil and gas reservoir zones lying between 4,000 and 6,100ft, and a similar multiplicity of separate "pay" sandstones can be found in many Californian fields.*

thick zones of closely interbedded reservoir rock and non-reservoir rock,
with wells being completed so as to produce from a considerable thickness of
such a sequence. When the reservoir rocks are reasonably thick and well
separated vertically, a well may be completed so as to produce from only one,
or it may be able to produce from two and even (rarely) from three reservoir
rocks separately.

The simple concept of a homogeneous oil reservoir is rarely found in
Nature. Even if a reservoir bed is a geologically distinct and separate unit,
it is likely to vary considerably through its vertical section and also laterally
as regards its properties of porosity and permeability. Argillaceous partings,
stylolites, bedding planes, or minor faults filled with impermeable "gouge"
material may divide the bed into may sections of varying properties which in
extreme cases can even form distinct sub-reservoirs—i.e. units separated by
impermeable zones which allow them to be individually productive without
mutual interference. Many limestone reservoir beds which at some time have
undergone subaerial erosion have marked secondary permeability in the top
few feet of the bed, while sandstones are often affected by local fissuring and
cementation.

The depths at which most of the early discoveries were made were only a
few hundred or at most two or three thousand feet. Inevitably, the average
depth of exploration wells has increased steadily throughout this century; the
deepest such well yet drilled reached 31,141ft. The deepest *producing* oil-well
is 21,465ft (Lake Washington, South Louisiana) and the deepest gas production is at 24,065–24,584ft, in Beckham County, Oklahoma. However, most of
the oil and gas accumulations so far found lie at depths of 2,000–10,000ft.
There seems to be a tendency for gas occurrences in a particular area to
extend to greater depths than oil accumulations.

Because of the great range in depth at which the accumulations occur, there
is a wide spread in the reservoir pressures and temperatures. Pressures as
high as 14,625 psi (at 16,000–16,018ft) and temperatures as high as 520°F (at
about 24,000 ft) have been recorded. However, most oil reservoirs have
temperatures lower than 225°F, and the continued presence of porphyrins in
crude oils shows that temperatures have generally never exceeded about
390°F in the past, since the porphyrins would have been destroyed at higher
temperatures.

The Sizes of Accumulations

A single accumulation is referred to as a *pool*. When a reservoir rock is
broken up into several sectors by faulting, each with an independent accumulation, every sector is a pool. A *field* consists of one or more pools; in the
latter case the pools are adjacent or they overlap in plan view and are associated
with a localised geological feature, such as an anticline, salt dome, or faulted
uplift. (The hyphenated field names often encountered in the USA show how
what initially seemed to be two or more separate fields have proved to be a
single field on further drilling). A series of oil- and gasfields occurring in
geographical contiguity and having similar or related geological features
constitute a *petroleum province*.

The "size" of an oilfield really refers to the volume cf its recoverable oil,
i.e. the "reserves" (to be considered more fully in Chapter 10), rather than its

areal extent. Individual oil and gas accumulations may in fact cover areas from a few tens of acres to tens of thousands of acres; most oilfields cover several square miles.

However, the mean area of the 26,000 oilfields of the United States is less than one square mile apiece. The areally smallest major US field is *Sante Fe Springs* in California, which covers only 2·3 sq miles; by contrast, *Pembina* in Canada covers some 1,170 sq miles. Several of the world's largest known oil accumulations—*East Texas, Kirkuk* (Iraq) and *Burgan* (Kuwait) have, by coincidence, areas of 120–130 sq miles apiece, although their plan shapes differ greatly.

The amounts of recoverable oil in commercial accumulations are usually in the range of tens of thousands to thousands of millions of barrels. Likewise, the recoverable gas varies widely in volume, with the larger fields containing millions of millions (i.e. trillions) of standard cubic feet.*

The volume of oil produced each year by an oilfield varies with the local technical and economic circumstances; clearly there is a minimum acceptable output in every case which justifies the expenditure involved. Small oilfields usually produce some millions of barrels per year, large ones may produce scores or even hundreds of millions of barrels of oil annually. The most productive oilfield in the world (*Burgan*) produces 800–1,000 MM brl p.a.

In the Los Angeles Basin of California, more than 6 B brl of oil has been produced since 1900 from a very limited area. This is one of the most prolific oil-producing regions of the world, with recoveries around *Signal Hill* of more than 500,000 brl/acre. In the Middle East, the phenomenal *Burgan* field in Kuwait had already produced more than 10 B brl by the middle of 1969, the first and only oilfield in the world to have yielded so great a volume of oil: the ultimate output of this field may well exceed 1 MM brl/acre. Very high per-acre productions have also been obtained in the past in the Baku area, in Mexico along the "Golden Lane" and in some of the Maracaibo Basin oilfields of Venezuela. Such high per-acre oil output figures are related to great thicknesses of excellent reservoir beds—in the case of the *Burgan* field, for example, at least 1,300 ft of very permeable sandstones.

The American Association of Petroleum Geologists uses a reporting system in which new oil- or gasfields are classified in the following groups, according to their estimated ultimate recoverable reserves of oil or gas (Table 1):

TABLE 1

CLASSIFICATION OF ACCUMULATIONS BY SIZE (ULTIMATE RECOVERY)[1]

	Oilfield MM brl	Gasfield Bcf
A	over 50	over 300
B	25–50	150–300
C	10–25	60–150
D	1–10	6–60
E	less than 1	less than 6
F	non-profitable	non-profitable

---

* For brevity, we shall use brl to denote barrels, MM for million (10^6), B for billion (10^9) and T for trillion (10^{12}) throughout this work. The volume of a (US) barrel is 5·61 cu ft or 0·159 cu m. Depending on the specific gravity of the particular crude oil, between 6·5 and 7·5 US barrels weigh one metric ton.

In this classification, a recoverable volume of at least 1 MM brl of oil is the qualification for the smallest normal commercial field. Accumulations of this size have sometimes been described in American practice as "profitable", but the term "significant" is nowadays preferred. About 26,000 "significant" oilfields have been discovered to date in the United States, and some 4,000 in the rest of the world.

A "major" discovery has generally been defined in the USA as one with at least 50 MM brl of recoverable oil reserves; in recent years, this criterion has been reduced to 25 MM brl, and it is noteworthy that only one in about 220 wells drilled in the United States each year nowadays discovers an oilfield of even these relatively modest dimensions. At the other end of the scale, the classification of "giant" in American practice has in recent years been used to describe fields with at least 100 MM brl of recoverable oil, or 1 Tcf of recoverable gas; 310 such accumulations are currently known in the United States.[13, 15]

It takes several years to evaluate an oilfield, although recent improvements in technology have allowed relatively rapid and accurate forecasts to be made on the basis of early drilling results coupled with seismic delineations, electric logs and reservoir studies. The interval has therefore shortened between initial discovery and the allocation of a field to its size group, and this usually occurs nowadays within three to four years. Upgrading of accumulations also sometimes occurs as the result of the discovery of new and deeper reservoir horizons, as for example in the case of the *La Paz* field in Venezuela, which was originally found in 1927 as a Tertiary producer, but was subsequently recognised as a "giant" field with the discovery of deeper Cretaceous production in 1945. (Sometimes, also, an accumulation must be downgraded as more information about it becomes available.)

The time required for the actual production of 100 MM brl of oil will vary with local economic as well as technical conditions. In Venezuela, for example, where about 40 of these US-type "giant" oilfields have been recognised, only 20 had actually produced 100 MM brl of oil apiece by the end of 1967. The average period between discovery and production of the millionth barrel in each case was about 12 years, with the shortest period 5 years, and the longest 30 years.

Outside the United States, the definition of a "giant" quoted above is not generally adequate, since some accumulations may be much larger than the American-type "giants", while still being unattractive to develop for economic reasons. For this reason, a larger "unit" is nowadays generally adopted—500 MM brl of recoverable oil reserves. (The importance of these international-type "giants" is discussed on p. 256)[13].

TABLE 2

USA—RATE OF DISCOVERY OF "GIANT" OILFIELDS*

	discoveries per year
1920–30	7·8
1930–40	5·0
1950–60	3·3
1960–70	1·7 (estimated)

* *Source of data: M. Halbouty, Bull. AAPG, 52(7) 1115, July, 1968. (In this case the term "giant" is applied to oilfields holding at least 100 MM brl of recoverable oil.)*

TABLE 3

WORLD'S LARGEST "GIANT" OILFIELDS

Field	Country	Ultimate Recoverable Oil in B brl
1 Ghawar Fields	Saudi Arabia	75
2 Greater Burgan	Kuwait	66
3 Bolivar Coastal Fields*	Venezuela	30
4 Safaniya-Khafji*	Saudi Arabia—Neutral Zone	25
5 Prudhoe Bay	Alaska, USA	20
6 Samotlor	USSR	15·1
7 Kirkuk	Iraq	15
8 Romashchkino	USSR	14·3
9 Rumaila	Iraq	13·6
10 Abqaiq	Saudi Arabia	12
11 Manifa*	Saudi Arabia	11
12 Marun	Iran	10·5
13 Fereidoon-Marjan*	Iran—Saudi Arabia	10

* Offshore or partly offshore.

The decline in the discovery rate of "giant" accumulations is most noticeable in the United States, as is shown by the statistics in Table 2; while some of the world's largest "giant" oilfields are listed in Table 3.

It is disturbing that this decline should have revealed itself first in the United States, where the application of scientific techniques to petroleum exploration operation has been widespread and exploratory drilling has been most intense. It may very well be repeated in other parts of the world as exploration for petroleum in such countries begins to approach the United States level. If indeed large new anticlines are becoming harder to find, this may indicate that the future search may have to be concentrated on the more obscure or "subtle" types of trap, in which accumulations of petroleum are due more to stratigraphic or hydrodynamic conditions than to straightforward structural effects.

The Life of Accumulations

Once a discovery is made and an oil accumulation has been located, the appraisal and development phases begin. Wells are then drilled to ascertain the extent of the reservoir and to find the edge-water boundary, while the reservoir formation is tested for its physical parameters, and samples of the fluid contents taken for chemical and market analysis.

When facilities are available, flow tests may be made, whereby the oil or gas is allowed to flow for a measured period through orifices of known diameters in order to give an idea of the productive capacity of the well and of the field.

There is no minimum rate of production for a viable oil-well. Wells inevitably decline in their ability to produce oil as they grow older, but if the crude produced can be marketed profitably, mean flow rates of a few barrels a day per well may still be acceptable in the latter stages of an oilfield's history.

Most oil-wells produce oil at average rates of between a few hundred and a few thousand barrels daily; some very productive wells in the Middle East yield over 20,000 brl/day apiece—and exceptional wells in Baku, Mexico and Iran are said to have had a productive capacity of more than 100,000 brl/day.

Once a field has been put on production it enters the period of its economic life—first a growing production as more wells are drilled and more oil and gas extracted, up to a peak, which may in fact be a plateau extending over a number of years; then a period of decline and senescence, finishing with economic death—a point which is as hard to define as clinical death. It may simply be taken at that point at which it is no longer commercially profitable to operate the field in all the circumstances prevailing.

It usually takes anything from five to eight years to develop the full potential productivity of a newly-discovered oil- or gasfield, depending on its location and other physical and economic factors; but commercial production may start much sooner, provided outlets are available.

The "payout" on the direct costs of drilling may come after between three and eight years of production, depending on the depth of the reservoir and the time it has taken to appraise and develop. The economic life of the field may be 20 years or more, depending on the production rate applied and any secondary recovery operations. Non-associated gasfields may have economically viable lives of 35 years or more.

A study of 74 oil- and 38 gasfields in the United States showed the average percentage extraction in the first ten years of the field's life to be 29% for oil, 30% for gas; the average time to the peak output averaged 11 years for oil, 12 years for gas; the half-life (i.e. time for production of half the ultimate output) for both oil and gas was 16 years; the total economic life averaged 46 years for oil, 39 years for gas. Only 11% of the fields had produced more than half their reserves in the first 10 years; 13% reached their peaks in under 5 years; 15% had half-lives of over 20 years; the fields overall had lives ranging from 14 to 87 years[7].

These statistics are of course only applicable in the economic conditions prevailing in a particular country and at a particular period, since the rate of production of an oilfield is essentially controlled by economic factors. In Chapter 10, reference is made to the concept of "maximum efficient rate" of production, which is related to "good oilfield practice", but it will be evident that such terms are relative only and therefore inevitably imprecise.

THE RESERVOIR FLUIDS

1. CRUDE OIL

Crude (i.e. unrefined or "natural") oils vary widely in their physical and chemical properties. They may be straw-coloured, green, brown or black, the last three being the common colours. Their densities generally lie in the range 0·79 to 0·95 g/ml, under surface conditions. Often, however, the gravity is expressed in terms of °API, which is defined by

$$°API = \frac{141·5}{Sp\ gr\ 60/60°F} - 131·5$$

On this arbitrary scale, values from 5° to 61°API have been reported. Verbal descriptions mentioning "high-gravity crude oil" without giving numerical values can be ambiguous, because a high gravity in °API is a low-density oil, whereas high gravity in terms of specific gravity is a high-density oil.

Viscosities range widely, from about 0·7 centipoise to more than 42,000 centipoises under surface conditions. Under subsurface conditions of higher pressure and temperature, with much gas in solution, the viscosities are markedly less than the values under surface conditions.

Fractions of some crude oils show optical activity which is thought to be indicative of biogenic origin.

In elemental composition, crude oils are mainly carbon and hydrogen, with varying amounts of oxygen (possibly up to 2%), nitrogen (up to 1·7%) and sulphur (up to 8%). Table 4 shows the elemental composition of some crude oils and related substances, in terms of their major components. In addition, a very wide range of elements have been found in trace amounts in the ash from crude oils (Table 5). Their source and mode of occurrence have been matters of much discussion. Some may be present in organic molecules, others may occur as inorganic molecules—oxides, sulphides, chlorides, for example —suspended in the crude oil, and presumably derived from the rocks or formation waters. Two of these trace elements, nickel and vanadium, are remarkable in the amounts in which they occur, in "asphaltic" crude oils especially. They are known to be present in the chlorophyll porphyrins (see Fig. 5, p. 36).

Although no crude oil has been completely analysed in terms of its constituent compounds, the oils, although complex in their make-up, are basically decidedly less complex than was once thought to be the case. Thus, in the course of work on one American example—the Ponca City (Oklahoma) crude —over a period of nearly forty years, only 230 hydrocarbons and four sulphur compounds were actually detected and determined, and these accounted for 45% by weight of the oil. The boiling points of these compounds ranged from −161·48°C to 475°C. Investigation of the rest of the crude raises special difficulties, because of the greater complexity of the components and the problems of separating them without breakdown taking place.

In the more volatile parts of crude oils, certain compounds are comparatively abundant. Thus n-heptane accounts for 2·3% of the Ponca City oil, and some ten years ago at least thirty-two hydrocarbons were known which were present to the extent of more than 0·5% in one or another crude oil. These were paraffins, naphthenes and aromatics with up to eighteen carbon atoms in their molecules. In this range, sixteen hydrocarbons accounted for 17·55% of the Ponca City crude; benzene amounted to 1·26% of another oil, and methylcyclopentane to 2·14% of a third.

The hydrocarbons in crude oils include paraffins, naphthenes and aromatics, as well as various hybrids. Some of the paraffins have as many as 78 carbon atoms in their molecules. At least 260 hydrocarbons, 50 oxygen-bearing compounds, 98 sulphur-bearing compounds and 33 nitrogen-bearing compounds have been reported. Unsaturated hydrocarbons of olefin type, if present at all, occur only in very small amounts.

The paraffins (alkanes or saturated hydrocarbons) have the general formula C_nH_{2n+2}. For values of n up to 4, the compounds are gaseous at normal pressure and temperature; from n = 5 to n = 15 they are liquids; the higher members are solids. When the value of n exceeds 3, the molecules can be straight-chain (normal, *n*-) or branched (*iso*-). The former have a continuous chain of carbon atoms with surrounding hydrogen atoms; the latter have one or more carbon atoms as side-chains.

The naphthenes or cyclo-paraffins have the general formula C_nH_{2n}, and are saturated ring compounds, i.e. there are only single bonds connecting adjacent carbon atoms. Those with five carbon atoms in a ring are the cyclopentanes; and those with six-carbon rings are the cyclohexanes.

The aromatics are unsaturated hydrocarbons with six carbon atoms in each ring, and a general formula of C_nH_{2n-6}. It will be evident that there are insufficient hydrogen atoms to leave only two unsatisfied bonds for every carbon atom, and therefore the bonding in the ring must differ from that of the cyclohexanes.

The compounds with the hetero atoms (O, N, S) in addition to carbon and hydrogen, occur mainly in the heavier fractions of crude oils and in the non-volatile residues left on distillation of an oil.

The oxygenated compounds include fatty, naphthenic and aromatic acids, as well as ketones, fluorenones and isoprenoid acids. Among the nitrogen-bearing compounds are pyridines, quinolines, pyrroles and carbazoles, while the sulphur-bearing compounds include thiols, thioalkanes, thiophenes and cyclo-alkylthiaalkanes. Quinolones (oxygen and nitrogen) and thio-quinolines (sulphur and nitrogen) have also been recognised. Porphyrin aggregates, free or complexed with nickel or vanadium, may account for 1–1,000 ppm in a crude oil. The porphyrin molecule includes oxygen and nitrogen atoms[19].

The peak amounts of normal paraffins in crudes are usually in the C_6 to C_8 range. Three isoprenoid hydrocarbons—phytane, pristane and farnesane—have been detected in crude oils, and as much as 0·5% by weight of pristane has been reported in one sample. In the lighter fractions, hydrocarbons with the simpler structures are more abundant than those which are highly branched. The simpler paraffins, naphthenes and aromatics are dominant; the many isomers are present in only small proportions. As the boiling point rises, the amounts of polynuclear aromatics and polycyclo-paraffins increase, whereas the monocyclo-paraffins, normal and branched-chain paraffins decrease in abundance.

Free sulphur occurs in some crude oils, and this has been taken to imply a low-temperature history, because sulphur can react with hydrocarbons to form hydrogen sulphide at temperatures as low as 100°C.

A broad grouping of the constituents of crude oils leads to a division into paraffinic, paraffinic-naphthenic, naphthenic, naphthenic-aromatic and mixed asphaltic types. Table 17, p. 39 gives examples with the amounts of paraffins, naphthenes, aromatics, and resins and asphaltenes. The asphaltic types of crude oils tend to contain more of the hetero atoms than the others.

In the asphaltic fraction of a crude oil are asphaltenes, resins and various so-called "oily constituents". It is believed that the asphaltenes are held in colloidal suspension in crude oils. In the laboratory, asphaltenes can be precipitated by the addition of light hydrocarbons such as pentane, and it is thought that the natural processes which increase the light-hydrocarbon content of a crude oil may also precipitate these substances[24].

A "sweet" or "low-sulphur" crude oil is one which contains less than 0·5% of sulphur; with more sulphur the crude is termed "sour" or "high-sulphur", and special refining operations are needed to remove the sulphur compounds which would give rise to objectionable corrosion and pollution effects.

TABLE 4

ELEMENTAL COMPOSITION OF SOME CRUDE OILS, MINERAL WAX, AND ASPHALTIC SUBSTANCES

	Source	C	H	S	N	N+O	O
				Percentage composition			
Crude oils	Humboldt, Kans.	85·6	12·4	0·37	—	—	—
	Healdton, Okla.	85·0	12·9	0·76	—	—	—
	Coalinga, Cal.	86·4	11·7	0·6	—	—	—
	Beaumont, Tex.	85·7	11·0	0·7	—	2·61	—
Wax	Ozokerite	84–86	14–16	0–1·5	0–0·5	—	—
Natural Asphalts	Athabaska Tar	84·4	11·2	2·73	0·04	—	—
	Bermudez Lake	82·9	10·8	5·87	0·75	—	—
	Trinidad Pitch Lake	80–82	10–11	6–8	0·6–0·8	—	—
	Asphalt from Utah limestone	89·9	9	0	—	—	—
	Native asphalt	82	11	2	2–2·5	—	—
Asphaltites	Gilsonite	85–86	8·5–10	0·3–0·5	2–2·8	—	0–2
	Glance Pitch	80–85	7–12	2–8	—	0–2	—
Asphaltic pyrobitumens	Wurtzilite	79·5–80	10·5–12·3	4–6	1·8–2·2	—	—
	Albertite	83·4–87·2	8·9–13·2	Tr–1·2	0·4–3·1	—	2·0–2·2

TABLE 5

TRACE ELEMENTS PRESENT IN CRUDE OILS

Source of oil	Elements identified
Canada .	Fe, Al, Ca, Mg, traces of Au and Ag
Ohio .	Fe, Al, Ca, Mg, traces of Au, Ag, Cu
Mexico .	Si, Fe, Al, Ti, Mg, Na, V, N, Sn, Pb, Co, Au
Argentina .	Fe, Al, Ni, V, P, Cu, N, K
Japan .	Si, Fe, Ca
Caucasus .	Si, Fe, Ca, Al, Cu, S, P, As, traces of Ag, Au, Mn, Pb
Egypt .	Fe, Ca, Ni, V
Venezuela .	Fe, Ni, V
Iraq .	Fe, Ni, V
Texas .	Si, Fe, Al, Ti, Ca, Mg, V, Ni, Ba, Sr, Mn, Pb, Cu, and traces of Cr and Ag

TABLE 6

RATIOS OF AMOUNTS OF CERTAIN HYDROCARBONS IN A NUMBER OF CRUDE OILS
(After Baker, 1962)[2]

Hydrocarbons	Volume ratio of hydrocarbons in crude oils	
	Average for 21 US crudes	*Average for 11 non-US crudes*
$\dfrac{\text{Methylcyclohexane}}{\text{Cyclohexane}}$	2·13	2·47
$\dfrac{\text{Toluene}}{\text{Benzene}}$	4·27	4·78
$\dfrac{\text{n-Hexane}}{\text{n-Heptane}}$	1·08	1·29

The *n*-paraffins in crude oils have been the subject of detailed chromatographic studies. In the higher molecular-weight range, there is little or no dominance of the odd-carbon-number molecules over the even-carbon-number molecules. However, in the lower-molecular weight range marked odd-carbon-number dominance is seen in some cases (Fig. 6, p. 36). There are also instances of greater abundance of the lower-molecular weight members in the crude oils of greater age or maturity, in particular.

Petroleum has isotopically lighter carbon than is found in limestones, while natural gas is isotopically lighter than the associated crude oil. Indeed, when a crude oil is fractionated by distillation, the fractions differ in their carbon isotopic composition. Broadly, the lower-boiling fractions show a diminution in the heavier carbon isotope (^{13}C) as the boiling point falls. In the higher boiling range, there is evidence of a small irregularity.

Certain substances found in crude oils have definitely biological aspects or affinities, so far as their structures are concerned. These include the chlorophyll porphyrins and the isoprenoid hydrocarbons.

In the carbon "skeleton" diagrams, a kind of "shorthand" for depicting the structures of the organic compounds, only the hetero atoms and special groups are shown by symbols. The carbon atoms are assumed to be at the free ends and junctions of the short straight lines representing the bonds, while hydrogen atoms are assumed to be present to satisfy the remaining valencies (bonds) of the carbon atoms, not drawn. The zig-zag form shown for the carbon chains should not be interpreted rigidly, while the ring structures are not necessarily planar; a three-dimensional arrangement must be involved, but the "shorthand" on paper cannot cater for this (Fig. 5).

2. NATURAL GAS[11, 21]

Crude oil is always associated in the subsurface with a mixture of gases, which for convenience is termed "natural gas". The composition of the gaseous mixture is variable, and it can contain both hydrocarbons (Table 7) and non-hydrocarbons.

The principal component of natural gas is normally methane, and small amounts of the immediately higher members of the paraffin series are usually also present. Benzene and several naphthenes have also been detected. In addition, natural gases may contain appreciable amounts of carbon dioxide, nitrogen and hydrogen sulphide; exceptionally, a natural gas can be almost entirely made up of carbon dioxide or hydrogen sulphide. Mercaptans have also been reported, and small amounts of helium; in a few cases, the concentration of the latter gas may justify processing to segregate the helium for industrial applications, if a sufficient volume of gas is available.

Helium and, more rarely, carbon dioxide, may be of economic value, but usually the non-hydrocarbon gases are a commercial drawback, since they may be actively harmful (e.g. hydrogen sulphide) and thus require removal, or at the least will lower the calorific value of the natural gas when used as a fuel.

Practically all natural gases also contain water vapour derived from the connate water in the pores of the reservoir rock, which must be removed by desiccants or refrigeration before the gas can be utilised.

TABLE 7

TYPICAL ANALYSES SHOWING HYDROCARBONS IN NATURAL GAS

	"Wet" Gas %	"Dry" Gas %		"Wet" Gas %	"Dry" Gas %
Methane	84·6	96·0	iso-Pentane	0·4	0·14
Ethane	6·4	2·0	n-Pentane	0·2	0·06
Propane	5·3	0·60	Hexanes	0·4	0·10
iso-Butane	1·2	0·18	Heptanes	0·1	0·80
n-Butane	1·4	0·12			

Source: D. Heron, Paper 93, 8th *Comm. Min. & Met. Congress*, 1964.

Under certain conditions of temperature and pressure, hydrocarbons with quite high boiling points may be present in the gaseous phase in the subsurface reservoir. These hydrocarbons separate as a liquid on pressure and temperature drop, and are then known as *condensate*. The ratio of gas to condensate may range roughly from 10,000 to 200,000 cubic feet per barrel (cf/brl). Condensates are variable in properties, but resemble motor gasolines. Indeed, in one case reported, a gasoline was of 59·1° API gravity and water white, with an initial boiling point of 102°F and a final boiling point of 406°F; while a condensate was of 59·8° API gravity, also water white, and with initial and final boiling points of 98°F and 406°F, respectively. Some laboratory tests have shown that even heavier coloured compounds can go into the vapour phase in natural gas under certain conditions.

The natural gas which occurs in subsurface petroleum accumulations, in solution in the crude oil or as a "free" gas cap, is termed "associated gas". Sometimes, non-associated gas accumulations occur in which virtually no liquid oil is found, but only gas and salt water. (The probable causes of the absence of oil in such cases are examined in Chapter 4).

The producing gas/oil ratios (GOR) of crude oils may vary from less than 100 to several thousand cubic feet per barrel of oil. The GOR varies with the methods of well control used, with the solution gas/oil ratio, and also with the position of a well on a particular structure and its production history.

(a) Methane

Methane is the dominant component of most natural gases, generally occurring in proportions of 90% or more by volume. It is a gas which apparently has both abiogenic and biogenic origins. Thus, methane forms part at least of the atmosphere of the larger planets and of their satellites, and of some terrestrial volcanic gases, where in part it could possibly have been formed by the combination of carbon derived from carbonaceous rocks with magmatic water vapour. The gas is most commonly produced, however, by the decomposition of the cellulose of vegetable matter by bacterial processes.

Methane occurs in coal mines, forming with air the dangerously explosive mixture known as "firedamp". It is exuded from freshly-cut coal faces and also occurs (usually with nitrogen, carbon dioxide and some carbon monoxide) in high-pressure pockets and fissures in coal seams.

The association of methane with coal deposits is presumably due to gradual organic metamorphism, whereby peats and lignites are converted into bituminous coals and anthracites as a result of rises in temperature and

pressure. This is thought to be a slow process, the earlier stages being bacterially controlled, while at the higher temperature associated with rises of
50° to 150°C the control is chemical. The rapid metamorphism of coal which
occurs when intrusive igneous rocks initiate local "baking" and destructive
distillation also produces methane and carbon dioxide. Both slow and rapid
coal metamorphism have been thought to account for the presence of methane
in some natural gases.

Attempts have been made to differentiate between methanes of biogenic
and abiogenic origin, without much success. However, analyses of the carbon
isotope ratios of various natural gases suggest that these gases are a mixture
of an isotopically lighter "bacterial" type of methane with a heavier
"chemical" methane. It may well be therefore that natural gas accumulations
in fact include methanes of different origins[20].

Enormous quantities of methane are produced from the oilfields of the
world as an unavoidable by-product (or more correctly a "co-product") of
crude oil production—perhaps as much as 10 trillion cubic feet every year. To
this output can be added the even larger volume of "dry" natural gas (again
principally methane) produced from the world's "non-associated" gasfields;
the annual world output of methane is probably not less than 25 trillion cubic
feet.

(b) Carbon Dioxide

Many natural gases contain small amounts of carbon dioxide, while in some
there are relatively high proportions of this component, particularly in
gases occurring in the western United States, Mexico and Canada.

As much as 92% CO_2 has been found in a gas from Upper Cretaceous
sandstones in the *Walden* field in Colorado, and gas containing 57% CO_2
comes from a Cambrian reservoir in Montana. Sufficient carbon dioxide was
at one time produced from the *Santa Maria* oilfield in California to justify
the operation of a local "dry-ice" plant on a commercial scale.

In Canada, carbon dioxide contents of more than 15% have been reported
from Cretaceous rocks in Alberta, while in New Zealand gases from Pliocene
strata have been found with up to 49% CO_2. Other carbon dioxide-rich gases
are known from Sicily, France, Australia, Japan, and in particular from
Mexico, where gases with 50–98% CO_2 have been produced from Cretaceous
limestones.

Much carbon dioxide is clearly of magmatic origin, and is related to areas
of recent vulcanism. The gas is produced as a common accessory of igneous
activity, and is also the product of contact metamorphism through the intrusion of igneous material into limestones or dolomites, as seems to be the case
in Mexico.

Carbon dioxide is also produced by the breakdown or oxidation of many
organic substances, and also by the oxidation of crude oils. Presumably,
therefore, some of the carbon dioxide present in natural gases may be the
product of biogenic and other processes acting on organic matter, while the
large proportions found in gases from areas where vulcanism is common are
the result of local igneous activity and the metamorphism of limestones.

(c) Hydrogen Sulphide

Hydrogen sulphide is produced in very large quantities by magmatic action; as much as 300,000 tons p.a. of hydrogen sulphide are believed to be produced from one fumarole area in Alaska alone.

In its non-volcanic occurrences, hydrogen sulphide is a frequent constituent in gases from swamps, peat bogs and stagnant waters, and it occurs in natural gases in varying proportions. One natural gas from Wyoming contains 63%, and another from Devonian strata at Panther River, Alberta, contains as much as 88% H_2S. Proportions of 10–15% H_2S are quite common.

Hydrogen sulphide is being formed at the present time as a result of the bacterial reduction of sea-water sulphates. This typically occurs in the Black Sea embayments and in North American lakes in which the oxygen of the deeper water is exhausted during summer stagnation, or as a result of prolonged ice cover in winter. The conditions required are the absence of oxygen, i.e. lack of circulation in the overlying waters, and the presence of marine or lacustrine organic material. Hydrogen sulphide produced by bacterial processes reacts with colloidal hydrated iron oxides in the sediments to form finely divided ferrous sulphide, which tends to blacken the sediments. (Similar environmental conditions can be shown to have existed at various times back to the early Pre-Cambrian; some of the world's major sulphide ore deposits occur, in fact, in this type of black shale).

In general, the proportion of hydrogen sulphide found in natural gases may be related to several factors, which include the type of organic source material involved, the presence of sulphates, and their bacterial breakdown.

In the United States, natural gases rich in hydrogen sulphide are found in the Texas Panhandle, western Texas, southeastern New Mexico, and in the Gulf Coast oilfields. In the latter cases, hydrogen sulphide-bearing gases are often associated with the calcium sulphate cap rocks of local salt domes; the famous *Spindletop* field is a notable example of this.

Many Middle East crude oils and associated gases contain appreciable proportions of hydrogen sulphide, and here again the gas seems to be related to the gypsum or anhydrite in the rocks. In Mexico, a high hydrogen sulphide content is typical of most natural gases, and it has been suggested that the gas may have been formed by the reducing action of liquid bitumens on gypsum.

When it contains an appreciable amount of hydrogen sulphide, a gas is termed "sour"; free of hydrogen sulphide it is termed "sweet"*. Apart from its noxious properties, hydrogen sulphide is actively corrosive, and so is generally removed by a chemical desulphurization process. A combined plant is often used to carry out dehydration and desulphurization simultaneously, while the same aqueous amine treatment can be used to remove the mildly corrosive carbon dioxide together with the hydrogen sulphide.

(d) Nitrogen

Nitrogen, the dominant element in the Earth's atmosphere, is evolved in volcanic gases as the result of deep-seated magmatic action. Some of this gas

---

* A "sweet" gas is more specifically one in which there is no "detectable" odour of H_2S present, i.e. where there is less than one part of H_2S in 10 million.

may be trapped in the course of migration upwards through sedimentary
formations, and this may be one explanation for the frequent occurrence of
nitrogen in natural gases.

In some oilfield gases, particularly in the southwestern USA, remarkably
high nitrogen contents have been recorded. Thus, 50% N_2 is not uncommon
in gas from shallow Pennsylvanian reservoirs in Oklahoma, while $84–86\%$ N_2
has been reported in gases from Permian reservoirs in the Westbrook field in
Texas. In the large gasfields of the Texas Panhandle a N_2 content of more than
11% is common.

In Holland, the gas from the Slochteren field contains $14\cdot4\%$ N_2, and it has
been suggested that it is the remaining constituent of an ancient atmosphere
trapped by rapid burial and subsequently released after the loss of oxygen by
bacterial or chemical action.

The nitrogen in natural gases may also have resulted from the bacterial
breakdown of nitrates in the organic sediments from which petroleum has
been derived. It has been suggested that ammonia formed as a result of the
decay of organic matter under certain types of bacterial attack, may act as an
intermediate product, subsequently being oxidized to nitrogen.

The nitrogen in some natural gases may also have been derived from the
metamorphism of organic matter in contiguous carbonate rocks. Thus, in
southern Alberta and southwestern Saskatchewan, very high nitrogen
contents (up to 85%) are found in gases from the Upper Devonian Beaverhill
Lake formation, in close proximity to carbon dioxide-rich gases, which
probably originated from the metamorphism of bituminous carbonates—the nitrogen presumably coming from the breakdown of the bituminous
material.

No satisfactory method has been found which can economically remove
the nitrogen which occurs in significant proportions in a number of natural
gases except by the liquefaction of the other components of the gas by a
refrigeration cycle, leaving the inert gas to be subsequently vented.

(e) Helium

Helium in significant proportions is frequently associated with nitrogen in
oilfield gases in southeastern Colorado, in the Texas Panhandle, in New
Mexico and in eastern Utah. In the richest of these gases, many of which
occur in quantities too small to be of commercial interest, helium percentages
of between 1% and 8% are found, often with high nitrogen contents—as much
as 37%.

The obvious explanation of the occurrences of helium in these natural gases
as the product of radioactive disintegration processes in shallow igneous
rocks containing uranium and thorium minerals is not always substantiated
by the evidence, since although in some areas helium- and nitrogen-bearing
gases occur in reservoirs relatively close to basement rocks, e.g. the buried
"granite ridge" of the Nemaha Mountains, in other areas, such as in southeastern Kansas, the gas from shallower zones contains a larger percentage of
helium than that from deeper strata. This may, of course, be due to local
migration conditions; but in general, there is no clear correlation between
helium content and local radioactivity, although it has been suggested that

the large proportion of helium in Panhandle gases is due to the occurrence of radioactive residual petroleum deposits in the neighbourhood.

In Canada, there tend to be the greatest concentrations of helium in the natural gases produced from regions of structural tension, such as the Peace River Arch, the Sweetgrass Arch, and the hinge belt of the Alberta basin. Tension and fracturing in the strata in these areas has presumably accelerated the release of occluded helium from mineral surfaces. There is, furthermore, a demonstrable relationship in Alberta between the helium content of the natural gases and the nearness of the gas-bearing rock horizons to the Pre-Cambrian basement.

In Europe, since most natural gases are produced from relatively shallow Tertiary or Mesozoic rocks, it is not to be expected that helium will occur in any quantity, and to date only a few examples of helium-containing gases come from Poland, Alsace and some of the North Sea wells. Helium is also produced from wells at Roma in Queensland, Australia, and traces of helium have been reported in the gases from a number of shallow wells in Japan.

(f) Other Gases

Argon is one of the rarer gases in the atmosphere, occurring in the proportion of 9,300 parts per million by volume (0·93%); however, even smaller amounts of argon are found in natural gases. Thus, in the natural gas from the *Panhandle* field at Amarillo, Texas, the concentration of argon is less than 0·1% of that of the helium present, while only a very few other natural gases show any argon at all. Most notable among these are some gases from Japan.

Very small traces of *radon* have been reported in the helium from certain wells in the United States—the highest concentrations being in some Texas Panhandle gases. This radon probably comes from the radioactive decay of very small quantities of radium present in the surrounding rocks.

3. OILFIELD WATERS[23]

Oilfield waters in general are termed "brines". The water found in petroleum reservoir rocks is usually referred to as "interstitial water", although the term "connate water" has also been used. Connate means "born with", but it is improbable that the water which now exists in the pores of a reservoir rock is the same water which was in the same group of pores at the time when the sediment was first laid down. The expressions "bottom water" and "edge water" are also in use, meaning water underlying the oil in the one case, and water lying down dip and therefore laterally with respect to the oil zone, in the other. Clearly, both terms are concerned with the same body of water in a given reservoir rock.

As noted earlier, water occupies part of the individual pores in the oil and gas zones, either at the so-called irreducible minimum saturation, or in increasingly greater proportions of the pore space with increase in depth in the oil/water or gas/water transition zones, before the bottom, fully water-saturated zone is reached.

The interstitial water is saline, and the total solids in solution usually exceed in amount those of sea-water. Since oil and gas accumulations commonly occur in rocks of marine origin the rocks can be expected to have incorporated sea-water when they were deposited. Hence, the water found in most oilfields should have affinities with sea-water.

Rocks laid down in water in non-marine areas would initially contain low-salinity water, and dune sands, originally with air in their pores, would be invaded by water when seas advanced to cover them with other sediments. Because of water flow induced by the compaction of fine-grained deposits, the reservoir types of rocks will transmit compaction fluids, so that the fluids in such rocks would be displaced and replaced. Furthermore, if laterally extensive reservoir horizons are exposed at the surface by uplift and erosion, meteoric waters can enter and mix with the water present previously in them, diluting it and modifying it in other ways.

It has been noted that the interstitial water within the hydrocarbon zone can differ in salinity from the edge or bottom water in the reservoir rock. This suggests isolation of the former water from some of the factors which have affected the latter, and is consistent with the concept of an irreducible minimum saturation at which flow and diffusion have been precluded.

Salinity gradients have been observed in formation waters, a condition to be expected when one water is displacing another and diffusion is unable to even-out the solute distribution.

Table 34 on p. 189 lists the amounts of the more abundant ions in a few oilfield waters, together with those of average ocean water. It will be seen that sodium, chlorine, calcium and magnesium are the most abundant ions, although other elements detected in oilfield brines in smaller amounts are Al, Fe, Mn, Zn, Ba, Cu, Ag, Rb, F, B and Ni; SiO_2 has also been recorded. Total salinities of more than 600,000 ppm have been measured[23].

It is interesting to note that in a few exceptional oilfields, (e.g. *Bay City*, Michigan, and in California) calcium chloride is the principal brine constituent, derived presumably by the interaction between the original sodium chloride brine and constituents of the local sediments. A few oilfields also are known to be associated with fresh water (*Quiriquire* in Venezuela and some Rocky Mountain pools, e.g. *Lander,* Wyoming), but in such cases the oil has apparently remigrated from its original reservoir.

Brines can also contain organic substances such as naphthenates and salts of aliphatic acids, phenols and benzene. Indeed, it has been proposed that a lateral increase in the benzene content of the brine in a given stratum may point towards an oil accumulation in that stratum[25].

Apart from their usually greater salinity, oilfield waters differ from modern sea-water in their low sulphate contents and their calcium/magnesium ratios. (In making comparisons with modern sea-water there is the assumption that ancient seas had waters of similar composition and concentration.) The brines are characteristically deficient in sulphate, and there is evidence of a reduced sulphate content in the water in marine sediments only a small distance below the sea floor. This results from the action of sulphate-reducing bacteria, which break down dissolved sulphates to satisfy their oxygen requirements when acting on organic matter in the sediments. Hydrogen sulphide gas, pyrites, and even sulphur, are then formed in the deposits.

In sea-water, magnesium is more abundant than calcium, whereas in oilfield waters the reverse is generally true. The relative decrease in magnesium is probably related to the formation of chlorite rather than dolomite, or to its slow incorporation in fine-grained, mixed-layer micas in the sediments on progressive burial[23].

Salinities less than that of average sea-water may represent an original condition, or be the result of dilution by meteoric waters. On the other hand, the increased salinity of most oilfield brines may have several possible explanations: usually the deeper the brine the greater is its salinity, and in some cases the water may have been concentrated before burial, by evaporation of water in a barred basin or land-locked gulf, or concentrated within the sediments under "sabkha" conditions. In such cases of high initial salinity, comparatively high atmospheric temperatures would be required. Alternatively, normal marine waters moving through the rocks may encounter in their passage pre-existing saline deposits in an evaporite series, and by solution thereby increase their own salinity.

It has been concluded that the hot brine pools which occur in the axial rift region of the Red Sea, which have temperatures as high as $56°C$ and layered salinities rising as high as $7\frac{1}{2}$ times that of normal sea-water, probably represent the re-solution of previously deposited halite and gypsum. Some oilfield brines may therefore have been formed by admixture of very concentrated solutions of this type with more dilute upper waters; and it is particularly interesting to note that the element enrichment is similar to that found in salt domes.

Experiments have shown that when a salt solution is forced to flow through compressed bentonite or shale membranes, the ratio of the upstream and downstream salinities appreciably exceeds unity. The bentonite and shale act as semi-permeable membranes; the ions in solution are impeded more than the water molecules. The natural operation of this mechanism could lead to a high salinity in water in coarse-pored rocks such as reservoir rocks underlying clays or shales, the cap rocks of oilfields. Compaction processes set water in motion in the sediments (as described on p. 64). In addition, when extensive permeable rocks are exposed by erosion and form elevated ground on the margins of a broad basinal structure, so that the central area of the basin is at a generally lower level, water entering the outcrops will move down dip to deeper levels, and where the pressure conditions are suitable some will escape upwards across the bedding. Where this trans-formational flow is taking place there will be an opportunity for the action of the "filtration" mechanism mentioned above, leading to increased concentration of solutes in the permeable rocks[3, 16, 17].

It appears that in the "filtration" mechanism Ca^{++} is less mobile than Na^+, and that Cl^- is less mobile than molecular water. Mg^{++} is slightly less mobile than Cl^-, which is more mobile than Ca^{++} and SO_4^{--}.[23]

Patterns of water flow undoubtedly change over time, with structural evolution, compaction of the rocks and the formation of faults, as well as the onset of erosion. Hence, the history of the water in the rocks is complex, and so far as salinities and composition are concerned the developments will not be the same in all areas. Some water may be essentially stagnant after a time, being "trapped" at some point, whereas other water may be in slow and continuous movement[12].

The late release of structural water from clay minerals may take place at a considerable depth. On movement into reservoir horizons, such water could cause dilution of salt waters already contained therein.

THE HISTORY OF OIL PRODUCTION[22]

Although petroleum had been obtained in small quantities for many centuries from surface seepages, and put to a wide variety of uses—ranging from medicinal applications to boat caulking—it was not until the first "intentional" oil-well was drilled in Pennsylvania in 1859 that the modern industry was born.

At that time, and in that area, wells were customarily drilled to obtain brine, which was used to produce solid salt by evaporation, and traces of oil had often been noted on the surface of the water recovered. This first true oil-well, drilled by the celebrated "Colonel" Drake, produced only a small volume of oil from a depth of 69ft, but its success rapidly brought about profound and world-wide changes.

Within a year, 175 further oil-wells had been drilled in Pennsylvania. The search for petroleum spread rapidly to other parts of North America and also to other countries.

By the year 1900, the world annual output of crude oil was 141 MM brl, from 11 countries. By 1920, in spite of the devastation caused by World War I, world output had increased to 695 MM brl. By 1946, just after the completion of World War II, world output had increased to 2,750 MM brl. In the years that have elapsed since then, the output of crude oil has continued to increase at an average annual rate of growth of more than 7%, equivalent to the doubling of production every ten years (Figs. 3, 4).

By 1973, commercial oil production was being obtained from 62 countries and totalled 19·9 B brl—roughly *140 times* the output in 1900. However, it is interesting to note that, since man-made frontiers have little relationship to the geological factors which control the size and productivity of individual oilfields, 85% of the total output comes from only 12 countries (Tables 8, 9).

TABLE 8

GROWTH OF WORLD OIL OUTPUT, 1890–1973
(MM brl)

	East Asia	Austral- asia	Middle East	Africa	E. Europe + USSR	West Europe	North America	South America+ W. Indies	World
1890	0·2	—	—	—	25·6	0·1	46·6	—	72·5
1900	6·0	—	—	—	69·9	0·3	64·5	0·3	141·0
1910	20·1	—	—	—	88·4	1·1	213·5	1·4	324·5
1920	29·1	—	12·2	1·0	44·8	0·6	600·2	7·1	695·0
1930	56·9	—	46·8	2·0	172·2	1·8	939·1	189·5	1408·3
1940	81·7	—	102·7	6·5	269·2	10·9	1,405·8	268·8	2,145·6
1950	87·7	—	640·9	16·7	307·0	19·6	2,075·1	644·1	3,791·1
1960	227·3	—	1,923·3	105·3	1,189·7	98·8	2,863·6	1,266·5	7,674·4
1970	589·3	65·2	5,154·1	2,208·4	2,717·8	114·7	4,112·1	1,729·9	16,691·5
1973	937·7	142·5	7,444·1	2,176·2	3,094·6	100·6	4,270·0	1,700·3	19,966·0

TABLE 9

PRINCIPAL OIL-PRODUCING COUNTRIES, 1973

	MM Brl	% of total
USA	3,359·9	16·8
USSR	3,031·2	15·2
Saudi Arabia	2,427·2	12·2
Iran	2,152·6	10·8
Venezuela	1,231·0	6·2
Kuwait	1,007·5	5·0
Libya	792·3	4·0
Nigeria	747·9	3·7
Canada	720·5	3·6
Iraq	695·1	3·5
Indonesia	482·3	2·4
Abu Dhabi	476·8	2·4
12 leading countries	17,124·3	85·8
50 other countries	2,841·7	14·2
World total	19,966·0	100·0

The United States produced in aggregate nearly half the total volume of crude oil output of the whole world for the first century of the oil industry's existence. The US proportion of annual world oil output was usually not less than 60% and sometimes more than 70% in most of the years from 1860 to 1946. However, since then the proportion has fallen with each succeeding year—it was less than 17% in 1973—and although more than three-quarters of all the wells drilled anywhere in the non-Communist world are still drilled in the USA†, the average production per well is now less than in any other oil-producing country, due to the great number of old wells still functioning.

Outside the USA, pre-revolutionary Russia was initially predominant as an oil producer, due mainly to the huge individual outputs of a relatively small number of Baku wells. Subsequently, the Russian oil industry declined, and its focus shifted eastwards from the old oilfields of the Caucasus to the newer oil areas of Ural-Emba and Siberia. Production from these newer areas has rapidly increased to make the USSR once more the world's second largest oil producer.

In the 1920's, Mexico, briefly, and subsequently Venezuela, occupied this position. Mexican production, based on a relatively few enormously productive wells drilled along the prolific limestone reef known as the "Golden Lane", soared for a few years but fell rapidly as these wells were exhausted.

After 1930, Venezuela was for many years the world's greatest oil exporting country. Production reached a peak in 1964, much of the output being sent to the United States and Europe. Since then, however, oil production has increased only very slowly.

The most notable increases of oil production in the post-war years have come from the newer oil-producing states of the Middle East. Oil was found in Iran as long ago as 1908, in Iraq in 1927, and in Saudi Arabia in 1938; but it was not until after the end of the last war that the remarkable expansion of

† *The total number of oil- and gas-wells that have been drilled in the USA is at least 1½ million, far more than in all the rest of the world; and 25,000–30,000 new wells are still being drilled every year, again far more than in all the other countries of the world together.*

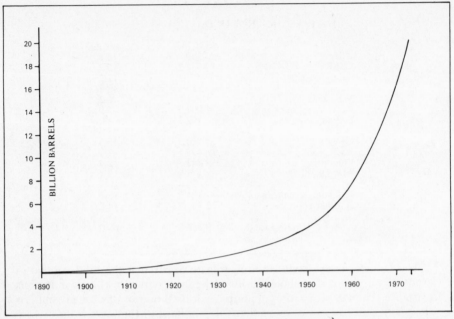

Fig. 3. World Annual Oil Output 1890–1973.
(Note: To obtain a smooth curve, only totals for every tenth year have been plotted up to 1970)

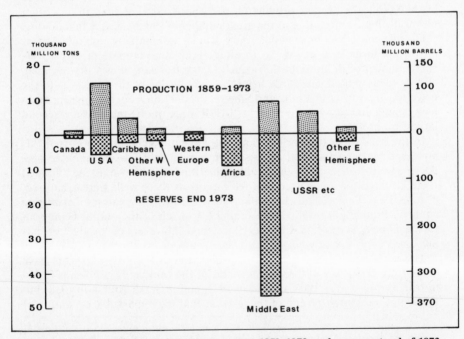

Fig. 4. World cumulative oil production by areas, 1859–1973; and reserves at end of 1973.
(*Source: BP Statistical Review, 1973.*)

TABLE 10

CRUDE OIL PRODUCTION OF MAJOR AREAS, 1973

	MM Brl	% of total
Middle East	7,444·1	37·3
USA	3,359·9	16·8
USSR	3,031·2	15·2
Africa (North and West)	2,176·2	10·9
South America and Trinidad	1,700·3	8·5
Rest of the World	2,254·3	11·3
World total	19,966·0	100·0

TABLE 11

CHANGING PROPORTIONS OF WORLD OIL OUTPUT, 1946–1973

	1946[1] %	1973[2] %
United States	63·2	16·8
USSR	5·9	15·2
Middle East	8·9	37·3
Venezuela	14·1	6·2
Rest of World	7·9	24·5
	100·0	100·0

[1] World output 2·75 B brl.
[2] World output 19·97 B brl.

the Middle East oil industry really began. Today, the belt of oilfields lying between Turkey and Oman produces more than a third of the world's total oil output and is still capable of considerable further development (Table 10).

The post-war years have also seen the rapid emergence of North and West Africa as important oil-producing areas. In North Africa, large new accumulations of oil and gas have been discovered and developed in Algeria, Libya, and Egypt; while in Nigeria development has also been remarkably rapid, in spite of the dislocation caused by the Biafran war.

Table 11 illustrates the remarkable changes that have taken place since 1946 in the proportions of the world's oil output that are obtained from the major producing areas. Over this post-war period, the proportion of the total provided by the USA has fallen to about a quarter of the 1946 figure, while the proportion due to the Middle East countries has more than quadrupled. The shares provided by the Soviet Union and the "Rest of the World"—the latter largely due to the developments in North and West Africa—have also greatly increased, whereas the contribution of Venezuela has approximately halved.

A notable feature of the petroleum industry in these post-war years has been the remarkable acceleration that has taken place in its tempo of development. The time that elapses between an initial oil discovery and the full development of a previously unexplored area has shortened from decades to only a few years.

Another very important recent trend has been the increasing importance of the offshore oilfields which have been discovered as techniques have been developed for carrying out exploration and development operations in waters hundreds of feet deep. In 1973, it was estimated (Table 41, p. 258) that more than 17% of the world's crude oil production came from offshore oil accumulations.

THE DEVELOPMENT OF PETROLEUM GEOLOGY[4, 6, 14]

Historically, petroleum exploration falls into several fairly well-defined phases. Initially, in the years which immediately followed the discovery of oil in Pennsylvania in 1859, proximity to seepages and earlier successful wells largely controlled the selection of drilling locations. Originally, it was believed that hydrocarbons occurred in subterranean lakes and caverns, but it was soon noted that the oil and gas accumulations in this area lay along distinct linear trends, which were actually lines of structural uplift, and that the hydrocarbons filled the pores and interstices of certain types of rock.

The first theory of anticlinal accumulation evolved in the 1860's, was based on the belief that gas, oil and water separated according to their densities so that oil and gas tended to be found at the culminations of anticlinal folds, which were thus soon recognised as targets for exploration. However, it was generally believed that this anticlinal concentration of hydrocarbons was due to the fissuring of the rocks towards the crests of the folds, rather than to any other structural factor.

Between 1883 and 1885, the techniques of field geology were used for the first time in petroleum exploration to produce structure contour maps, and thus to locate the crestal portions of the folds which it had now been accepted would be zones of maximum oil productivity. Early examples of successful exploration based on the delineation of anticlines in this way in the eastern United States were the *Grapeville-Belleveron* and *Volcano-Burning Springs* fields. Other discoveries were made in California and Mexico by the same techniques, and the discovery well on the *Salt Creek* dome in the Rockies was also drilled on the basis of structural contours.

In spite of these early successes, little use was made of geology in the search for oil in the United States during the next 20 or 30 years. At that time there were such a large number of oilfields awaiting discovery at shallow depths that methods of exploration other than "hit or miss" were scarcely necessary. Drilling near surface seepages was a simple and adequate method of exploration, and it was not until well into the 20th century that a need developed for improved methods of exploration, not only in America but also in other parts of the world, notably the (then) Dutch East Indies. As the demand for petroleum mounted and the shallow oilfields become more difficult to find, the second phase of petroleum geology history began, initiated by the discovery of the *Cushing* oilfield on geological advice in 1910, which led to a great increase of geological activity in the search for anticlines.

The initial simple levelling instruments previously used for surface surveys were replaced by the plane-table and alidade or theodolite, and it became common practice to use augers and light drills for obtaining rock samples in areas obscured by surface vegetation or alluvium. Exploration drilling was carried to greater depths, and the results obtained frequently showed that the structures deduced from evidence of surface exposures did not coincide with the structures which existed in the subsurface. The need for correlation between the formations penetrated by succeeding exploration wells became increasingly important, and methods were needed for relating drill samples where surface outcrops were rare or did not exist.

Methods of correlation were evolved to meet the requirements of particular areas; for example, in Trinidad the use of heavy mineral residues was used as a means of identifying and correlating the confusing number of different sandstone beds penetrated by the early wells, while on the Gulf Coast micropalaeontology was shown to be a powerful tool for separating and correlating beds which were lithologically indistinguishable.

It was also recognised in this period that every well drilled, whether it found oil or not, should be treated as a valuable source of information. In order to collect this information, various techniques of sampling, coring and instrumental well-surveying were developed (described in Chapter 7).

As the more easily found shallow oil accumulations were exhausted, the need was increasingly felt for a method of investigating the subsurface at depth, and of locating and delineating subsurface structures. Geophysics was first used as an aid to geology in petroleum exploration in the years 1920–22, when torsion balance surveys were carried out in Egypt, Germany and the USA. The first oilfield to be discovered by a gravity survey was the *Nash* salt dome field in Texas, in 1926. Seismic surveying methods were successful over the next few years in locating structural hydrocarbon accumulations in several parts of the world—notably in Iran and the East Indies. Particularly successful results were obtained by seismic surveys in the US Gulf Coast area, where many further salt dome accumulations were discovered by this means. As a result, seismic surveys (mainly reflection, less frequently refraction) became very widely used in exploration activities all over the world, and the resultant successes brought about what has been called the "Golden Age" of petroleum exploration. (Geophysical surveying techniques are described in Chapter 6.)

The search for anticlinal accumulations still continues, so that this phase of exploration history has not really closed. However, in recent years it has become increasingly apparent that additional structural traps detectable by reversals of dip are becoming harder to find, and in particular fewer and fewer "giant" oilfields are being discovered, as was pointed out on p. 9.

In this context, it must be remembered that the methods by which oil has been found have remained basically the same for many years, although a number of refinements have been added. Thus, developments in geophysics have been principally along the lines of transmitting greater proportions of seismic energy into the subsurface for subsequent reflection, or of developing techniques which eliminate multiple reflections when processing the records. Magnetic tape recordings and digital computer presentation have enabled much greater sensitivity to be obtained in interpreting seismic results. Progress has been made also in remote-sensing methods of exploration, using aerial surveying coupled with sensitive magnetometers and geochemical sensors. However, no method has yet shown signs of replacing the well-tried gravity and seismic methods of survey as standard methods of subsurface investigation.

The development of air photography has cut down very largely the time and effort needed to make an initial surface geological map of an unknown area. New techniques have also been evolved to help determine prospectivity; thus, palynology, the study of ancient spores and pollens, has been introduced as a standard technique of stratigraphic correlation, while the geochemical examination of sediments can give valuable indications of likely source beds.

The modern trend is for the petroleum geologist to take the broadest possible view of the sedimentary history of any region under investigation, and to use all possible sources of information to determine the palaeogeographical setting and the changes which have occurred in time, the environmental and facies patterns, and the tectonic history. On this basis he is able to estimate the likelihood of oil being found at all and, if found, of it existing in quantities large enough to be "commercial"—a term which was discussed above. However, when it comes to selecting the exact point at which to drill, the decision still has to be made on a limited amount of information. The many improvements made in the techniques used are inevitably negatived to some degree as time goes on by the difficulty experienced in finding hydrocarbon accumulations at greater depths and in less easily identifiable traps. The more intense the degree of exploration has been in a particular country, the greater the difficulty will there be in making further discoveries on the same scale as has been achieved in the past. Thus, in the United States, where exploration has been most intense, the percentage of successful exploratory wells was only 17 in 1972. This "success ratio" of about 1 in 6 was considerably worse than the 1 in 4·4 that was achieved in 1945, indicating that the search for oil becomes more complex and difficult with the passage of time. Every scientific means has to be employed as effectively as possible for the discovery of new fields and the methodical development of old ones. It is estimated, for example, that if the discovery to dry hole ratio in the United States could be improved only to 1 to 5, current exploration costs would be cut almost in half.

Petroleum geology, coupled with applied geophysics, still provides the best available means of evaluating the economic prospects of an area in terms of its likely oil and gas content per unit of investment capital deployed. Although until recently such judgments made before drilling commenced had inevitably to be largely subjective, more objective methods have recently been devised for translating geological evaluations into numbers and making appraisals on the basis of the probabilities of the optimum developments of a number of geological parameters[10, 18]. The versatility and speed of computer analysis provides a method by which such statistical analyses can be extended and refined. Various assumptions can be combined, using combinations of the best data available and different operating practices, economic factors and constraints, to arrive at informed assessments of profit/investment ratios, "pay-out" times, etc.[9]

Many other computer-based techniques are now applied in petroleum geology. Thus, computerised data can often be plotted directly onto maps, while geophysical data are very suitable for high-speed computer treatment, thus facilitating interpretation[5].

The success achieved by modern techniques of petroleum geology allied with applied geophysics over the past quarter-century or so may perhaps be best illustrated by reference to Table 45 on p. 285, which compares world "published proved" oil reserves at the ends of 1945 and 1973. The totals were about 59 and 600 B brl respectively—an approximate *tenfold* increase. If to the total of 541 B brl of "new" oil we add the volume of crude oil actually *consumed* in the 28 years—about 253 B brl—we see that some 794 B brl or say 28 B brl p.a. of oil must have been developed during this time—an average annual increase of about half the total world "published proved" reserves at the beginning of the period.

REFERENCES

1. "AAPG Classification", *Bull. Amer. Assoc. Petrol. Geol.,* in Appendix F, **58** (8), 1504, (1974).
2. E. G. Baker, "Distribution of hydrocarbons in petroleum", *Bull. Amer. Assoc. Petrol. Geol.,* **46** (1), 76–84 (1962).
3. J. D. Bredehoeft, C. R. Blyth, W. A. White and G. B. Maxey, "Possible mechanism for concentration of brines in subsurface formations", *Bull. Amer. Assoc. Petrol. Geol.,* **47** (2), 257–269 (1963).
4. R. J. Cordell, "Future of geology in petroleum exploration", *Bull. Amer. Assoc. Petrol. Geol.,* **52** (2), 475, (1968).
5. E. L. Dillon, "Expanding uses of computers in geology", *Bull. Amer. Assoc. Petrol. Geol.,* **51** (7), 1185, (1967).
6. R. H. Dott and M. J. Reynolds, "Sourcebook for Petroleum Geology", Mem. 5, Amer. Assoc. Petrol. Geol. (1969).
7. W. Eggleston, "Economic life of oil- and gasfields in the United States", *Journ. Petrol. Technol.,* **18** (6), 661–664, (1966).
8. P. H. Frankel and W. L. Newton, "Comparative evaluation of crude oils", *Journ. Inst. Petrol.,* **56**, 547, 1–9 (1970).
9. J. M. Forgotson and C. F. Iglehart, "Current uses of computers by exploration geologists", *Bull. Amer. Assoc. Petrol. Geol.,* **51** (7), 1202–12, (1967).
10. V. A. Gotautas, "Quantitative analysis of prospect to determine whether it is drillable", *Bull. Amer. Assoc. Petrol. Geol.,* **47**, 1794–1812 (1968).
11. B. Hitchon, "Geochemical studies of natural gas" Pt. I: "Hydrocarbons in Western Canadian natural gases", *J. Can. Petrol. Tech.,* **2** (2), 60–76 (1963); Pt. II: "Acid gases in Western Canadian natural gases, *ibid.,* **2** (2), 100–116 (1963); Pt. III: "Inert gases in Western Canadian natural gases", *ibid.,* **2** (4), 165–174 (1963).
12. B. Hitchon and J. Hays, "Hydrodynamics and hydrocarbon occurrences, Surat Basin, Queensland, Australia". *Water Resources Res.,* **7** (3), 658–676 (1971).
13. M. T. Halbouty, A. A. Meyerhoff, R. E. King, *et al.,* "World's giant oil and gasfields", in "Geology of Giant Petroleum Fields", Amer. Assoc. Petrol. Geol., Tulsa (1970).
14. M. King Hubbert, "History of petroleum geology and its bearing upon present and future exploration", *Bull. Amer. Assoc. Petrol. Geol.,* **50**, 2504–2518, (1966).
15. G. M. Knebel and G. Rodriguez-Eraso, "Habitat of some oil", *Bull. Amer. Assoc. Petrol. Geol.,* **40** (4), 547–561 (1956).
16. J. G. McKelvey and I. H. Milne, "The flow of salt solutions through compacted clays", Pp 248–259, in: "Clays and Clay Minerals" (ed. A. Swineford). *Proc. 9th. Nat. Confer. on Clay Minerals,* Lafayette, Ind. (1960).
17. L. U. de Sitter, "Diagenesis of oilfield brines", *Bull Amer. Assoc. Petrol. Geol.,* **31** (11), 2030–2040 (1947).
18. I. T. Schwade, "Geologic quantification", *Bull. Amer. Assoc. Petrol. Geol.,* **51** (7), 1225 (1967).
19. G. C. Speers and E. V. Whitehead, "Crude Petroleum", pp. 638–675, in: "Organic Geochemistry" (eds. G. Eglinton and M. T. J. Murphy), Springer-Verlag, New York (1969).
20. W. Stahl, "Carbon isotope fractionations in natural gas", *Nature,* 251, 134, (1964).
21. E. N. Tiratsoo, "Natural Gas—A Study", 2nd Ed. Scientific Press, Beaconsfield, (1972).
22. E. N. Tiratsoo, "Oilfields of the World", Chapter 1, Scientific Press, Beaconsfield, (1973).
23. D. E. White, "Saline waters in sedimentary rocks", pp. 342–366 in: "Fluids in Subsurface Environments" (eds. A. Young and J. E. Galley), Mem. 4, Amer. Assoc. Petrol. Geol., Tulsa, Okla. (1965).
24. P. A. Witherspoon and R. S. Winniford, "The asphaltic components of petroleum", pp. 261–297, in: "Fundamental Aspects of Petroleum Geochemistry" (eds. B. Nagy and U. Colombo), Elsevier Publishing Co., (1967).
25. W. M. Zarella, R. J. Mousseau, N. D. Coggeshall *et al.,* "Analysis and significance of hydrocarbons in subsurface brines", *Geochim. Cosmochim. Acta,* 31 (7), 1155–1166 (1967).

CHAPTER 2

The Origin of Petroleum

THE problem of the origin of petroleum has been debated for many years. Most of the broader geological and chemical ideas are of long standing, although in some cases there has been refinement and the presentation of more detail as knowledge of the composition of different petroleums has increased, and laboratory experiments and industrial processes have provided new information. However, the fact that crude oil and natural gas are essentially mobile substances, and hence are not necessarily found at the place where they were formed, adds to the difficulties; thus, the environment for their formaton is not immediately apparent, in contrast with the case with coal, the other major organic fossil fuel.

The hypotheses that have been made about petroleum origin fall into two groups: the first of these (chronologically) has assumed that the source material was inorganic; the second has argued that petroleum was derived from former living organisms. The respective labels *abiogenic* and *biogenic* have been attached to these groups, and there has also been much discussion as to whether a biogenic source material was in fact animal, vegetable or both, and whether the transformation process was essentially biochemical, thermal or radioactive. It is important to stress that our real concern must be with the formation of hydrocarbons and near-hydrocarbons in quantities sufficient to form accumulations of commercial interest. There may well be other conditions and processes which also result in the formation of hydrocarbons, yet these hydrocarbons do not provide large accumulations and are in some senses rarities.

Because a petroleum accumulation is a geological entity, any acceptable hypothesis about the origin of oil and natural gas must satisfy geological requirements with respect to the source material, the transformation processes involved, and the possibility of producing an accumulation. The universal association of petroleum accumulations with sedimentary rocks must also be explained, and also the chemical composition and other properties of crude oils[13].

Over a hundred years ago, it was suggested that petroleum had been formed in the interior of the Earth by the action of water on metallic carbides such as those of calcium and iron. Another mechanism proposed involved the interaction of alkali metals with carbon dioxide and water. In the laboratory, hydrocarbons have in fact been formed in this way; however, metallic carbides and alkali metals are not known to exist in the Earth's crust, and no acceptable conditions can be envisaged which would lead to their occurrence in sedimentary rocks[4, 30].

Methane is an important component of the atmospheres of some planets and satellites in the solar system, and small amounts of organic matter have been reported in certain types of meteorites the origin of which is not clear: it may be abiogenic, extra-terrestrial biogenic, or due to terrestrial contamination. Consequently, the possibility cannot, at present, be excluded that some abiogenic terrestrial hydrocarbons are linked with the Earth's cosmic history.

A "duplex" origin has also been proposed for ancient crude oils: an abiologically-formed oil might have resulted from processes involving Fischer-Tropsch-type reactions occurring long before the first sediments were formed. The first living organisms then used this hydrocarbon mixture as a source of carbon and hydrogen, and possibly also of nitrogen and sulphur. The organisms were associated with this early oil, their biochemistry evolved, and their remains contributed to the total crude oil[34].

Although in view of the obviously organic origin of coal, there is a strong temptation to accept an organic origin of petroleum, it should be noted that, in recent years there has been some renewed support for an abiogenic origin, particularly in the USSR[33a].

Organic matter in sediments

Among living organisms, only members of the plant kingdom are able to use inorganic substances for the formation of organic compounds. In addition to carbon dioxide or carbonate ions and water, certain other substances are required in small amounts, and restrictions in the supply of these nutrients limit the development of the organisms concerned. Unusual conditions can lead to the proportions of certain substances in the organism differing considerably from the amounts developed under what may be considered to be normal conditions. The proportions can also depend on the general temperature level, and there may be greater or lesser numbers of individuals per unit volume or area, depending on circumstances.

Petroleum is commonly associated with marine rocks. Erosion of the land is the principal source of the inorganic nutrients for plant life in marine areas, and these are available in solution for use by the plant organisms. Nutrient-bearing water will generally be most plentiful near to land, although it is known that some nutrient-bearing currents travel along the sea floor and rise to the ocean surface in zones of upwelling at the continental slope.* The organisms also require sunlight for their activities, and hence they thrive in relatively near-shore near-surface waters, and also in the near-surface zones in areas of upwelling.

*The nature of the continental margins is discussed in Chapter 9.

Plant organisms provide food for animals of many types, and thus the areas of the seas and oceans mentioned above will understandably be the regions of most ample development of organic matter. Some plant and animal organisms will develop on the sea floor and in the inter-tidal areas, and relics of these, together with excrement and unconsumed organic matter from the overlying water, will be available for incorporation in the sediments.

Particulate organic matter is unlikely to be much denser than sea-water, so its sedimentation from the water will be slow and take place only where there is little or no movement of the water. Such areas will also allow the sedimentation of fine-grained mineral matter—e.g. clay particles or very fine carbonate precipitates. Lack of motion in the water leads to depletion of its dissolved oxygen content, and anaerobic conditions thereby created will favour the preservation of relatively more organic matter on the sea floor and in the new deposits than would be the case if aerobic conditions prevailed. Water in which there is significant motion will allow the sedimentation only of the coarser mineral particles, and will be comparatively rich in dissolved oxygen. Such water will be less favourable for the deposition of organic matter, especially the smaller fragments, and will favour its extensive decomposition through the activities of the more destructive aerobic bacteria.

The proportions of organic matter in sediments must depend on the relative rates of deposition or incorporation of organic matter and inorganic mineral matter, and in sands, silts and clays the relative amounts have been reported to be roughly in the ratios of 1:2:4, respectively[15, 49].

Organic matter in the sediments need not be restricted to that which developed in the water body in which deposition took place. It can include contributions of particulate and "dissolved" organic matter directly derived from land organisms and organic matter adsorbed on inorganic mineral matter; in addition there can be some re-cycled ancient organic matter, much of it intimately mixed with inorganic mineral matter, obtained from the erosion of coasts and more distant areas inland.

Delta plain deposits are dominated by organic matter from land plants; delta front, pro-delta and continental shelf deposits have a mixture of land and marine organic matter. Beyond the continental shelf, marine organic matter occurs, although it may be noted that land plant material has been found 1000km offshore in the Atlantic. The organic matter production is of the order of tens of grams/m^2/day in the delta plain area, about a tenth of that in the delta front and continental shelf area, and some hundredth of the rate in the area beyond the continental shelf. The delta plain has deposits ranging from barren sands to organically-rich swamp deposits. The sediments of the delta front, pro-delta and continental shelf are organically fairly rich, and this area is favourable for transgressions and regressions. Beyond the continental shelf, the deposits are organically lean.

Studies of the ratios of the carbon isotopes, ^{13}C and ^{12}C, in the organic matter in recent sediments afford evidence of changes in the proportions of land-derived and marine-derived organic matter in moving from fresh-water to marine environments. Marine organic matter is isotopically heavier than terrestrial organic matter[40].

Lakes have organic matter of mixed origin, aquatic and terrestrial, and some, like swamps, can have organically rich deposits. Many, especially the smaller ones, are transient, their deposits being eroded while they are still youthful, only to be deposited elsewhere after a phase of transport in which oxidative and destructive processes are likely to be active.

The organic matter incorporated in the sediments will differ in some measure from that in the living complexes. It will also include bacterial and fungal products and residues, which at times constitute an important proportion of the total. In the youthful sediments, biochemical activity will continue until conditions inhibit further bacterial action. Certain types of compound will be greatly reduced in amount, if not completely destroyed, and new substances will appear. In some cases, aerobic conditions may exist to a small depth in the sediment, being replaced by anaerobic conditions at greater depths; in other cases the bottom waters and the sediments will be anaerobic.

Rapid changes take place in plankton after death, and the same is true for other organisms. Increase in cell membrane permeability allows loss of soluble body substances, access of oxygen with subsequent cross-linking, as well as hydrolysis of proteins and carbohydrates to give amino acids and aldehydes[1].

The main types of substance formed in organisms are the lipids, proteins and carbohydrates. The proportions of these differ considerably between the plant and animal kingdoms; carbohydrates are much more abundant in the former than in the latter. There are also differences in the amount and members of a given type in different organisms, and also in the various parts of single organisms (Table 12). Some organisms or parts of organisms have a better chance than others of being entombed in the sediments, with or without much alteration at the time of inclusion. Soluble decay products formed before sedimentation may, however, be included in the sediments: (a) in interstitial water; (b) by becoming adsorbed on inorganic mineral matter; and (c) by being adsorbed on organic débris or by becoming involved in polymerization reactions[42, 43].

The term lipid has not been used consistently, but has implied a "fatty" appearance, solubility in the so-called fat solvents, not in water. Lipids range in amount from about 1% in many organisms to 90% in *Botryococcus* and other boghead-type algae. Vertebrate fats have under 2% of unsaponifiable matter, whereas a series of invertebrates had 13% to 37% of unsaponifiable matter in their lipids. Generally, the more primitive the animal the higher the proportion of unsaponifiable matter. The unsaponifiable components include waxes and sterols[2].

TABLE 12

COMPOSITION OF ORGANIC MATTER IN ORGANISMS AND
MARINE SEDIMENTS (*After* Brandt and Trask)[49]

	Ether Extract %	Crude Protein %	Carbohydrates %	Crude Fibre %
Peridineans	1·5	14	85?	?
Diatoms (algae)	8	29	63	0
Copepods (small crustacea)	8	65	22	0
Higher invertebrates	10	70	20	0
Organic matter in marine sediments	1	40	47	0

Numerous studies have been made on recent and ancient sediments to ascertain the types of substances present (Tables 13 and 14), but these studies do not necessarily reveal the exact condition in which the substances occur. In addition, the recovery technique may lead to the isolation of a substance which is a modification of that actually in the sediment, so that related rather than parental substances, or fragments are identified. For example, some of the organic acids may be derived from simple esters, salts of the acids or

TABLE 13
ORGANIC MATTER IN RECENT SEDIMENTS
(*After* Erdman, 1965)[11]

	Percentage of dry sediment		Percentage of organic carbon				$10^3 \times$ percentage of organic carbon		
	Organic carbon		Amino acids	Carbohydrates		Lipids	Sterols		Carotenoids
	1–15cm	120–135	120–135	0–15cm	120–135	120–135	0–15cm	120–135	0–15cm
Semi-marine: Sound	1·54	1·32	3·0	8·1	7·9	1·7	17	10	15
	1·13	0·80	2·0	9·6	5·9	2·5	16		29
Semi-marine: Shallow continental shelf	0·89	0·85	2·5	11·6	4·7	2·2	11	14	
	0·59	0·53	3·0	10·7	8·1	2·9	30	21	25
Marine: Basin— 2000ft	2·88	3·00		6·4	6·9		17	10	79
Basin— 6000ft	4·91	3·14		10·0	4·8		9·9		10
Continental slope 13,500ft	0·97	0·84		13·7	19·5				3

TABLE 14
FATTY ACIDS IN ORGANISMS, SEA WATER AND SEDIMENTS
(*Based on* Parker)[31a]

		Sea water			Recent sediment		Ancient sediment	
Fatty acid carbon number	Mixed plankton, total fatty acids %	Pacific µg/l	Gulf of Mexico µg/l	Van-couver Island µg/l	Baffin Bay† µg/g	San Nicholas* µg/g	Eagle Ford* µg/g	Green River, total fatty acids %
12	0·77	0·4	329		0·73			2·5
13	0·06				0·21			7·0
14	10·72	1·8	21	1·6	4·07	1·05	0·18	3·0
15	1·17	6·6			4·4	0·74	0·2	4·0
16	39·14	14·6	252	2·44	17·8	17·02	0·5	7·0
17	3·0	15·5			3·0	0·7	0·32	5·0
18	22·0	22·2	49	0·8	11·59	3·83	0·38	7·5
19		14·1				0·7	0·18	5·0
20	5·8					2·36	0·14	5·2
21	16·0					0·65	0·12	4·0
22						2·49	0·10	6·5
23						0·75	0·08	5·0
24						4·00	0·08	5·6
25						0·81	0·08	3·0
26						3·71	0·08	4·1
27						0·47	0·06	2·0
28						3·11	0·06	3·7
29						0·24	0·04	2·3
30						1·39	0·08	4·6

† *Method did not detect molecules over* C_{20}. * *Method excludes unsaturated and branched acids.*

from complex polymeric mixtures, e.g. *kerogen,* the general name given to the material in sediments which does not dissolve in the solvents used for recovering hydrocarbons and similar substances. Similarly, the sugars found are not necessarily free in the sediments, but may be produced by hydrolysis in the extraction processes. Differences in technique may result in differences in the amounts recovered, and may even lead to the appearance of small quantities, at least, of substances which represent the breakdown products of materials initially present. In some cases, the techniques make possible the recognition of even small amounts of certain kinds of substances without characterising them completely[16].

Palmitic and palmitoleic acids are stated to be the dominant fatty acids in animal tissues, and stearic, oleic and other C_{18} acids the principal acids in plant tissues. The unsaturated fatty acids are more plentiful than the corresponding saturated acids (Table 15). The most abundant biologically-formed hydrocarbons are the terpenes. *n*-Paraffins are the alkanes most abundantly produced biologically, with the C_1 and C_7 to C_{36} homologues found commonly in organisms, although members as high as C_{62} have been reported. n-C_{17} is the principal paraffin in red and green benthonic algae, whereas n-C_{15} is dominant in brown benthonic algae. Pristane, an isoprenoid-type alkane, is common in various unicellular and multicellular marine and terrestrial animals.*[29]

TABLE 15

C_{18} FATTY ACIDS IN ORGANISMS AND SEDIMENTS
(*Based on* Parker, van Baalen and Maurer; and Parker and Leo)[23, 31]

	% of total fatty acids in C_{10} to C_{20} range			
	18:0	18:1	18:2	18:3
Blue-green algae				
Trichodesmium erythaeum	2·6	2·8	4·2	19
Coccochloris elabens	1·2	13	17	2·6
Microcoleus elabens	3·7	14	5	18
Nostoc muscorum G	3·1	16	14	4
Anabena variabilis	4·4	14	14	17
Agmenellum quadruplicatum	2·6	16	14	5·2
Bacteria				
K_2	3·2	5		
E	2·0	4·1		
L	<0·5	21		
G	2	12		
	Parts per million			
Living algal mat	130	1200		
First mud (under mat)	180	330		
Second mud (under mat)	97	n.d.		
Baffin Bay core: 0–10cm	53	89		
37–41cm	35	50		
60–64cm	58	20		

* C_{27}, C_{29} and C_{31} *waxes occur in land plants, raising the possibility of their being present in coals, near-shore sediments and oils.* C_{15}, C_{17} *and* C_{19} *hydrocarbons which are formed in plankton, may be incorporated directly or indirectly in offshore sediments, and thence in petroleum. It has been noted that waxy crude oils are largely in sand-shale sequences; they are commonly associated with coals and other highly carbonaceous strata, and are low in sulphur. The parental organic matter could have a terrigenous contribution, and there can be evidence of deposition of the sediments in waters of lower than normal salinities*[14].

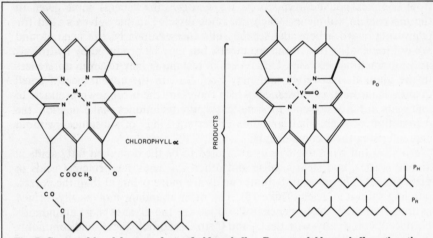

Fig. 5. Suggested breakdown products of chlorophyll α. P_O=vanadyldeoxophylloerythroetio-porphyrin; P_H=phytane; P_R=pristane.

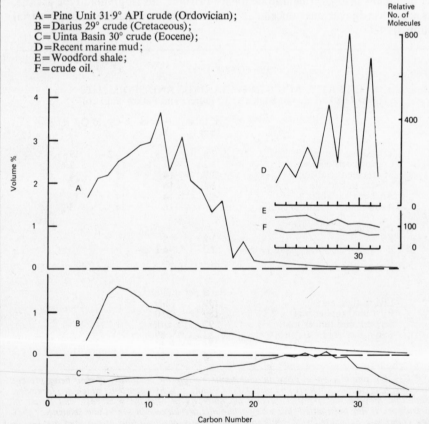

A=Pine Unit 31·9° API crude (Ordovician);
B=Darius 29° crude (Cretaceous);
C=Uinta Basin 30° crude (Eocene);
D=Recent marine mud;
E=Woodford shale;
F=crude oil.

Fig. 6. Distribution of *n*-alkanes in crude oils, and in extracts from a recent marine mud and an ancient shale [5,27].

Hentriacontane ($C_{31}H_{64}$) has been found in a brown alga, a red alga and several species of green algae, and pentacosane ($C_{25}H_{62}$) in a red alga. The plant waxes and seed lipids include non-isoprenoid branched-chain alkanes, as well as olefinic and acetylenic hydrocarbons. Sponges have lipids with substantial amounts of sterols, and solid hydrocarbons such as $C_{27}H_{56}$, $C_{29}H_{60}$ and $C_{31}H_{64}$ have been indicated.

Although there are instances of odd-carbon-number dominance, examples of no marked dominance have been recorded among the Porifera and Coelenterata. Thus values of 1·0 to 1·4 were found for the carbon-preference indices, using the C_{24} to C_{34} normal paraffins. The carbon-preference index was the mean of

$$\frac{\Sigma\ C_{25}\ \ldots\ C_{33}}{\Sigma\ C_{26}\ \ldots\ C_{34}}\quad \text{and} \quad \frac{\Sigma\ C_{25}\ \ldots\ C_{33}}{\Sigma\ C_{24}\ \ldots\ C_{32}}.$$

About 25% of the lipid material in the bacterium *Desulfovibrio desulfuricans,* some 1–2% of the cell weight, consists of saturated hydrocarbons, including a prominent series of normal paraffins in the C_{25} to C_{35} range, with a carbon-preference index of about 1·0.[19]

Algal lipids have sterols in their unsaponifiable fraction, including cholesterol, which was for a long time considered to be a typical product of animals. Sterols are steroid alcohols. They are optically active, and occur widely in Nature, either free or as esters of the higher aliphatic acids.

Triglycerides and waxes, especially cetyl palmitate, are present in the lipids of coelenterates, such as sea anemones, reef-building corals and gorgonias. It has been suggested that if the organic matter of reefs, built of corals, bryozoa or coralline algae, is not completely broken down or consumed by other organisms, the triglycerides and waxes might be hydrolysed, with conversion of the acids thus freed to their calcium salts[3].

The carotenoids are widespread as red and yellow pigments in Nature, one group, the carotenes, being hydrocarbons and the other group, the xanthophylls, having oxygen in the end parts of very similar molecules. The proportions of the major carotenoids in freshwater and marine algae range from about 2 to 50 mg/100g for ß-carotene, and from 8 to 550mg/100g for the xanthophylls. Carotenoids generally disappear comparatively quickly in sediments, the oldest occurrence of ß-carotene probably being about 20,000 years in a lake sediment, except for a case in which low temperatures associated with glaciation may have favoured preservation. Chlorophyll is a source of porphyrins in the sediments, and also one of the sources of the isoprenoids (Fig. 5)[37].

Ocean water has yielded organic extracts of 4·58 to 13·8mg/l. Fractionation showed the presence of at least 25 paraffin hydrocarbons including C_{10}, C_{12}, C_{14}, C_{16} and C_{18} up to at least C_{24}. Free fatty acids, triglycerides, diglycerides, monoglycerides, free sterols and sterol esters of fatty acids were also indicated to be present.

Small amounts of normal paraffins have been detected in lake and river water, in addition to sea-water; they have also been observed in oilfield brines (Table 16), as well as in water expressed from a lake mud by centrifuging. In the lake water, and possibly also in the other surface waters at least, there is some question about the precise way in which these substances occur.

TABLE 16

n-PARAFFIN CONTENTS OF WATERS
(*After* Peake and Hodgson)[32]

	Amount μg/l	Odd-carbon- number Preference
Surface Waters		
Lake: Cooking Lake (October)	3·8 (C_{20} to C_{33})	Strong
Cooking Lake (March)	0·3 ,,	Weak
Lake Wabamun	0·2 ,,	None
Sea: Victoria, B.C.	0·2 ,,	,,
Dartmouth, N.S.	< 0·2 ,,	,,
River: North Saskatchewan River	1·6 ,,	,,
Subsurface Waters		
Joffre oilfield (Nisku)	5·1 (C_{16} to C_{33})	None
Judy Creek oilfield (Beaverhill Lake)	14 ,,	,,
Countess gasfield (Bow Island)	1·1 ,,	,,

As much as 10ppm of benzene has been found in waters in oil-bearing horizons, and a lateral decrease in concentration with increased distance from the oil accumulation has been observed (2·6ppm adjacent to the accumulation to 0·9 ppm at a distance of 5 miles in one instance). Acetic acid has also been found in petroleum and in the associated waters. The quantities of fatty acids appear to fall off with increase in molecular size[54].

Although certain hydrocarbons have been detected in living organisms, and these are known to be present in petroleum, there are other components in petroleum which have not been found in such organisms. Naphthenes, the lower members of the aromatics, and asphaltic substances are in the latter category, while certain of the aliphatic hydrocarbons and acids fall into the former category. There has been a suggestion that petroleum is not strictly formed in the sediments, but that it is the result of a process which collects together hydrocarbons and closely-related substances which were formed in organisms. However, if the substances mentioned above do not occur in living organisms, they must be developed in the sediments from other organic matter. Nevertheless, this does not preclude the possibility of hydrocarbons produced by organisms in suitable settings contributing directly to petroleum, without necessarily being the sole source of those particular compounds[53].

Modern investigations

Numerous investigations had been made over many years on recent and ancient sediments in order to throw light on the origin of petroleum, and a considerable amount of information had been collected on some points. However, as new techniques became available in the early 1950's, there began a new and rapid advance in the knowledge of the organic matter in sediments apparently related to petroleum, as well as of some details concerning the composition of crude oil.

For example, a series of rock samples were taken from a well drilled offshore in the Gulf of Mexico, the samples coming from depths down to about 100ft below the sea floor. These samples were extracted using a mixture of solvents capable of dissolving hydrocarbons and other petroleum components. The

TABLE 17

COMPOSITION OF SOME CRUDE OILS IN TERMS OF THE MAIN HYDROCARBON GROUPS, RESINS AND ASPHALTENES
(*After* Sachanen, 1950)[36]

Type of crude	Source	Paraffins %	Naphthenes %	Aromatics %	Resins and asphaltenes %
Paraffinic		40	48	10	2
Paraffinic-naphthenic	Oklahoma City	36	45	14	5
Naphthenic	Emba-Dossor	12	75	10	3
Naphthenic-aromatic	Santa-Fe	20	45	23	12
Mixed asphaltic	Inglewood	8	42	27	23
Bermudez asphalt	Bermudez Lake	5	15	20	60

TABLE 18

RESULTS OF CHROMATOGRAPHIC ANALYSES OF ORGANIC SOLVENT EXTRACTS FROM RECENT SEDIMENTS
(*After* Smith, 1954)[54]

Source of samples	Depth below sea floor (ft)	Chromatographic analysis of extracted organic matter (%)			
		Paraffin-naphthene	Aromatic	Asphaltic	Remaining on alumina
7 miles off Grande Island, Louisiana	3–4	6·0	1·5	14·0	78·5
	18–22	17·9	2·5	12·1	67·5
	102–103	25·0	5·7	10·6	58·7

extracted material, amounting to about 0·031gm/100gm of dried rock, was then transferred to a column packed with alumina in order to fractionate it by means of chromatography. In this manner, paraffin-naphthene, aromatic and asphaltic fractions were eluted by the use of a series of progressively more polar solvents, although the major part of the extract remained on the alumina. The relative amounts of these fractions are given in Table 18 for rock samples taken at several different distances below the sea floor. The fractions eluted represent assemblages of substances which have, at least broadly, counterparts in crude oils (Table 17). Furthermore, if it is legitimate to assume that the organic matter originally included in the deposits was substantially the same in composition for each level, then there appears to be progressive development of paraffin-naphthenes and aromatics at the expense of the asphaltic material and that which remains on the alumina[44].

Similar studies followed on samples obtained from recent and ancient deposits from various environments. The details of the extraction techniques varied, as did the solvents, and the chromatographic columns differed in some cases. Hence, exact comparisons cannot be made between the various samples, but the findings had much in common, namely that hydrocarbons were detected as well as asphaltic (O-N-S) substances.

Tables 19 and 20 give data for recent deposits from marine and non-marine areas; Table 20 shows the results of investigations of a series of ancient deposits. The organic matter contents of the samples were commonly smaller for the marine deposits than for some of the non-marine deposits. The amount of the solvent extract from unit weight of sediment was small, and commonly very small; the amounts of hydrocarbons detected were correspondingly small. Both quantities showed considerable variation in value, the figures for the Broxburn Curly Oil Shale being unusually high[17].

The techniques and solvents employed in many of the studies automatically meant that low-boiling-point materials in the sediments would be entirely or largely lost in the recovery process, yet these are known to represent an important fraction of many crude oils. Consequently, special methods were developed in order to recover some at least of these substances. The results of one series of such investigations are presented in Table 21. Significant amounts of C_4 to C_8 hydrocarbons were detected in ancient sediments, and C_1, C_2 and C_3 hydrocarbons were known to be present, although they were not reported quantitatively because of losses in the procedure. With the exception of methane, these low-boiling-point hydrocarbons were not recorded in recent sediments[8].

In some investigations, the hydrocarbons have been further fractionated to segregate the normal paraffins, and the molecular distribution of these paraffins has been determined. For the C_{20} to C_{30} range, the normal paraffins from recent sediments commonly showed a distinct odd-carbon-number dominance, whereas extracts from ancient sediments and crude oils showed little or no dominance of the odd-carbon-number molecules over the even-carbon-number molecules (Fig. 6). The heavy n-paraffins from 77 samples of non-reservoir rock type recent sediments from the Gulf of Mexico had carbon-preference indices of 2·4 to 5·5; a series of 241 shales and mudstones of Miocene to Mississippian age had values of 0·7 to 4·5, but principally 0·9 to 2·5, and 40 crude oils had values of 0·9 to 1·15 (for carbon numbers of about 24 to 34). Nevertheless, recent sediment samples from the Cariaco trench showed no marked dominance[5].

It may further be noted that organic extracts from depths of 10 to 810ft below the sea floor on the upper continental slope of the northern part of the Gulf of Mexico had normal paraffins with carbon-preference indices of 0·85 to 1·39. Cores from the Mississippi cone area of the upper continental slope from depths of 150 to 768ft below the sea floor had, however, indices of 2·4 to 4·2. Perhaps organic matter contributions from terrestrial sources in the latter cores influenced the values, while the former group may have contained only marine organic matter. Furthermore, cores from progressively greater depths in a single hole in the first area all had carbon-preference indices of about 1·0 for the n-paraffins, but there was a shift of the peak value of the n-paraffin distribution to a lower molecular weight with increased depth, and the appearance of n-paraffins between C_{15} and C_{18}.[35]

The normal fatty acids recovered from rocks and recent sediments have also been examined for molecular distribution. Commonly the even-carbon-number molecules are dominant in recent sediments, and the dominance appears to diminish in older rocks (Table 22).

TABLE 19

ORGANIC MATTER AND HYDROCARBON CONTENTS OF RECENT DEPOSITS
(After P. V. Smith, 1954)[44]

			Chromatographic Analysis of Extract	
Non-marine Sample	*Hydrocarbon*	*Organic Matter*	*Paraffin-Naphthene*	*Aromatic*
	ppm	*%*	*%*	*%*
Stony Lake, N.J.	224	11·2	1·4	0·6
Lake Wapalanne, N.J.	105	2·23	1·6	3·1
Mirror Lake, N.J.	52	2·3	1·1	1·1
Marsh sample, N.J.	116	6·81	0·9	0·8
Peat, Minnesota	986	10·5	3·8	5·6

TABLE 20

ORGANIC EXTRACTS, HYDROCARBONS, AND CHROMATOGRAPHIC ANALYSES OF EXTRACTS FROM SOME RECENT AND ANCIENT SEDIMENTS
(After Bonnett, 1958)[17]

		Percentage of extract			*Hydro-carbons (ppm)*	*Organic extract as % of sediment*
		Paraffin-naphthenes	*Aromatics*	*Asphaltic*		
Recent	Silt ⎫ from Wash	18·7	19·1	25·1	201	0·0503
	Clay ⎭	15·2	—	13·6	36	0·0235
Cretaceous	Gault Clay	26·3	10·8	28·8	56	0·0151
Jurassic	Kimmeridge Clay	7·8	12·2	26·7	39	0·0197
	Oxford Clay	9·1	20·5	32·7	156	0·0529
Triassic	Rhaetic Shale	12·3	49·5	15·1	426	0·0689
Carboniferous	Coal Measure Shale (Caerphilly)	5·9	22·9	42·6	67	0·0233
	Broxburn Curly Oil Shale	47·7	11·0	12·4	3,290	0·583
	S₂ Shale	9·4	29·6	38·2	43	0·011

TABLE 21

LIGHT HYDROCARBONS IN SEDIMENTS
(After Dunton and Hunt, 1962)[8]

Hydrocarbon content in parts per million*

	C_4	C_5	C_6	C_7	C_8	*Cyclo-pentanes*	*Cyclo-hexanes*	*Total*	*% Organic matter*
B-7 (Eocene)	3·77	4·66	4·33	2·87	2·54	5·33	2·19	25·63	2·2
Lias (Jurassic)	55·70	87·80	66·74	80·25	134·60	129·04	37·67	591·67	11·44
Wolfcamp (Permian)	80·45	106·30	81·10	60·95	58·95	284·37	76·70	748·82	4·08
Woodford (Mississippian)	358·31	165·42	111·02	58·88	46·30	50·29	69·28	859·70	8·9
Salina (Silurian)	11·44	7·38	5·02	3·19	2·86	2·86	1·83	34·58	0·61

* *All of the samples contained C_1, C_2 and C_3 hydrocarbons, but these were not reported because of losses in the procedure.*

TABLE 22

NORMAL FATTY ACIDS AND NORMAL PARAFFINS IN EXTRACTS
FROM SEDIMENTS
(*Based on* Kvenvolden, 1966)[21]

	n-fatty acids		*n-paraffins*		*Paraffins* / Acids
	µg/g	CPI*	µg/g	CPI	
Modern					
Algal ooze 1500 yr.	570	6·57	35	5·19	0·06
Carbonate 2000 yr.	77	5·58	0·96	1·80	0·01
Mud 10,000 yr.	28	9·00	0·61	2·57	0·02
Mud 30,000 yr.	26	8·4	2·3	2·13	0·09
Ancient					
Repetto, Pliocene	1·9	3·18	1·5	1·16	0·79
Mowry, Cretaceous	32	1·06	43	1·10	1·34
Heath Shale, Mississippian	54	1·10	32	1·13	0·59
Chattanooga, Devonian	21	1·23	52	1·05	2·48

* *CPI=Carbon-preference index.*

TABLE 23

ORGANIC CARBON AND HYDROCARBONS IN SEDIMENTS
(*Based on* G. T. Philippi, 1965)[33]

Sediment	*Organic carbon* %	*Hydrocarbons boiling above 325°C* ppm	*Hydrocarbons* / Organic carbon
Recent	0·3–3·4	20–140	0·0014–0·0137
Ancient: Los Angeles Basin*	0·53–5·8	67–3032	0·009–0·113
Ventura Basin**	0·76–4·14	116–1215	0·015–0·055
Other areas ***	0·53–3·67	267–2360	0·034–0·101

* *Pliocene and Miocene.* ** *Pleistocene, Pliocene and Miocene.* *** *Miocene to Ordovician.*

As noted earlier, there are exceptions to what seem to be general patterns; early conditions, including the nature of the original organic matter, may account for some of them. *n*-Fatty acids and *n*-paraffins in the C_{23} to C_{33} range from the Skull Creek and some younger rocks alone showed even- or odd-carbon-number dominance, respectively. A Mowry shale extract had *n*-paraffins with slight even-carbon-number dominance in the C_{20} to C_{26} range and odd dominance in the C_{29} to C_{33} range; a Big Muddy crude oil had *n*-paraffins with even dominance in the C_{20} to C_{24} range, and odd-carbon-number-dominance in the C_{27} to C_{33} range[21, 22].

The determinations of the normal paraffin and normal fatty acid contents of extracts from rocks have permitted the calculation of their ratios (Table 22). It is evident that these ratios vary widely, and that for recent sediments the ratios are decidedly lower than for ancient sediments.

Comparisons have also been made between the amounts of hydrocarbons boiling above 325°C and the organic carbon content of the rocks from which these hydrocarbons were recovered. Table 23 shows that the ratio of such hydrocarbons to the organic carbon content is generally markedly lower for recent than for ancient deposits. In the Los Angeles Basin and the Ventura Basin this ratio increased markedly below 10,000ft and 14,000ft, respectively.

In investigations on the sediments and crude oils in the Los Angeles and Ventura Basins, it was found that there were changes with depth in the distribution of naphthenes with various numbers of rings per molecule, in the 325–370°, 370–420° and 420–470°C fractions of the *iso*paraffin-naphthene concentrate, changes which were similar to those observed in the associated crude oils[33].

The Toarcian shales of the Lias in the Paris Basin are generally 10m to 20m thick, of fairly constant mineralogical composition, and with indications of relatively uniform conditions of deposition over a substantial area, leading to the assumption that the original organic matter was probably reasonably uniform. Their age is about 180×10^6 years, and geological data point to maximum depths of burial of 400m to 2540m. Examination of the organic extracts from samples which had been buried to different maximum depths showed the quantity of naphthenes with 4, 5 and 6 rings to be high in the shales which had not been deeply buried, but it was smaller in the samples which had suffered deeper burial. There was a diminution in optical activity in the deeper samples, for this phenomenon is associated with the more complex structures of the above type. As the depth of the maximum burial increased, so did the relative amounts of alkanes and monocyclic naphthenes. The naphthene-aromatic molecules, i.e. those which have at least one aromatic ring associated with naphthene rings, showed an increase in the ratio of aromatic to naphthene rings with increase in the maximum depth of burial[46].

The four- and five-ring naphthenes in the C_{27}–C_{29} range may be linked with the steranes and triterpanes.

On progressively deeper burial, the light hydrocarbons were found to increase relatively more rapidly than the medium and heavy hydrocarbons. Hydrocarbons with fewer than fifteen carbon atoms were very scarce at shallow depths, but could be as much as a third of the total at a depth of 2500m. It was recognised that methane. and to a less extent other gaseous hydrocarbons, could be lost during the coring, handling and storage of the samples.

The pristane/phytane ratio was well under unity at small depths of burial, and markedly over unity for the deeper samples: the n-C_{17} and n-C_{18} paraffins were decidedly less abundant than the isoprenoids at small depths of burial, but more abundant than the isoprenoids in the deepest samples. Generation of the normal paraffins must account for some of these features.

Generation of petroleum components

On the preceding pages attention has been drawn to a number of points:

(a) The composition and some other characteristics of crude oils and natural gas.

(b) Some of the main types of organic substances in organisms.

(c) The nature of the organic substances which have been detected in recent sediments, sea-water and in ancient sediments.

(d) Differences in the nature and relative amounts of the substances present as the rocks are older or have been more deeply buried.

(e) Assumptions which have to be adopted when attempts are made to explain certain patterns of occurrence of the organic compounds, and in particular to deal with the problem of the origin of petroleum.

One of the important tenets of geology is that "the present is the key to the past". This applies especially to processes, even though it may apply more accurately to the type of process than to the rate, which may well have varied in some cases between very wide limits over geological time, while its action currently may sometimes be inferred without necessarily being observable. In addition, although the same processes may be proceeding at different periods the materials being affected may show some differences.

Organisms are known to differ in many respects and also to have evolved. They may, however, be made up broadly of the same types of compounds in spite of evolution with, however, possible differences in the proportions or complexity of these types. Yet the basic units from which these compounds are constructed are likely to be the same now as in the very distant past. If this postulate is true, then the organic matter included in the sediments of given environments over time may well show a considerable degree of uniformity in broad terms. Transformations, which are likely to take note of the basic units involved in the construction of the various compounds, may therefore be expected to yield products which, so far as the dominant types of compound are concerned, are generally similar, in spite of organic evolution.

It was argued earlier, that the formation of petroleum called for something more than the collection together of substances formed by living organisms; and that it involved changes of greater or lesser extent in substances originally formed by such organisms. Hence, there must be an agency or a series of agencies for effecting the changes. Three have commonly been considered: bacteria or similar minute organisms, radioactivity, and heat. Because petroleum characteristically occurs in rocks not associated with igneous activity, and which do not show signs of the baking caused by igneous activity, it has been concluded that if heat is the agent, high temperatures are not required and that the necessary temperatures are attained in the course of burial of the rocks.

Bacteria

There seems little doubt that bacteria play some part in the formation of petroleum, but their role probably does not extend beyond modifying the organic matter at the time of, or soon after, its inclusion in the sediments, with the addition of their own products, including body substances. The biomass of bacteria in soils, sediments and water ranges from a very small percentage to 15% or more of the organic matter present. There are indications that anaerobic bacteria may take part in the production of certain aromatic hydrocarbons in recent marine sediments, these hydrocarbons most commonly being detected at depths in the sediments exceeding a few feet.

Radioactivity[15, 24]

Low-level radioactivity is widespread in sedimentary rocks. Clays and shales normally show a higher degree of radioactivity than do sandstones and limestones. The general association of this higher degree of radioactivity with the type of rock usually best endowed with organic matter is at least favourable for the significant action of radioactivity as a transforming agent. Experiments have been made on the effects of radioactivity, on fatty acids, for example. The products included hydrogen, carbon dioxide, carbon monoxide and methane, as well as, in the case of palmitic acid, a liquid similar in properties to *n*-pentadecane. This last compound would, with carbon dioxide, be expected to result from decarboxylation[52].

Calculations based on observed rates of transformation in the laboratory and the radioactive and assumed initial organic contents of a Palaeozoic formation, have led to the extremely tentative conclusion that the rates of hydrocarbon generation in the rock indicated that the amounts of oil formed might attain a level thought to be sufficient to result in an oilfield.

The intensity of radioactivity would decline over time, and hence, the transformation activity would fall off as time passed. As will become apparent later, the opposite is to be expected for thermal transformation.

Heat

Measurements made in wells show that the temperature increases with depth, and that the mean rate of increase changes significantly from one area to another. For a typical series of wells of substantial depths the gradient ranged from 0·44 °C/ft to 1·0 °C/ft, and these are not the limiting values.

The time rate at which the temperature of a layer of rock increases will be a function of the rate at which it is being buried and of the local rate of heat flow. Mean rates of burial vary widely (Table 24), and hence the times required for a layer of rock to attain a certain temperature can vary very widely. Furthermore, because of the complexities in the geological histories of some areas—burial, uplift and erosion, a further phase of sedimentation, etc.—the time-temperature history of a given layer of rock and its contained organic matter can be far from simple, while its present temperature need not be the highest temperature to which it has been subjected.

TABLE 24

RATES OF SEDIMENTATION

	Thickness (Ft)	Rate (cm/100 years)
Gulf Coast Area, USA		
Recent plus Pleistocene	8,000	12·2 (maximum)
Pliocene	6,000+	1·8+(maximum)
Miocene	40,000+	8·1+(maximum)
Oligocene	16,000+	4·9+(maximum)
Eocene plus Paleocene	24,000	2·0 (maximum)
Cretaceous	13,500	1·0 (maximum)
Other Areas		
Pliocene (Los Angeles Basin)		2·25+
Mesozoic (Dukhan)		0·25

Many experiments have been made on the effect of heating organic matter, simple or complex, or heating rocks containing organic matter, with a view to throwing light on the origin of petroleum. The principal parameters varied have been temperature and time, while pressure has been generated when heating has taken place in sealed capsules. In some cases, when single compounds have been heated, the effects of the presence of clay minerals and/or water have also been investigated. Indeed, if the intention is in some respects

TABLE 25

FORMATION OF HYDROCARBONS, RESINS AND
ASPHALTENES BY HEATING LOWER TOARCIAN
SAMPLES FROM THE PARIS BASIN
(*After* B. Durand)[47]

Time	Amount formed (*mg/g of original organic carbon*)		
Days	180°C	200°C	220°C
1	5·9	6·8	9·2
10	6·6	9·2	14·0
90	10·4	12·2	30·0
270	12·8	21·2	36·2

to match natural conditions, the presence of inorganic mineral matter and liquid water is essential.

In view of the obvious inability to match geological time, relationships between rate of reaction, time and temperature have been sought. Starting with some of the earliest work, it has been customary to consider the process as being dominated by a first-order reaction in which:[25]

Rate = collision frequency × energy factor × probability factor.

Particle size, concentration and speed determine the collision frequency, and it is evident that particle weight and the temperature affect two of those factors. The energy factor also depends on the temperature; there is, moreover, dependence on the activation energy E, which is a measure of the proportion of the collisions involving sufficient energy for the reaction to take place. The fraction is given by $e^{-E/RT}$, where $e = 2·817$, $R = 1·986$ and T is the absolute temperature. The proportion of the collisions for which the particles have the appropriate orientation determines the probability factor. In mathematical terms, the rate is given by: $dc/dt = -kc = -cAe^{-E/RT}$, where c is the concentration of the parent substance, and A is the frequency constant. The rate decreases with increase in the value of E; it increases with increase in the temperature.

The generalised findings in the experimental work have been that at a given temperature the amount of petroleum-like substances formed increased with the length of time of heating, whereas for fixed periods of heating, the amount formed increased with increase in the temperature used (Table 25). Clearly, if the natural process is dependent on heat provided by progressive burial of the rock with its organic matter, the rate of transformation will accelerate as burial proceeds and the temperature rises, provided that the reduction in the concentration of the parent substance does not offset the burial factor.

Catalysis

Behenic acid ($C_{21}H_{43}$ COOH) has been heated in the presence of clay (kaolinite) and in the presence of clay and water for lengths of time as much as 1848 hours, at temperatures in the range 200–300°C. Low-molecular-weight hydrocarbons were formed, as well as normal paraffins in the C_{15} to C_{34} range and fatty acids in the C_{15} to C_{24} range. There were branched-chain and unsaturated hydrocarbons, but the main *n*-paraffin formed was $C_{21}H_{44}$, a compound to be expected via a simple decarboxylation process. For reactions which have not gone to completion, it is to be expected that the compounds found, and their amounts, will depend on the stage at which the experiment is stopped in order to sample the products. For the experiments described above, the absolute quantity of unsaturated compounds detected generally decreased with increase in the time of heating. Hence, these compounds appear to be intermediates[9].

When the behenic acid was heated at 275°C for 330 hours in the presence of the clay, the yields of the individual hydrocarbons ethane to pentane were three to six times greater than in the absence of clay. Clay influenced the reaction, even in the presence of large amounts of water. For the uncatalysed reactions, calculations of the rates of reaction for 100°C yielded values which were very low, and were considered to be geologically unacceptable for the formation of petroleum. Unfortunately, for the clay-catalysed reactions it did not prove possible to assign probable values to the kinetic parameters needed for making projections[18].

In other experiments with behenic acid, Ca-montmorillonite, temperatures of 200 and 250°C, and times of 50 to 500 hours were employed. Most of the acid was converted to an insoluble kerogen-like substance, less than 1% of the acid consumed appearing as n-paraffins. $C_{21}H_{44}$ was the principal n-paraffin formed, and its amount passed through a maximum at about 50–150 hours. The analytical technique permitted the recognition of other n-paraffins in the C_{10} to C_{20} range. When powdered calcite was used instead of the clay, the most prominent n-paraffin produced was $C_{20}H_{42}$. Heating n-$C_{18}H_{37}COOH$ with powdered calcite gave $C_{17}H_{36}$ as the main n-paraffin produced. In both cases, as for the clay, there were also larger and smaller n-paraffins in the products[39].

Other experimental work

It has been found possible to break down ß-carotene in dilute solution in hydrocarbon solvents, rapidly at 188°C, more slowly at 150°C, and in several months at 110°C. Heating ß-carotene dispersed in water-wet mud also led to its degradation. The products, a yield of about 5%, comprised toluene, m-xylene, 2,6-dimethylnaphthalene and ionene[11].

The amino acid, phenylalanine, is broken down in aqueous solution on heating, to give small amounts of benzene, toluene and vinyl-benzene. The last of these compounds could be an intermediate in the formation of ethyl-benzene, a characteristic petroleum compound. Examination of the formulae for other amino acids—alanine, valine, leucine, *iso*leucine, *D*-amino acid and α-amino butyric acid—suggests that cleavages comparable with those indicated for phenylalanine would yield hydrocarbons, among them methane, ethane, propane, n- and *iso*-butane, and n- and *iso*-pentane, in addition to the aromatics listed for phenylalanine. Bacteria and other organisms form *D*-amino acid, while acid hydrolysis of proteins would yield the other amino acids.

It is of interest to note that pure alanine in a dilute solution had a half-life of 2×10^7 sec when heated at 190°C, but when heated at the same temperature as part of the protein of algae the half-life was 10^6 sec. Evidently the behaviour of a simple system can be an imperfect guide to the behaviour of a more complex system. Catalysts generally act by providing a reaction path of lower activation energy than that for the uncatalysed reaction.

Some thirty years ago, a study of petroleum occurrences led to the suggestion that a minimum temperature of 65°C, burial to 5000ft and a time of 10^6 years were required for the formation of petroleum. However, this conclusion was undoubtedly based on the occurrence of oilfields, which call not only for the formation of the appropriate compounds but also for their segregation in a trap. In addition, it does not show which are the controlling factors: the time, the temperature or the pressures (perhaps 2000–5000 psi) associated with burial to 5000ft. The depth of burial automatically means that the other factors must have values of the order of magnitude quoted, and it may well be that so far as the generation of the components of petroleum is concerned only the temperature-time function is important[7].

It has been suggested that most petroleum was formed at temperatures well below 100°C, although it has been claimed that in the Los Angeles Basin and the Ventura Basin the bulk of the oil was generated above 115°C. Even though the second statement may be valid, it does not necessarily mean that the temperature in question was necessary for the formation of the accumulated oil.

The tendency for even-carbon-number dominance in the fatty acids in organisms would cause a simple decarboxylation reaction to give odd-carbon-number paraffins in the rocks and crude oils. Such a feature is observed in the younger rocks and crude oils, but disappears in the heavier molecules with increased age. This suggests that something more than simple decarboxylation is involved, and, indeed, some of the experimental work shows that there are other reactions. Various reaction schemes have been proposed to explain the changes in the distribution of the normal paraffins and normal fatty acids, and their development from a limited number of parental fatty acids. One detailed scheme is shown on p. 49.

Further changes[38]

Many years ago, it was argued that with increasing age and temperature, hydrocarbons would change so as to yield a heavier and a lighter group, ultimately leaving only a gas, methane, and a solid natural petroleum, coke. In recent years, investigations have been made on the composition of the gas from fine-grained cuttings and that from reservoirs in Eastern Alberta. This showed a marked increase in the proportions of ethane, propane and butane relative to methane in going downwards from levels at which the temperature was about 40°C. In the Devonian of Western Canada, comparable investigations on similar rocks revealed a rapid decrease in the proportions of the same hydrocarbons relative to methane at points below those with temperatures of about 90°C, to leave little more than methane in the gas at greater depths. It appeared that these rocks could have had maximum temperatures some 55–80°C higher than the present values. Devonian rocks in a well in the Northwest Territories showed similar features for the gas to those observed in the Devonian of Western Canada, and at the same time the amounts of C_4 to C_7 hydrocarbons fell off, as did the amounts of $C_{15}+$ hydrocarbons. It was suggested, therefore, that immature rocks have dry gas and asphaltic or naphthenic oils, the quantities being small relative to the organic matter

---

REACTION SCHEME[18]

Key: **Original material in bold type**
Intermediates in italic type
Final products in roman type.

RCOOH $\longrightarrow$ $*R + CO_2 + *H$

$*R +$ **RCOOH** $\longrightarrow$ $RH + *(R\text{-}H)COOH$

$*(R\text{-}H)COOH$ $\longrightarrow$ $R_{\overline{1}}^{=} + *R_2COOH$
$\searrow$ $*R_3 + R_{\overline{4}}^{=} COOH$

$*R_3 +$ **RCOOH** $\longrightarrow$ $R_3H + *(R\text{-}H)COOH$

$*R + R_{\overline{1}}^{=}$ $\longrightarrow$ $*(R\text{-}H + R_1)$

$*(R\text{-}H + R_1) +$ **RCOOH** $\rightarrow$ $(R\text{-}H + R_1)H + *(R\text{-}H)COOH$

$R_{\overline{1}}^{=} + *R_3$ $\longrightarrow$ $*(R_1 + R_3)$

$*(R_1 + R_3) +$ **RCOOH** $\rightarrow$ $(R_1 + R_3)H + *(R\text{-}H)COOH$

$*R + *R_3$ $\longrightarrow$ $(R + R_3)$

$*R_2COOH +$ **RCOOH** $\longrightarrow$ $R_2HCOOH + *(R\text{-}H)COOH$

$*R + R_{\overline{4}}^{=} COOH$ $\longrightarrow$ $*(R + R_4)COOH$

$*(R + R_4)COOH +$ **RCOOH** $\longrightarrow$ $(R + R_4)HCOOH + *(R\text{-}H)COOH$

$*R + *R_2COOH$ $\longrightarrow$ $(R + R_2)COOH$

$*H + *R_3$ $\longrightarrow$ R_3H

$*H + *R$ $\longrightarrow$ RH

$*H + *R_2COOH$ $\longrightarrow$ R_2HCOOH

$*H + *(R + R_4)COOH$ $\rightarrow$ $(R + R_4)HCOOH$

$R_{\overline{1}}^{=}$ olefin

$R_{\overline{4}}^{=} COOH$ unsaturated acid

$*R$ radical

$*(R\text{-}H)COOH$ secondary radical, i.e. R has lost one hydrogen atom (two valencies have to be satisfied for R-H; only one for R).

---

content. At 30° to 60°C, labile molecules in reservoirs and shales began to break down to give wet gas and gasoline, while $C_{15}+$ saturated hydrocarbons continued to form. A less dense and more paraffinic petroleum develops as the temperature is further increased and condensates appear. However, at about 180°C the changes cease to be beneficial from the point of view of oil, and crude oil components begin to move rapidly to the lowest free-energy state, namely methane and graphite[12, 45].

Support for the earlier stages of this series of changes is provided by the studies of the hydrocarbons in the Toarcian shale of the Paris Basin. Fig. 7 summarises in simple form the overall behaviour, while Fig.8 supplements it in some ways and draws attention to a number of important points. Fig. 7, so far as depths are concerned, refers to a particular c. . The diagram can be considered to represent:[46, 47]

(a) the behaviour of a specific sample of source rock as it became more deeply buried, showing the hydrocarbons "associated" with it during its burial history, or

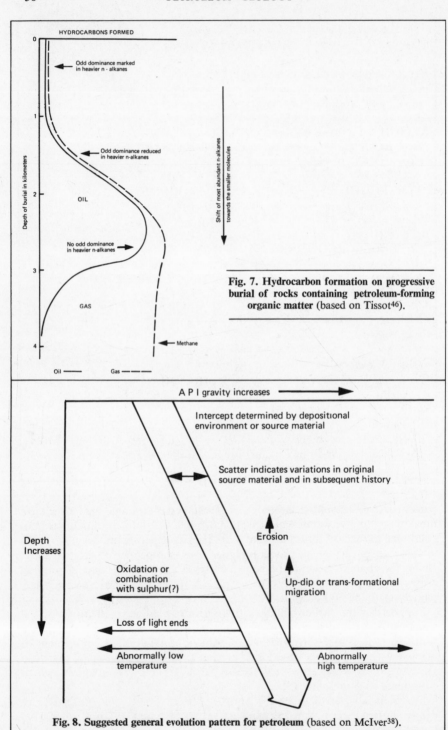

Fig. 7. **Hydrocarbon formation on progressive burial of rocks containing petroleum-forming organic matter** (based on Tissot[46]).

Fig. 8. **Suggested general evolution pattern for petroleum** (based on McIver[38]).

(b) the vertical distribution of the hydrocarbons "associated" with a series of source rock horizons in a thick series.

In the latter case, however, vertical variations in the amount of organic matter suitable for yielding oil would affect the detailed form of the oil curve as would variations in any catalytic agent associated with the organic matter. The detailed burial history and the integrated temperature-time function would affect the shape of the curve in both cases. The form of the curve is such as to suggest a depth range over which oil will be developed and appear in significant amounts, its base being marked by an oil "extinction" zone.

Changes in the organic matter in the rocks depend on the temperature-time function. The maximum temperature reached, although important, does not alone determine the composition of the products developed. Young rocks rapidly buried to considerable depths in an area of high overall temperatures could yield a more mature hydrocarbon mixture than the same kind of parental organic matter of much greater age buried more slowly and to a lesser maximum depth in an area of low overall temperatures. For areas of comparable temperature distribution characteristics, and for rocks in the same depth range, a thick sequence of limited age range would have a thicker "main oil zone" than a sequence of similar thickness yet wider age range and including substantially older deposits. In the second case, the bottom dry gas zone ("oil extinction zone") would be shallower (relative to the top of the "main oil zone"). As was observed in the Toarcian of the Paris Basin, a rock sequence of given age and similar constitution, buried to different depths in a major area, would normally have the more mature indigenous hydrocarbons where the depth of burial was greatest, in the absence of locally exceptional temperatures. Nevertheless, irregular basinal subsidence could lead to exceptions; rocks of a given limited age which have suffered comparatively little burial over a long period and are then quickly buried to a greater depth than equivalent rocks which had become deeply buried at a relatively early stage, although not to as great a depth as the first-mentioned rocks[46].

When rocks have been eroded down to the dry gas (oil extinction) levels and then re-buried it would appear that they cannot be expected to generate more oil, whatever the new depth reached. Hence, in some cases, major unconformities can impose limits on the vertical distribution of the heavier hydrocarbons in source rocks.

When large molecules are disrupted, there is redistribution of hydrogen. The formation of a light saturated molecule leads to ring structures or unsaturated bonds in the remainder of the parent molecule. Thus, aromatics can be a product, while polymerisation may yield asphaltenes. The simultaneous formation of low-molecular-weight hydrocarbons and complex products such as asphaltenes may be considered to be a normal feature as petroleum increases in age and maturity. Light paraffins and asphaltenes are incompatible in solution and eventually the asphaltene solubility limits are exceeded, leading to the precipitation of the asphaltenes. Since increase in burial and hence reservoir pressure will force more gas into solution, this also can cause precipitation of the asphaltenes. Some of this gas can have resulted from the disruption process. Evidently, the characteristics of the crude oil are dependent on a complex set of mechanisms[10, 41].

Source rock potential

Irrespective of the correctness of the values for the temperatures suggested as being required for the significant formation of petroleum or for the marked degradation of petroleum to leave gas only, there is a strong interest in obtaining guides to these conditions. Although some of the data may effectively be of the nature of the results of a post-mortem examination, there is always the hope of being able to obtain information which will (a) avoid the waste of money and effort, or (b) indicate where effort has a good prospect of leading to success. The cost of drilling exploratory wells is high, and hence it is important to extract the maximum amount of information from such wells.

A variety of techniques have been employed for trying to assess the capacity of rocks to generate petroleum, or to indicate whether conditions have reduced the chances of finding oil, but left prospects for finding gas. (It should be noted, however, that there are reasons for believing that natural gas accumulations do not always originate in the same way). Some aspects of the problem can be linked with the idea of early or "eometamorphism", without getting into the realm of the metamorphic rocks. Other aspects are concerned with depositional conditions, including the amount and nature of the organic matter incorporated in the rocks.

Many years ago, it was suggested that in the Pennsylvania area of USA the "rank" of the coals, as measured by the fixed-carbon content (on a dry, ash-free basis), was a guide to the occurrence of heavy oils, light oils or only natural gas. When lines of equal fixed-carbon content (isocarbs) were drawn on a map, they had a northeast-southwest trend, parallel to the trends of the folds in the Appalachians, and their values increased to the southeast, i.e. towards the Appalachians. The oil generally decreased in density to the southeast, where there were gas accumulations only. It was argued that the increase in fixed-carbon content was an index of the degree of metamorphism of the coals, and that this increased in going towards the zone of Appalachian folding. The petroliferous horizons were inferred to have been affected by comparable conditions[51].

This may be an over-simplification of the picture, which is in any case somewhat generalised, and features other than the modification of the petroliferous content of the rocks may well be involved. It should also be noted that Hilt's Law, which relates to coals, states that the fixed-carbon content of coals increases, with simultaneous decrease in the volatile content of the coals, as the depth of burial increases. The claimed relationship between the fixed-carbon content of coals and the quality of oil in associated strata was referred to as the carbon ratio theory.

The standard technique for measuring the fixed-carbon content of a coal involves heating the coal, after drying, at a high temperature under precisely controlled conditions. However, coals are complex substances: they are made up of several different main organic components (macerals) whose proportions vary from one coal to another, and whose responses to burial differ. Thus, for example, some German coal samples had vitrinite contents of 47·2% to 86·8%, exinite contents of 4·9% to 24·3%, micrinite contents of 0·9% to 5·1% and inertinite contents of 3·0% to 27·6%, in terms of volume, on an ash-free basis. These are not the limits known for coals. The fixed carbon contents of the macerals of one coal were 31%, 63% and 77%, for exinite, vitrinite and micrinite, respectively, whereas for another coal the corresponding values were 75%, 74% and 83%. Each maceral showed an increase in fixed-carbon content with increase in coal rank. Almost certainly variations in the proportions of the macerals account for the changes in fixed-carbon contents in a vertical sequence of coals in a borehole sometimes failing to agree strictly with Hilt's Law. There are, in addition, other complicating factors. The influence on the fixed-carbon content of differences in the proportions of the macerals diminishes with increase in coal rank[6].

In using coals as a measure of the alteration of the organic matter in sediments, it is necessary, therefore, to consider the behaviour of a single maceral, vitrinite. Relationships have been developed between its reflectivity and its fixed-carbon content. Measurements of reflectivity can be made on small particles of the maceral, and where this is available in the sediments it has been extensively used as a guide to the degree of alteration of any associated organic matter and in particular to the depth of burial or the temperature attained.

The ratio of the residual (C_R) and total organic carbon (C_T) has also been employed as a measure of the alteration. In essence, this is equivalent to the use of the fixed carbon in coals, but can be applied to rocks containing little organic matter. First, the soluble organic matter is removed, then on part of the sample the total organic carbon is determined by combustion. The remainder is heated as for obtaining the fixed carbon of coals, after which its carbon content (residual carbon) is determined by combustion.

Spores and pollens in the Toarcian shale showed a progressive change in colour from light yellow to dark brown as the depth of burial increased. In Canada, similar colour darkening of the organic matter in Cretaceous, Mississippian and Devonian rocks was observed in each formation with increase in the depth of burial, whilst for single wells there was progressive darkening with depth, a consequence of the greater age and the attainment of higher temperatures by the deeper samples. The colour of the cuticular plant débris in the organic matter was used to build up a thermal-alteration index, the nature of any associated hydrocarbons being noted. In brief, yellow spores and kerogen were associated with dry gas and asphaltic or naphthenic oil (10–35° API); yellow to brown kerogen with wet gas and paraffinic oil; black kerogen or graphite with dry gas, hydrogen sulphide and carbon dioxide.

Yet another approach depends on changes in the clay minerals in the sediments. This seems to be of more limited value[45, 46].

Finally, measurements of electron-spin resonance (ESR) have been made on the organic matter in rocks, and interpreted in terms of temperatures. Disruption of the kerogenous material on heating creates free radicals. Unpaired electrons are preserved, and there are more unpaired electrons the higher the temperature to which the organic matter has been subjected. In the US Gulf Coast region, it has been possible to build up relationships between the ESR data and the subsurface temperatures, in areas where the geological indications are that the present temperature is likely to be the maximum temperature to which the rocks have been heated. Measurements on other samples can then be interpreted in terms of maximum palaeotemperatures. Obviously, the opportunities for building up suitable calibration curves tend to diminish as older and older rocks are involved, and for them inferences about former depths of burial and temperatures may become necessary in order to build up a depth-temperature relationship. Again, some measure of basic uniformity of the organic substance incorporated in rocks over periods of hundreds of millions of years seems to be a necessary assumption.

Although for some degree of analytical or processing simplicity short-cuts to indicative parameters have attractions, the more tedious and elaborate, and seemingly more definite determinations of hydrocarbons, etc., have value, even though they must measure what is still in the source rocks, not what has moved out to form petroleum accumulations. With certain possible limitations, the nature of the source-rock petroleum-like components may be some guide to the present condition of any accumulated oil in areas where the wells have, for one reason or another, not penetrated the relevant rocks at a position where a potential trap for hydrocarbons exists, with an appropriate geological history.

It was noted earlier, that in detail what is recovered from the rocks by means of solvents depends on the analytical technique employed. Both the amounts and the detailed make-up may be affected, and for the very light components there can also be losses both before and during the laboratory measurements. Hence, statements about the "richness" of source rocks must involve quantities which are to some extent a function of the analytical details. At the same time, it is necessary to bear in mind the possibility of contamination under certain circumstances. Hence, the data *as a whole* will commonly be of more value than if attention is concentrated upon a single parameter. Measurements of the organic carbon content of source rock samples and of the content of the heavier "hydrocarbons", present fewer manipulative and, in some senses, analytical difficulties than do determinations of the light hydrocarbons. While the ratio of the solvent extract to the organic carbon content seems to have a characteristic range of values for typical source rocks, it appears that the total organic extract should exceed several hundred parts per million. However, if the latter is high absolutely and also relative to the organic carbon, then contamination may be suspected. The hydrocarbon content of the extract may be expected to increase with advance in maturity of the source rocks, as evidenced by the Toarcian shale studies quoted and other data, and again high values in some circumstances may indicate contamination. Rocks classed as "good source rocks" commonly have from about 1% to less than 10% of organic carbon.

REFERENCES

1. P. H. Abelson, "Organic Geochemistry and the Formation of Petroleum", *Proc. 6th World Petroleum Congress,* Frankfurt, 1963, Section 1, pp. 397–407.
2. W. Bergman, "Geochemistry of Lipids", in *Organic Geochemistry* (Ed., I. A. Breger). Pergamon: Oxford, 1963, pp. 503–542.
3. W. Bergman and D. Lester, "The Occurrence of Cetyl Palmitate in Corals", *J. Org. Chem.,* **6**, 120–122 (1941).
4. M. Berthelot, "Sur l'Origine de Carbures et des Combustibles Minéraux". *Compt. Rend.,* **63**, 949–951 (1866).
5. E. B. Bray and E. D. Evans, "Hydrocarbons in Non-reservoir Rock Source Beds", *Bull. Amer. Assoc. Petrol. Geologists,* **49** (3), 246–257 (1965).

6. U. Colombo, F. Gazzarrini, R. Gonfiantini, G. Kneuper, M. Teichmuller and R. Teichmuller, "Carbon Isotope Study on Methane from German Coal Deposits", in *Advances in Organic Geochemistry*, 1966 (Eds. G. D. Hobson and G. C. Speers). Pergamon: Oxford, 1970, pp. 1–26.
7. B. B. Cox, "Transformation of Organic Material into Petroleum Under Geological Conditions", *Bull. Amer. Assoc. Petrol. Geologists*, 30 (5), 645–659 (1946).
8. M. L. Dunton and J. M. Hunt, "Distribution of Low Molecular Weight Hydrocarbons in Recent and Ancient Sediments", *Bull. Amer. Assoc. Petrol. Geologists*, 46 (12), 2246–2248 (1962).
9. E. Eisma and J. W. Jurg, "Fundamental Aspects of the Generation of Petroleum", in *Organic Geochemistry* (Eds., G. Eglinton and M. T. J. Murphy). Longman, Springer-Verlag. 1969, pp. 676–698.
10. J. G. Erdman, "Geochemistry of the High Molecular Weight Non-hydrocarbon Fraction of Petroleum", in *Advances in Organic Geochemistry* (Eds., U. Colombo and G. D. Hobson). Pergamon: Oxford, 1964), pp. 215–237.
11. J. G. Erdman, "Petroleum—Its Origin in the Earth", in *Fluids in Sub-surface Environments* (Eds., A. Young and J. E. Galley). Amer. Assoc. Petrol. Geologists: Tulsa, Oklahoma, U.S.A., 1965, pp. 20–52.
12. C. R. Evans, M. A. Rogers and N. J. L. Bailey, "Evolution and alteration of petroleum in Western Canada", *Chemical Geology*, 8 (3), 147–170 (1971).
13. H. D. Hedberg. "Geologic Aspects of Origin of Petroleum", *Bull. Amer. Assoc. Petrol. Geologists*, 48 (11), 1755–1803 (1964).
14. H. D. Hedberg, "Significance of High-wax Oils with Respect to Genesis of Petroleum", *Bull. Amer. Assoc. Petrol. Geologists*, 52, (5), 736–750 (1968).
15. G. D. Hobson, *Some Fundamentals of Petroleum Geology*. Oxford University Press: London, 1954.
16. G. D. Hobson, "The Organic Geochemistry of Petroleum", *Earth Science Reviews*, 2, 257–276 (1966).
17. G. D. Hobson, "Oil and Gas Accumulations, and Some Allied Deposits", in *Fundamental Aspects of Petroleum Geochemistry* (Eds., B. Nagy and U. Colombo). Elsevier: Amsterdam, 1967, pp. 1–36.
18. J. W. Jurg "On the Mechanisms of the Generation of Petroleum". Thesis (1967).
19. C. B. Koons, G. W. Jamieson and L. S. Cieresko, "Normal Alkane Distributions in Marine Organisms: Possible Significance in Petroleum Origin", *Bull. Amer. Assoc. Petrol. Geologists*, 49 (3), 301–304 (1965).
20. K. A. Kvenvolden, "Normal Paraffins in Sediments from San Francisco Bay, California", *Bull. Amer. Assoc. Petrol. Geologists*, 46 (9), 1643–1652 (1962).
21. K. A. Kvenvolden, "Molecular Distributions of Normal Fatty Acids and Paraffins in Some Lower Cretaceous Sediments", *Nature*, 209, 573–577 (1966).
22. K. A. Kvenvolden and D. Weiser, "A Mathematical Model of a Geochemical Process: Normal Paraffin Formation from Normal Fatty Acids", *Geochim. et Cosmochim. Acta*, 37 (8), 1281–1309 (1967).
23. R. F. Leo and P. L. Parker, "Branched-chain Fatty Acids in Sediments", *Science*, 152, 649–650 (1966).
24. S. C. Lind "On the Origin of Petroleum", *The Science of Petroleum* (Eds., A. E. Dunstan *et al.*). Oxford University Press, 1938, Vol. 1, pp. 39–41.
25. C. G. Maier and S. R. Zimmerly. "The Chemical Dynamics of the Transformation of the Organic Matter to Bitumen in Oil Shale", *Bull. Univ. Utah*, 14 (7), 62–81 (1924).
26. B. J. Mair, "Hydrocarbons Isolated from Petroleum", *Oil Gas J.*, 14 September 1964, 131–134.
27. R. L. Martin, J. C. Winters and J. A. Williams, "Composition of Crude Oils by Gas Chromatography: Geological Significance of Hydrocarbon Distribution", *Proc. 6th World Petrol. Cong.*, Frankfurt, 1963, Section V, pp. 231–268.
28. R. D. McIver, "The Crude Oils of Wyoming—Product of Depositional Environment and Alteration", in *Symposium on Early Cretaceous Rocks*. 17th Annual Field Conference, Wyoming Geological Association, 1962, p. 248.
29. W. G. Meinschein, "Hydrocarbons—Saturated, Unsaturated and Aromatic", in *Organic Chemistry* (Eds., G. Eglinton and M. T. J. Murphy). Longman, Springer-Verlag, 1969, pp. 330–356.
30. D. Mendeleef, "Entstehung und Vorkommen des Mineralols". Abstract by G. Wagner, *Deut. Chem. Ges. Ber.*, 10, 229 (1877).
31. P. L. Parker, C. van Baalen and L. Maurer, "Fatty Acids in Eleven Species of Blue-green Algae: Geochemical Significance", *Science*, 155, 707–708 (1967).
31a. P. L. Parker. "Fatty Acids and Alcohols". In: *Organic Geochemistry* (Eds. G. Eglinton and M. T. J. Murphy). Longman, Springer-Verlag, 1969, pp. 357–373.

32. E. Peake and G. W. Hodgson, "Alkanes in Aqueous Systems: I. Exploratory Investigations on the Accommodation of C_{20}–C_{30} n-alkanes in Natural Water Systems",
American Oil Chemists' Society, **43** (4), 215–222.

33. G. T. Philippi, "On the Depth, Time and Mechanism of Petroleum Generation",
Geochim. et Cosmochim. Acta, **29** (9). 1021–1050 (1965).

33a. V. B. Porfirov, "Inorganic origin of Petroleum". *Bull. Amer. Assoc. Petrol. Geol.,*
58 (1), 3–33 (1974).

34. R. Robinson, "The Duplex Origins of Petroleum". pp. 7–10, in *Advances in Organic
Geochemistry* (Eds. U. Colombo and G. D. Hobson), Pergamon, Oxford: (1964).

35. M. A. Rogers and C. B. Koons, "Generation of Light Hydrocarbons and Establishment of Normal Paraffin Preference in Crude Oils", in *Origin and Refining of Petroleum*
(Ed., R. F. Gould). Amer. Chem. Soc.: Washington, D.C., 1971, pp. 67–80.

36. A. N. Sachanen, "Hydrocarbons in Petroleum", in *The Science of Petroleum* (Ed.,
A. E. Dunstan, A. W. Nash, B. T. Brooks and H. T. Tizard), **5** (1), 55–77. Oxford
University Press: London, 1950.

37. R. B. Schwendinger, "Carotenoids", in *Organic Geochemistry* (Eds., G. Eglinton and
M. T. J. Murphy). Longman, Springer-Verlag. 1969, pp. 425–437.

38. W. F. Seyer, "Conversion of Fatty and Waxy Substances into Petroleum Hydrocarbons", *Bull. Amer. Assoc. Petrol. Geologists,* **17**, 1251–1267 (1933).

39. A. Shimoyama and W. D. Johns, "Formation of alkanes from fatty acids in the presence
of $CaCO_3$". *Geochim. et Cosmochim. Acta,* **36** (1), 87–91 (1972).

40. S. R. Silverman, "Carbon Isotope Evidence for the Role of Lipids in Petroleum
Formation", *J. Amer. Oil Chemists' Soc.*

41. S. R. Silverman, *Influence of Petroleum Origin and Transformation on its Distribution
and Redistribution in Sedimentary Rocks.* Preprints for Panel Discussion 1 of 8th
World Petroleum Congress, Moscow, June 1971.

42. H. M. Smith, "The Hydrocarbon Constituents of Petroleum and Some Possible Lipid
Precursors", *J. Amer. Oil Chemists' Soc.,* **44** (12), 680–690.

43. H. M. Smith, *Crude Oil: Qualitative and Quantitative Aspects.* U.S. Bur. Mines I.C.
8268 (1966).

44. P. V. Smith, "Studies on Origin of Petroleum", *Bull. Amer. Assoc. Petrol. Geologists,*
38 (3), 377–404 (1954).

45. F. L. Staplin. "Sedimentary organic matter, organic metamorphism and oil and gas
occurrence", *Bull. Can. Petrol. Geology,* **17** (1), 47–66 (1969).

46. B. Tissot, Y. Califet-Debyser, G. Deroo and J. L. Oudin. "Origin and evolution of
hydrocarbons in early Toarcian shales". *Bull. Amer. Assoc. Petrol. Geologists,* **55** (12),
2177–2193 (1971).

47. B. Tissot and R. Pelet, *Nouvelles données sur les mécanismes de genèse et de migration
du pétrole: simulation mathématique et application à la prospection.* Preprints of Panel
Discussion 1 of 8th World Petroleum Congress, Moscow, June 1971.

48. B. Tissot, B. Durand, J. Espitalié and A. Combaz. "Influence of nature and diagenesis
of organic matter in formation of petroleum". *Bull. Amer. Assoc. Petrol. Geologists,*
58 (3), 499–506 (1974).

49. P. D. Trask, *Origin and Environment of Source Sediments of Petroleum.* Gulf Publishing
Co., Houston, Texas, U.S.A., 1932.

50. P. D. Trask and H. W. Patnode, *Source Beds of Petroleum.* Amer. Assoc. Petrol.
Geologists: Tulsa, Oklahoma, U.S.A., 1942.

51. D. White. "Metamorphism of organic sediments and derived oils", *Bull. Amer. Assoc.
Petrol. Geologists,* **19** (5), 589–617 (1935).

52. W. L. Whitehead and C. W. Sheppard. "Formation of Hydrocarbons from Fatty
Acids by Alpha Particle Bombardment". *Bull. Amer. Assoc. Petrol. Geologists,* **30** (1),
32–51 (1946).

53. F. C. Whitmore, "Transformation of Organic Matter into Petroleum-Chemical and
Biochemical Phases: Fundamental Research on Occurrence and Recovery of Petroleum", *Proc. Am. Petrol. Inst.,* **124** (1943).

54. W. M. Zarella, R. J. Mousseau, N. D. Coggeshall, M. S. Norris and G. J. Schrayer,
"Analysis and Significance of Hydrocarbons in Sub-surface Brines", *Geochim. et
Cosmochim. Acta,* **31** (7), 1155–1166 (1967).

CHAPTER 3

The Migration of Petroleum

GENERAL PRINCIPLES

THE types of rock in which subsurface hydrocarbon accumulations are characteristically found—the "reservoir rocks"—differ notably from the "source rocks" in which, as described in the previous Chapter, the generation of petroleum is thought to have occurred. The most obvious difference is in the sizes of the pores in the two kinds of rock; source rocks have very small pores, while reservoir rocks are essentially coarse-pored.

In order to account for the presence of petroleum in reservoir rocks, it becomes necessary to assume that there has been movement of fluid from the source rocks into the reservoir rocks, i.e. *migration* of the petroleum must have occurred in order that an accumulation could be formed.

Supporting evidence for this assumption may be summarized as follows:

1. The proportion of organic matter incorporated in the source rocks believed to give rise to petroleum, and the small fraction of this material evidently changed into oil and gas, are very much less than can provide, essentially *in situ,* for the saturations of oil and gas in subsurface accumulations[22].

2. The oil, gas and water in such accumulations are segregated according to their densities, provided that there are no marked variations in the detailed pore structure of the reservoir rock. It is hard to believe that the type and amount of the organic matter or the nature of the transforming mechanism could have anticipated this distribution. Moreover, the volume of the hydrocarbons is sensitive to the physical conditions.

3. The hydrocarbons generally occur in what is locally the highest available part of the reservoir rock. Again, it is difficult to envisage that organic matter in sediments would have been especially abundant just where an anticline, a fault or some other form of "trap" would develop at a later stage in the geological history of the area.

4. Some types of reservoir rock, e.g. fractured igneous and metamorphic rocks, or limestones with secondary porosity developed at unconformities, acquired the characteristics which enable them to serve as reservoir rocks relatively late in their geological histories[9].

There must therefore be some mechanism available which enables hydro-carbons and the other closely related substances found in petroleum to move from the places where they are formed to other places where they can be concentrated into an accumulation large enough to be of commercial signifi-cance. This entire process is known as *migration* and *accumulation,* the former term referring to the movement and the latter to the eventual segregation of the hydrocarbons in a highly concentrated form in a particular part of a rock with reservoir characteristics, i.e. a rock in which the spaces between the grains of solid matter are relatively large and extensively interconnected.

It has been found convenient to consider two stages in the migration pro-cess: (a) the transfer of petroleum or some closely-related ancestral material from the source rock to the reservoir rock; and (b) movements of this material *within* the reservoir rock to the locality in which an accumulation is created. The former process is known as "primary migration" and the latter as "secondary migration". (The term "tertiary migration" has been applied to any subsequent movements of the accumulated hydrocarbons. Such move-ments could represent adjustments in response to changed geological condi-tions, such as changes in depth of burial, tilting of the strata, or the development of avenues, by faulting or erosion, along which the hydrocarbons could escape to the surface to give seepages.)[4]

A fundamental problem encountered in considering primary migration is that of the state in which the hydrocarbons and related substances can be presumed to occur in the source rocks. The possibilities include being in "solution" in the water, being present as free globules, or being adsorbed on solid organic matter or on solid inorganic mineral matter; and, of course the hydrocarbons could be present in more than one of these states.

Clearly, primary migration requires that the substances must be free or become free at some stage to move from the source rock in which they were formed. There is no certainty as yet that the total material extracted by vigorous solvents in laboratory investigations on source rocks would be potentially able to move from the source rock under natural conditions. It is noteworthy, however, that the value of the ratio of organic-solvent extract to organic carbon in a source rock is now considered to be one of the criteria of significance in determining source rock quality. Nevertheless, it must be remembered that such measurements show what is still in the rock, and this may not necessarily be a satisfactory guide to the amount or nature of the material which has moved out of that rock.

As rocks become buried more and more deeply in the Earth's crust, the temperature and pressure of the fluids they contain will rise. At the same time, those rocks that are compactible will have their porosities reduced, with a corresponding loss of fluids from their pore space. The rocks undergoing most compaction will be those which are fine-grained and therefore fine-pored, commonly the clays. On most occasions, the expelled fluids will move up-wards relative to the mineral matter, although there can also be cases of downward movement; and it is possible that in some instances the direction of flow of the fluids set in motion by compaction will not always be the same throughout the clay's history. Furthermore, when there is relative upward movement of the fluids they will not necessarily be moving into zones of lower temperatures and pressures; however, uplift and erosion of the rocks can be expected to lead to a drop in the temperature of the fluids and also at some stage to a drop in the pressure.

Measurements have been made of the solubility of hydrocarbons in pure water (Table 27). For a given class of hydrocarbons, the solubility decreases as the size of the molecule increases: naphthenes are decidedly more soluble than paraffins, while the aromatics are far more soluble than either of these groups. Except in the cases of the gaseous members of the paraffin series, pressure has little or no effect on solubility. With the same exception, rise in temperature normally increases the solubility of hydrocarbons in water. Inorganic compounds in solution in the water decrease the solubilities of the hydrocarbons. The presence of soap-like substances in the water, the so-called colloidal electrolytes, greatly increases the "solubilities" of the hydrocarbons, provided that the concentration of the colloidal electrolyte exceeds a certain amount, the critical micelle concentration, above which the molecules of the colloidal electrolytes are believed to be present as aggregates of large numbers of molecules, and these aggregates (micelles) can take up molecules of hydrocarbons. This is known as *solubilization* of the hydrocarbons, a phenomenon which is influenced by the presence of inorganic compounds in the aqueous solution.[2, 5]

For the paraffin hydrocarbons, plots of solubility in pure water *versus* molar volume or the number of carbon atoms in the molecules show a discontinuity in the trend at about C_{10}, i.e. the "solubilities" of the larger molecules prove to be greater than would be predicted from the behaviour of the molecules up to about C_{10} in carbon number. From a study of the beha-

TABLE 27

SOLUBILITY OF HYDROCARBONS IN WATER

	Solubility in $g/10^6 g$ water
Methane	24·4
Ethane	60·4
Propane	62·4
n-butane	61·4
iso-butane	48·9
n-pentane	38·5
iso-pentane	47·8
n-hexane	9·5
2-methyl pentane	13·8
2,2-dimethyl butane	18·4
n-heptane	2·93
2,4-dimethyl pentane	3·62
n-octane	0·66
2,2,4-trimethyl pentane	2·44
Cyclo-pentane	150
Cyclo-hexane	55·0
Methylcyclo-pentane	42·6
Methylcyclo-hexane	14·0
Benzene	1780
Toluene	538
o-xylene	175
Ethyl benzene	159
*iso*propyl benzene	53

Source: C. McAuliffe.[17]

viour of normal paraffins in the $C_{20} - C_{33}$ range it has been inferred that there are aggregates of these molecules in the water; the hydrocarbons are said to be "accommodated" in the water. The use of a series of filters (0·05, 0·22, 0·45, 3 and 5μ pore sizes) afforded evidence of entities of different sizes in the water, the amount present after passing through a given filter pore size diminishing with increase in molecule size.[17, 18]

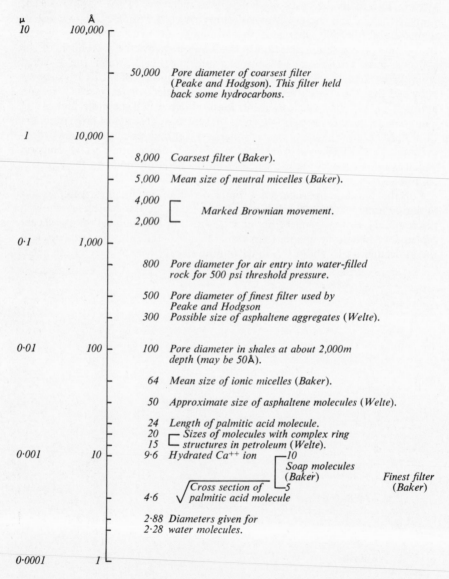

The fact that hydrocarbons can be "dissolved" in water, and that compaction and other processes lead to water movement through the sediments from source rocks to reservoir rocks is not, however, the full story. There must also be a release mechanism if migration of hydrocarbons in "solution" is involved in the primary migration process. Except in the case of gases, the solubility of hydrocarbons seems unlikely to be dependent on pressure; for gases the solubility decreases with rise in temperature. However, it seems that the solubility of the heavier hydrocarbons falls as the temperature is reduced. Thus, hydrocarbons would be released progressively from a saturated solution as the temperature fell, and hence the release could occur in any rock type where the temperature was lower than elsewhere along the flow path of the water. Consequently, unless the reservoir rock was very thick, little oil would be deposited in it from water in transit. Alternatively, reduction in the cover over the rocks, and the associated fall in temperature could also release hydrocarbons from "solution" in the water in the absence of flow. In addition, as noted earlier, water might flow downwards, or even relatively upwards into a reservoir rock, without suffering a drop in temperature. In neither case would there be release of oil as a result of temperature effects[10].

Release would create "free" oil. By temperature drop, "free" oil (unless it were immediately adsorbed) could appear in reservoir rock and non-reservoir rock alike, and because it would be only a part of the oil initially in "solution", the amounts freed could be expected to be small in any given rock layer. Consequently, change in temperature is unlikely to provide an important contribution to the formation of an oil accumulation.

If there is solubilization through the presence of colloidal electrolytes, circumstances can be visualised under which a substantial proportion of the oil could be released in a reservoir rock. This requires that the water in the reservoir rock be deficient in the colloidal electrolyte present in the source rock. Hence, when water from the latter enters the reservoir rock, initially at any rate, the concentration of colloidal electrolyte can be reduced to a value less than the critical micelle concentration, thereby releasing a substantial proportion of the solubilized oil. Nevertheless, even for a "solution" comparatively rich in oil from a source rock, this mechanism could not create a high concentration of oil in the reservoir rock as a whole: yet it could eventually give substantial local concentrations in limited parts of a very extensive reservoir rock[12].

A major difficulty in relation to primary migration of oil in "solution" is the quite small amount present even when there is solubilization. Concentrations of the order of 10 ppm have been suggested. Hence, very large volumes of water would be needed to carry substantial amounts of hydrocarbons from the source rock into a reservoir rock. Such volumes could well exceed those available from compaction[13].

There have been suggestions that clays can function as semi-permeable membranes holding back ions from aqueous solutions in transit. Thus, where a clay cap rock overlies a reservoir rock, this phenomenon could lead to an increase in the salinity of the water in the reservoir rock. Such an increase in salinity might cause salting-out of hydrocarbons from "solution" in the water, to accumulate in the reservoir rock. *

---

*The possibility of cap rocks acting as "filters" to retain hydrocarbons in reservoir rocks before there is significant aggregation as free oil cannot be ignored. There would still, however, be the problem of how the hydrocarbon aggregates or micelles undergoing primary migration were able to move through the fine-grained source rock, to enter the reservoir rock.

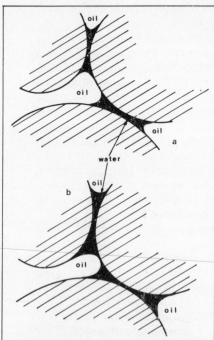

Pressures in water   Pressures in oil

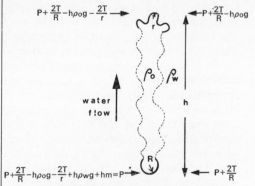

$$P + \frac{2T}{R} - h\rho og - \frac{2T}{r} \rightarrow$$

$$\leftarrow P + \frac{2T}{R} - h\rho og$$

water flow

$$P + \frac{2T}{R} - h\rho og - \frac{2T}{r} + h\rho wg + hm = P \cdot \rightarrow$$

$$\leftarrow P + \frac{2T}{R}$$

Fig. 10(a). Influence of height of oil mass within pores in a water-wet reservoir rock on the curvature of oil/water interfaces.

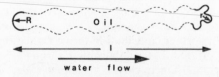

water flow

Fig. 10(b). Influence of horizontal water flow on a horizontally-extensive oil mass within the pores of a water-wet reservoir rock. The curvature of the oil/water interface is greater, and hence the radius of curvature is smaller, at the leading end than at the trailing end.

Fig. 9. Effect of height of oil column on curvature of oil/water interface and hence on proportion of the pore space occupied by oil. (a) and (b) are identical pore forms, but (a) is higher than (b) in the oil column, and therefore the oil occupies a greater fraction of its pore space.

Reservoir rock

Fig. 11(a). Small amount of oil has entered and stayed at base of a horizontal reservoir rock.

Reservoir rock

Oil movement

Fig. 11(b). Additional oil has entered base of reservoir rock, giving continuous connection; when there is sufficient vertical continuity, buoyant rise occurs.

Reservoir rock

Fig. 11(c). Oil build-up by entry at top of horizontal reservoir rock. (The convention has been used of large circles implying a greater fraction of pore space occupied by oil—not large radius of curvature of oil/water interface, see Fig. 9).

Oil movement

Reservoir rock

Fig. 11 (d). Oil entry at top of reservoir rock, or very small globules entering at base and being able individually to rise to top. Inclination and lateral continuity give greater fraction of pore space occupied by oil at upper end of oil mass.

Compaction-motivated water is available in greatest volume early in the history of a given compactible bed. On the other hand, it has been argued that the transformation of organic matter into oil and gas accelerates when the burial of the source rock becomes substantial, i.e. at a time when much of the compaction of the bed has taken place. The oil content of the rock at that stage appears in general to be far greater than could be contained in the water remaining in the rock. If capable of being mobilized, this oil would be available to go into solution in any water passing through the source rock. This state of affairs need be no barrier to the migration of oil in solution, provided that at any stage the amount of oil in the source rock is enough to saturate the water then in the rock.

There have been suggestions that late-formed hydrocarbons might be aided in their primary migration by structural water released from the clays when burial results in the attainment of temperatures at which certain mineral transformations take place. However, for transport in "solution", the volumes so released do not seem to be very important; they may represent only a fraction of the pore volume, and compaction will still be necessary for fluid expulsion to take place.

When there is a globule of one fluid in another immiscible fluid, there will be an excess pressure in that globule which is determined by the following relationship: $p = 2T/r$ where p is the excess pressure in dynes/cm², T is the interfacial tension between the two fluids, and r is the radius of the globule in centimetres. A more general form for the pressure across the interface is: $p = T \left(\dfrac{1}{r_1} + \dfrac{1}{r_2} \right)$ where r_1 and r_2 are the principal radii of curvature of the interface.

An oil globule of lower density than the surrounding water and smaller diameter than any throat connecting pores in a sediment would be able to rise by buoyancy. However, when its diameter exceeds the diameter of a throat the globule must be distorted, i.e. its curvature increased, in order for it to pass through the throat. Density differences and suitable vertical dimensions, as well as favourably directed water flow, would provide an increase in curvature.

Consider a complex globule of oil of height h and density ρ_o, in water of density ρ_w, inside the pore network of a rock (Fig. 10a). Suppose that there is a vertical hydrodynamic pressure gradient m in the water, causing upward flow. The water/oil interfacial tension is T, and the radius of curvature at the top of the globule is r, and at the bottom R. The units are centimetres, dynes/cm, and dynes/sq cm/cm. It can readily be shown that:

$$2T/r - 2T/R = hg \, (\rho_w - \rho_o) + hm$$

Now, $2T/r$ and $2T/R$ are respectively the capillary pressures at the top and bottom of the globule, and because $\rho_w > \rho_o$, $2T/r$ must exceed $2T/R$. In other words, $R > r$. Also, if r and R are fixed, it is evident that h will be greater when m is zero than when m is a positive quantity; it will be even greater when m is a negative quantity, i.e. when the water flow is downwards. A maximum value for R is fixed by the pore sizes in the rock. In the case of two rocks of similar pore geometry, but with one having pores and throats with sizes a fraction f of those of the other, then in the absence of water flow the two relationships will be:

$2T/r-2T/R=hg$ ($\rho_w-\rho_o$) and $2T/fr-2T/fR=h'g$ ($\rho_w-\rho_o$) where h' is the height of the complex globule in the finer-pored rock.

The second relationship can be re-written as:

$$2T/r-2T/R=fh'g\ (\rho_w-\rho_o).$$

Comparison with the first relationship shows that $fh'=h$, and hence for the same relative distortion of the top of the globule, the height of the globule must be greater in the fine-pored than in the coarse-pored rock. Experiments in sand packs confirm this point.

A further conclusion is that a single spherical globule marginally greater in diameter than a throat in a very coarse-pored rock might be able to squeeze through the throat by buoyancy, whereas a geometrically similar globule in a fine-pored rock could not do so.

Comparable arguments show that for horizontally extensive, complex oil globules water flowing past them will increase the curvature at the leading end relative to that of the rear end (Fig. 10b). If l is the length of the globule and m the pressure gradient in the water, $lm=2T/r-2T/R$, where r is the radius of the leading end, and R is the radius of the rear end.

Hence, for a given sand, oil and water, as well as pressure gradient, the length must exceed a certain value before the leading end can be squeezed through a throat of specified size. Again, considering geometrically similar pore systems, with one having pore sizes a fraction f of those in the other, it is clear that the critical lengths will be related by $l=fl'$, for a given pressure gradient, i.e. the critical length (l') is greater in the rocks with the finer pores. However, since the permeability is proportional to the square of the pore radius, for the same volume rates of flow, the pressure gradient in the finer sand will be m', such that $m'f^2=m$, f having the same meaning as above. In this instance, the critical lengths will be related by $lf=l'$, i.e. the critical length will be less in the rock with the finer pores.

The above theoretical conclusions suggest that a globule in a single pore in clays exceeding the pore throats in dimensions will be unable to move from pore to pore by virtue of its own properties or assisted by flow of the surrounding water, and therefore bodily translation cannot lead to aggregates sufficiently "tall" or "long", to be able to move by buoyancy or under the influence of flowing water. Indeed, any mechanism that could accomplish this initial aggregation would be capable of moving the oil through and out of the source rock without the aid of buoyancy or the flow of water.

It has been suggested that direct squeezing of oil globules by clay platelets during compaction would drive oil from pore to pore, but this does not seem to afford a particularly efficient mechanism for expelling oil from clays. The principal porosity reduction would result from the expulsion of the far more abundant and more easily moved water in the pore system.

The suggestion has also been made that it is not hydrocarbons but some more soluble parental substance which undergoes primary migration. The problem then is—why is this substance preferentially retained in the reservoir rock? Undoubtedly, some hydrocarbons are formed in the source rock; thermal effects could also lead to their being formed in the reservoir rock when suitable substances are present. Also, if the hydrocarbons undergo primary migration solubilized by colloidal electrolytes, the latter compounds might be broken down in the reservoir rock, as well as elsewhere, to give hydrocarbons. Irrespective of the precise mechanism involved in primary migration, it is essential that there should be preferential retention of hydrocarbons or related organic substances in the reservoir rock.

Although upward primary migration seems likely to be common, downward movement cannot be excluded, and the mechanism seems to be linked in some way with the process of *compaction*.

Compaction Effects

Freshly deposited, fine-grained sediments contain an excess of water, and volumetrically there can, in some cases, be more than 90 parts of water to less than 10 parts of mineral matter. As a particular section of such a sediment is progressively buried more deeply by further deposition, its water content is reduced as the centres of the constituent grains come closer and closer together, i.e. compaction takes place and the porosity is reduced. Thus, burial to 10,000ft may reduce the porosity to less than 10%. Empirical relationships between porosity and depth have been derived from porosity measurements made on such sediments. Relationships have also been derived theoretically.

Compaction is more extensive the finer the mineral grains. In detail, it is dependent on a number of factors: the nature of the constituent minerals, the shape, size and size distribution of the grains, the rate of sediment accretion, the thickness of the compactible beds, and the thickness and nature of any interbedded rocks present which undergo little compaction. Water entry at the base of a compactible series, or downward escape from the base, will also influence the compaction process. Although ultimately the water as a whole must move upwards relative to the mineral matter, locally and temporarily there can be downward relative flow. The expelled water carries both inorganic and organic material in solution. The formation of a fault can accelerate compaction in the adjacent compactible beds.

The volumes of water passing through a given layer of rock as a result of compaction depend on many factors, including its depth of burial, and the thickness and characteristics of any underlying compactible rock. However, for a "100m" zone of source rock underlain by "500m" of similar compactible clay-type rock, burial of its top to "500m" would cause about 9·4 litres/sq cm to enter its base and some 13·1 litres/sq cm to leave its top. Burial of the top to "1000m" would raise these volumes to about 12·2 litres/sq cm and 16·3 litres/sq cm, respectively[8].

Lateral variations in the thickness of a compactible series involve diff rent amounts of thinning on deeper burial, so that overlying beds will h e a structural form which is influenced by this differential compaction, e.g. a surface which was horizontal at the time of deposition can become inclined, the higher parts overlying the zones of least thickness of the compactible beds.

Because water is moving in a compactible formation, the increase in fluid pressure with depth will differ from that which occurs in a stationary column of water. For upward flow, the downward pressure increase will be greater

TABLE 28

SUMMARY OF COMPACTION PROCESSES

Mineral	Fine-grained mineral matter —clay, etc.	Organic matter, e.g. peat, algal mat.	Gypsum	Sandstones and limestones.
Degree	Very high.	Very high.	Considerable.	Small.
Timing	Progressive. Irreversible.	Progressive. Irreversible.	Delayed. Reversible.	Delayed. Irreversible.
Scale	Grain.	Complex.	Molecular.	Molecular.
Means	Grains re-arranged. Free water loss. Late loss of "bound" water. Late mineralogical changes.	Loss of water and breakdown of compounds.	Loss of constitutional water.	Re-arrangement of solid mineral matter via solution. Water loss. (Pressure solution, stylolites).

than hydrostatic, being a combination of a hydrostatic and a hydrodynamic component.

By changing into anhydrite, gypsum can, in a sense, undergo compaction, the water lost being water of crystallisation. However, this change requires a particular set of physical conditions (temperature and pressure) to be attained before it takes place, and the salinity of any free water present also influences the conditions required. Hence, this change calls for substantial burial of the gypsum before compaction and release of water take place. The transformation is reversible (Table 28).

All clays consist of a mixture of silt, platy "clay minerals" and colloids. The individual particle sizes lie between 0·005 and 0·001mm, or less. X-ray investigations show that clay minerals have a lattice structure of a layered type; the fundamental layers comprise one based on silica atoms and another based on aluminium atoms, with links provided by oxygen and hydroxyl groups. Some degree of substitution of the aluminium atoms in particular is possible, and there can also be incorporation of other metallic atoms to a limited extent. The platy structure includes inter-planar water which can be expelled step-wise under particular conditions attained by the rise in temperature associated with deep burial. At a given temperature, the fourth, third, second and first inter-planar layers of water are removed at progressively higher pressures, and as the temperature is increased the respective pressures required decrease. In addition, montmorillonite in clays can be converted into illite[19, 20].

As an example, a 150ft bore-hole in the young (Pliocene-Holocene) deltaic deposits of the Pedernales area of East Venezuela provided evidence of water flow in the compacting clays, as well as indications of the carriage of youthful oil in the moving water. An enclosed sand body showed water pressures in excess of hydrostatic. The sand lens had a hydrocarbon concentration four to five times greater than was found in the sands in free communication with the surface or in the clays. In this sand, the ratio of aromatic to paraffin-naphthene hydrocarbons was three to one, whereas the ratios were about one to one elsewhere. The sand was about 5,000 years old, and it was estimated that the rate of accumulation of the oil was 0·025 ppm per year. Analyses of the moving waters from the sands suggested hydrocarbon contents under 16 ppm. Carbon dating of the hydrocarbons which gave an apparent age of 13,400±1,200 years led to the conclusion that older hydrocarbons are moving up-dip from deeper parts of the sedimentary basin and mixing with the locally-formed hydrocarbons[14a].

Stylolites are the complicated partings sometimes found between blocks of limestones, dolomites and, less frequently, sandstones. They are formed of dark-coloured, clay-like residual material originally disseminated in rock which has since been dissolved. In essence, stylolites are *pressure-solution features* developed after the rock had become indurated and buried to a considerable depth.

There is evidence that burial to 2,000–3,000ft may be needed for vertical loading to cause the formation of stylolites. The boundaries between the rock blocks interpenetrate and show a stepped form, with the interpenetration being parallel to the pressure which gave rise to the stylolites, while the other sectors of the partings are normal to the pressure[5a].

Stylolites can be the result of vertical pressure or of lateral pressure. In the former case, there is evidence of reductions of the original bed thickness by as much as 40%. Deposition of the dissolved material elsewhere (presumably higher in the sequence when loading is the cause) in the rock mass is responsible for drastic reduction of the previous porosity of the rock. Should the reduction in bed thickness as a result of stylolitization not be uniformly distributed horizontally, overlying rocks may develop an arched form, thereby creating a trapping structure. Non-uniform settling of overlying beds, associated with the formation of stylolites, may not only lead to the tilting of such beds, but also may create near-vertical fractures which could allow trans-formational oil and gas migration, or, after sealing, provide fault-trapping. The stylolites may also act as impermeable barriers in nearly flat, fracture-free limestone reservoir rocks, as in Abu Dhabi.

The formation of stylolites is effectively a form of compaction. The reduction of the total pore space in the sequence calls for the expulsion of fluid from the zone affected.

SECONDARY MIGRATION

Because of the low concentration of hydrocarbons and related organic substances in the source rock and the slow rate at which they enter the reservoir rock, the initial concentration at the point of entry will be small. Should any hydrocarbon globules entering or released be smaller than the throats in the reservoir rock, and entry is at the bottom, these globules could rise by buoyancy to the top of the reservoir rock Exit into the cap rock, with its far smaller throats and pores, would be prevented by capillary pressure phenomena. For entry at the top of the reservoir rock (Fig. 11) the globules must remain there until displaced by other globules or by water flow, but water would move them only after adequate lateral continuity had developed. Continued entry of hydrocarbons would lead to greater continuity of the hydrocarbons, both horizontally and vertically, whether entry was at the top or at the bottom of the reservoir rock. Inclination

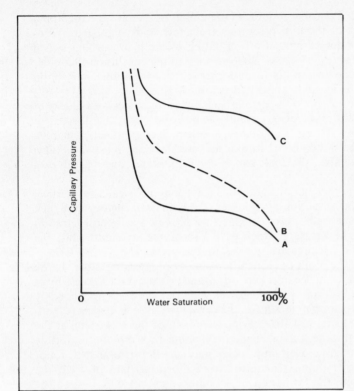

Fig. 12. Capillary pressure curves: A. Relatively coarse-pored rock with pore throat sizes within a narrow range. B. Rock with pore throats ranging widely in size (the larger throats are comparable in size with those of A). C. Comparatively fine-pored rock with a narrow range of pore throat sizes.

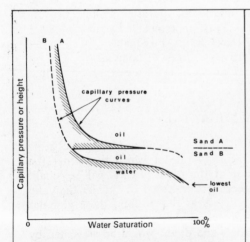

Fig. 13. Effect of sand layering on vertical distribution of water (and oil) saturation. Sand A has finer pores than those of the underlying sand B. This causes an increase in the water saturation in the former where they are in contact.

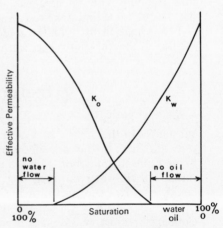

Fig. 14. Effective permeability/saturation relationships.

of the cap rock-reservoir rock boundary or the seat seal-reservoir rock boundary could, for a given amount of hydrocarbons, give greater effective height than for horizontal boundaries. Inclination might exist before or develop during or after some measure of primary migration. Buoyant movement of the hydrocarbons would take place once critical effective heights had been reached. Then, oil which had entered at the base of the reservoir rock would rise slowly to the top, while oil entering at the top, or which had risen to the top, would move laterally upwards under any inclined boundary to a "trap"—a position at which the direction of dip was reversed on where finer-pored rock was otherwise interposed on the up-dip side by faulting or by a lateral change in the reservoir rock itself. Any water flow in the reservoir rock, depending on its direction, could facilitate or impede this phase of segregation of the hydrocarbons in a trapping position.

Laboratory measurements of capillary pressure for a rock sample provide a relationship between capillary pressure and the saturation of the wetting fluid, usually brine, for the given pair of fluids. The relevant relationship, so far as the rock pores are concerned, is between throat sizes and the volume of fluid that is expelled from the pores as a consequence of the non-wetting fluid being able to pass through the given throat size, indicated by the capillary pressure value. The capillary pressure curve will be flat over a considerable range of saturations for a rock with throats, and by implication pores, of uniform size; the curve will be inclined when there is a spread in throat sizes, even in the high wetting-fluid saturation range (Fig. 12). At given saturations, a rock with small throats will show higher capillary pressures than one with large throats. A "cap rock" or a "seat seal" has a high capillary pressure, especially the entry pressure, whereas a reservoir rock has a low capillary pressure.* When two rock types are in contact, with fluid communication, in the absence of flow there must be the same pressure in a given fluid at a particular level. Under certain circumstances, differences in capillary properties will lead to differences in brine saturation at a given level (Fig. 13). Differences in brine saturation at a particular level will also exist when a reservoir rock varies laterally in pore throat sizes, a feature which is associated with a change in level of the fluid contacts, and also in the thicknesses of the fluid transition zones.

In a rock body containing immiscible fluids, there will be an inter-fluid meniscus curvature which is associated with the bottom of the transition zone of the fluids and a greater curvature which is linked with the top of that zone. These can be represented by two radii, R_B and and R_T, respectively, and two different wetting-fluid saturations. For a geometrically similar pore system, but with pore openings a fraction, f, in size of those for the first rock, the corresponding radii of the menisci at the bottom and top of its transition zone will be fR_B and fR_T. In both rocks, the differences in capillary pressure between the top and the bottom of the transition zones must be equal to the thickness of the zones multiplied by the difference in densities of the wetting and non-wetting fluids. The coarser rock gives:

$$h_c g (\rho_w - \rho_o) = 2T/R_T - 2T/R_B$$

---

* High and low capillary pressure are commonly used in relation to the value for initial entry in drainage measurements, or to the general level of the values for saturations markedly above the irreducible minimum saturation for the wetting fluid.

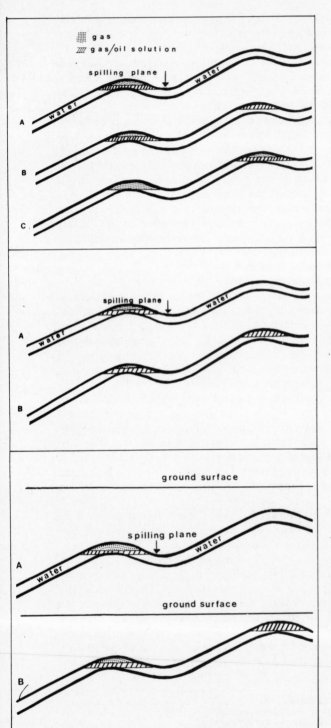

Fig. 15. Differential entrapment of oil and gas. Oil and gas are assumed to enter the reservoir rock low on the monocline, i.e. on the left. (A), (B) and (C) represent successive stages as progressively more oil and gas move upwards from the left.

Fig. 16. A. Oil and gas accumulation in a monoclinal dome. B. Tilting of the structures (down to the left) reduces the closure of the monoclinal dome, and oil spills under the spilling plane, moving to the next dome to the right.

Fig. 17. Oil and gas accumulation in a monoclinal dome. B. Erosion reduces the depth of burial and the reservoir pressure, causing the original gas cap to expand and gas to come out of solution in the oil. When the hydrocarbon volume exceeds the storage capacity of the dome, excess oil spills under the spilling plane to reach the next dome to the right.

The finer rock gives:
$$hg\,(\rho_w - \rho_o) = 2T/fR_T - 2T/fR_B = 1/f\,[2T/R_T - 2T/R_B]$$
Hence $h_c = fh$. Thus, for given pore geometries, the finer the pores, the thicker is the transition zone. The relationships also indicate that the thickness of the transition zone depends on the interfacial tension and the difference in density of the two fluids; for a given rock and interfacial tension the transition zone will be thicker the smaller the difference in density; for a given rock and difference in fluid density, the transition zone will be thicker the greater the interfacial tension.

Although there is a link of a reciprocal nature between capillary pressure and permeability, it is not the low permeability of a cap rock which determines its sealing property; it is its high *capillary pressure*. Low permeability is an impediment to flow; it does not provide a seal. Adequately *tall* hydrocarbon columns can result in hydrocarbons entering cap rocks and even inclined seat seals, as well as seals provided by non-reservoir rock faulted against reservoir rock. Fine-grained lenses within reservoir rocks may remain entirely water-filled, even though enveloped by oil- or gas-charged rock, provided that their overall height does not exceed a critical value which is dependent on the throat and pore sizes of the two rock types and the properties of the hydrocarbons and the water. It is, moreover possible for the upper part of the fine-grained lens to contain oil, with water in the lower part, when there is oil in the surrounding coarser rock.

Laboratory measurements have provided relationships between water and oil (or gas) saturations in a rock and the effective permeabilities of the rock for each of these fluids (Fig. 14). At low concentrations of a given fluid there is zero permeability for that fluid; for intermediate concentrations there is permeability to both fluids; at high concentrations of a fluid, that fluid only can flow. When the quantity of hydrocarbons which has been trapped is sufficient only to give a transition zone, and not an overlying zone in which the interstitial water is at the irreducible minimum saturation, cores will show some hydrocarbon content. Yet when a well is tested, either water and oil or only water will enter the well bore. As noted earlier, the thickness of the transition zones depends on the rock pore sizes, the fluid densities and interfacial tension. Fine-pored rocks containing dense oil could have thick transition zones and when the height of closure is less than the thickness of the transition zone the trap, although occupied by oil down to the spilling plane, can never produce water-free oil.

The existence of these fluid saturations for which no flow takes place raises a query with respect to ideas proposed earlier for secondary migration in order to form an accumulation. It was suggested that primary migration leads to widespread entry of oil or gas into the reservoir rock, and that subsequently the hydrocarbons were caused to move to give high concentrations at trapping points. Would this mechanism leave behind residual saturations of oil or gas at all points along the flow paths? In the case of gas, solution and diffusion effects offer a means for removing isolated globules, since globules with high curvature would tend to go into solution and there would be transfers to globules with low curvature; in a sense, big globules would grow at the expense of little ones. Although solution might also affect oil globules, for various reasons it would be less effective than for gas. There might also be transformation effects. It may also be wondered how often search has been made in seemingly water-saturated cores of reservoir-rock type for residual amounts of oil. Furthermore, extremely low rates of movement are being considered. Would these lead to the occurrence of residual oil concentrations such as arise for the far more rapid rates of flow involved in laboratory measurements and in oil production? Perhaps the very slow movement would preclude the disturbances which would favour the disruption of stringers of oil at the waists or necks in pore throats, thus suppressing the formation of the isolated globules that may account for residual oil saturations.

The accumulation of oil and gas in a series of domes on a regional monocline has been discussed in detail. If oil and gas were entering an extensive reservoir rock at points which were structurally lower than the lowest of the domes, then they would first accumulate in the lowest dome until this was filled to the spilling plane (Fig. 15). Thereafter, as more oil and gas reached the dome, only the gas would be retained within it, and an equivalent volume of accumulated oil, plus any incoming additional oil, would spill from the trap to move up the monocline into the next dome. This process could continue until the lowest dome was effectively filled with gas down to the spilling plane. (Presumably there would be a residual amount of oil.) In the meantime the next higher dome would have been collecting the oil, and because of its position some gas might have been released from solution to give a small gas cap. The continued entry of gas and oil into the reservoir rock would lead to a sequence of events in the second dome similar to those which occurred for the first dome; if the quantities were right, oil would eventually spill from the second dome to move further up dip, and it, too, might be filled with gas down to the spilling plane. It is evident that by this mechanism the dome or domes lowest on a monocline could be gas-bearing, while the domes immediately up dip are oil-bearing, and domes still higher on the monocline carry only water. This kind of distribution has been observed in some areas. The mechanism is known as *differential entrapment,* and should be looked upon as a principle rather than as a mechanism which applies only in the precise form described above. Thus, oil and gas need not be entering only at points further down dip than the lowest of the domes[6].

Other mechanisms could lead to comparable distributions. Should gas and oil have accumulated in one or more of a series of domes on a monocline, after which tilting occurred to increase the general steepness of this monocline, the tilt would reduce the closure of individual domes (Fig. 16). As a result, any dome would have its storage capacity reduced, and if full or nearly full of hydrocarbons before tilting took place some of these, the lowermost of which would be oil but might include some gas also, would spill out and move to the next dome higher on the monocline, initiating a series of events resembling some of those involved in differential entrapment.

When the pressure on a gas/oil solution is reduced gas is released, and the hydrocarbon system increases in volume (Fig. 75, p. 266). If erosion of the overlying rock cover takes place, the reservoir pressure may be expected to fall. Should gas release cause the volume of the hydrocarbons to become greater than the storage space of the trap the excess, which would be the lowest material —oil—would pass under the spilling plane to enter an adjacent trap (Fig. 17).

Lastly, any breakdown of oil to give gas and to increase the total volume of the hydrocarbons under the reservoir conditions, could also lead to expulsion of oil from a trap.

The processes listed above could operate alone or in combination. Evolutionary changes could preclude working out the exact mechanism in a particular case. Thus, deeper burial, gas leakage or the removal of gas by solution in moving water might raise oil/water contacts to levels above spilling planes, even though such planes had earlier exerted control[11].

Inclined fluid contacts have been attributed to water flow, or to lateral changes in pore sizes in the reservoir rocks. In laboratory studies of capillary pressure, it has been found that the capillary pressure/saturation curves exhibit hysteresis, i.e. the drainage (falling saturation of the wetting fluid) curves do not coincide with the imbibition (increasing saturation of wetting fluid) curves. Should there be hysteresis for the much slower natural processes, this could lead in certain circumstances to inclined fluid contacts[14].

There is the possibility that after an accumulation has formed with a horizontal fluid contact the area could undergo tilting. The fluid contact would tend to respond to the changed conditions. The end which was being relatively uplifted would try to move downwards, i.e. to force hydrocarbons into water-bearing rock (a drainage process), whereas the end which was being relatively lowered would try to move upwards, i.e. allow water to enter previously hydrocarbon-bearing space (an imbibition process), The existence of hysteresis would mean that stability could be attained without the fluid contacts being returned to horizontality after tilting ceased. Uneven additions of hydrocarbons to opposite sides of an anticlinal trap could lead to a similar state of affairs, for the side on which the hydrocarbon/water contact was tending to move down most rapidly by accretion would seek to rise and at the same time to force down the other side[9].

The likely storage volume of a trap at a given stage in its history has been compared with the volume which its present hydrocarbon content would have occupied at the estimated pressure and temperature corresponding with that stage. If the inferred hydrocarbon volume exceeded the storage volume, then accumulation must have taken place in part at least after that stage; if it was less than the storage volume, it could have taken place earlier than that stage. The reasoning involved assumes no late breakdown of oil to give gas; and no gas removal by seepage, or by solution in moving water. Any of these processes could vitiate the argument[15].

Since it seems that the formation of petroleum components is a progressive process, the components and their proportions developed may well vary with the depth of burial. Accordingly, the material available for primary migration will change over time. Moreover, there could be lateral variations in the organic matter present in a source rock or source rocks feeding a single extensive reservoir rock. Even though transformation may also take place in the reservoir rock, the products may differ in detail from those formed at corresponding times in the source rock, for the latter contains a different organic complex from the reservoir rock.

Earlier, it has been suggested that the directions of water flow even during compaction need not be constant at every locality. When consideration is given to the sedimentary history of a basin, its structural evolution (p. 236) and any phases of uplift and erosion, it is evident that water flow patterns are likely to change in the course of time. The very marked permeability contrasts between reservoir rocks and non-reservoir rocks are of importance. Equally, it must not be forgotten that the areas of contact between the different rock types are extremely large compared with the thicknesses of the rocks, a feature which tends to be masked by the widely-used convention of drawing geological cross-sections with exaggerated vertical scales. Undoubtedly, the volumes of water flowing across the bedding are far greater than have been supposed by many, quite apart from the water set in motion by compaction. The mere presence of hydrocarbon accumulations in a reservoir rock will exert some influence on flow patterns, and the interaction between flowing water and the accumulations will lead to adjustments of the positions of the accumulations.

The low solubility of common rock cements requires the passage of many pore volumes of water to provide significant amounts of cement, except where they result from local solution and re-deposition. In some cases, the hydro-carbon-bearing sector of a reservoir is stated to be less well cemented than other parts. This condition has been attributed to hydrocarbon accumulations largely pre-dating the cementation phase. If this inference is correct, there is evidence in support of hydrocarbon accumulations influencing the pattern of the water flow[9].

Extensive lateral flow in reservoir rock horizons seems more probable when uplift of the peripheral parts of a basin affords opportunities for meteoric waters to enter some outcrops, especially when lower outcrops provide suitable outlets for the water entering at higher levels. Thus, there is the likelihood that lateral movements of water may be more important in the later stages of the history of a basin, whereas compaction may have more influence in directing water flow during the depositional phase[7].

In particular cases, it may be feasible to indicate a limit to lateral migration, as a distance, in terms of the concept of a "gathering ground" for the accumulation, and to migration across the bedding from a consideration of the presumed source rock. More generally, it may be stated that migration, primary or secondary, is obviously possible *so long as the conditions needed for those processes to take place are satisfied*. Migration is halted when these conditions are no longer met, i.e. when there is a change in lithology, a reversal of the dip or the imposition of some other barrier to the movement of hydrocarbons that has hitherto been possible.

THE RESERVOIR ROCKS

By far the most important types of reservoir rocks are the sandstones and carbonates. Thus, one survey has shown that about 59% of the production from the world's major oilfields comes from "sandstones" and 40% from "carbonates"—these terms being loosely used to include, on the one hand, orthoquartzites, greywackes, arkoses and conglomerates, and on the other precipitated carbonates, muddy limestones, reef limestones, and many kinds of dolomites. Other rocks also occasionally provide petroleum reservoirs, but these occurrences are relatively rare; they include various igneous and meta-morphic rocks, fractured shales and cherts.

The porosity and the relatively high permeability characteristic of reservoir rocks may be *primary* or *secondary* in origin. Primary reservoir properties depend essentially on the size, shape, grading (size distribution), and packing of the constituent rock particles, and the manner and degree of initial consoli-dation of the sedimented particles. Secondary post-deposition factors may either enhance or diminish the primary porosity and permeability; these properties will be adversely affected by deposition of minerals from percola-ting fluids, and by changes associated with increases of temperature and pressure arising from deep burial or lateral compression. The development of joints and fractures by tectonic activity or other mechanisms will increase the bulk permeability without necessarily adding much to the porosity, while partial solution and recrystallization or replacement under certain circumstances, will generally enhance both properties. Weathering during temporary exposure is a significant agent in causing partial solution of carbonate grains or cements.

Sandstone Reservoir Rocks

Sandstones result from the breakdown of the so-called crystalline rocks as a result of weathering, followed by transport and eventual deposition of the comparatively large resultant particles, or from similar actions on previously formed sandstones. They are formed of grains of quartz, possibly embedded in a matrix of finer particles, with varying proportions of cementing material. The grains may be sub-angular, or more or less rounded, and the cemen-ting material is most commonly calcium carbonate, silica, or iron oxides. Sandstones may vary in make-up from those in which the larger particles are almost entirely quartz ("orthoquartzites" when they are also quartz-cemented) to muddy or dirty sandstones with appreciable amounts of other materials such as mineralogically-composite grains, volcanic ash, felspar, calcite or dolomite. The matrix may also include finer particles of these materials, as well as smaller quartz grains and clay minerals. A sandstone may be described as an arenite if its texture is essentially sand-grain supported, a wacke if clay-silt matrix supported, or a sub-wacke if between these extremes. The main grain sizes (usually between $\frac{1}{16}$mm and 2mm diameter), the dominant mineral, and the degree of sorting and consolidation are parameters which are used to classify sandstones.

The terms "sand" or "pay sand" are often used in oilfield parlance to describe an oil-saturated sandstone, although completely uncemented sands are rare. When a poorly-cemented sandstone is in fact an oil or gas reservoir, problems usually arise as a result of sand being carried into the well-bore during production. Such an unconsolidated sand may have over 40% porosity, depending on the degree of rounding and size distribution of the individual particles and the manner of packing. The presence of cement automatically reduces the porosity, and extensive cement deposition may reduce the porosity to less than 10%, with a relatively much greater reduction in the permeability. However, the effect of cement deposition on effective porosity and permea-bility is complex.

Before hydrocarbons move into them, reservoir rocks will inevitably have been water-bearing and almost certainly water-wet. When oil or gas enters, many reservoir rocks remain water-wet, and at the higher hydrocarbon saturations the residual interstitial water is present as wetting films of molecular thicknesses on the mineral surfaces, but mainly as collars round grain contacts and as complete fillings of pores surrounded by unusually small throats. Cement deposition and grain growth reduce the porosity, but the precise effect on hydrocarbon storage capacity depends on the sites of these mineral additions, especially when they are small in amount; additions at grain contacts could reduce the volume of interstitial water which could be held at grain contacts without materially affecting the maximum volume available for storing hydrocarbons.

Furthermore, for geometrically similar grains and grain arrangements, and hence, geometrically similar pore spaces, the permeability is proportional to the square of the characteristic grain size. Deposition of mineral matter in the pore spaces causes a reduction in porosity directly proportional to the amount deposited, whereas even when this mineral matter is uniformly distributed the relative reduction in permeability is decidedly greater.

When there are clay partings on the bedding planes of a sandstone, they will have little effect on the permeability of the sandstone parallel to the bedding planes, but they will markedly reduce the permeability in a direction normal to the bedding planes. Small discontinuous partings or wisps of clay will have comparable although somewhat smaller effects on the permeability across the bedding, and in this case a small sample used for making permeability measurements normal to the bedding will give a lower value than for the rock in bulk, when the parting extends right across the sample.

Deep burial with the resultant increase in pressure may lead to the fracturing of rock grains and some rearrangement to give increased packing density and a lower porosity. This is particularly noticeable in areas of great thicknesses of relatively soft Tertiary strata, as in the USA Gulf Coast area. If the initial porosity of a sandstone was say 35%, it will probably be reduced to 25% when the rock has been subjected to a pressure of about 20,000 psi—equivalent to burial to about 18,000ft.

Rise in temperature associated with increased pressure will result in a general deterioration of porosity and permeability in a sandstone. In addition to the effect already mentioned, there may be pressure solution at grain contacts, the growth of quartz grains at the expense of the matrix material, or the development of stylolites with the consequent segregation of insoluble clayey material along planes normal to the direction of the pressure (commonly, therefore, along bedding planes, for simple burial effects).

In the foregoing, the ultimate concern has been with the holes in the rock—their sizes, shapes and interconnections, for these determine the porosity, permeability and capillary pressure. However, study of the size and shape of the constituent particles, which is feasible by the use of drill cuttings, supplemented in some cases by examination of the so-called sedimentary structures seen in cores, makes it possible to determine the *environment of deposition,* i.e. whether the sand is aeolian, fluvial or marine in origin. Such determinations may be a guide to gross reservoir behaviour, and to the possibility of other prospects existing in the area.

The widespread, very uniform deposits which are sometimes found—called "blanket" or "sheet" sandstones—have been formed by seas transgressing over a wide area, the sands which were deposited at the end of the process being considerably younger than those formed at the start. Sandstones more usually take the form of comparatively small, lenticular bodies, grading relatively rapidly both laterally and vertically into clays and shales, typically deposited by regressing seas or in deltas.

Two of the largest oilfields in the world, *East Texas* and *Burgan,* have Cretaceous sandstone reservoirs many square miles in extent, whose permeability and uniformity are such that they contain enormous volumes of oil which can be produced at a very high rate. The productive area at East Texas covers at least 140,000 acres, and the productive thickness at Burgan exceeds 1300ft. Another notable "blanket"-type sandstone field is *Pembina* in Canada, covering some 755,000 acres. The deltaic Athabaska "tar sands" of Alberta have a thickness of more than 400ft and cover an area of many square miles.

The close geographical contiguity of an organic clay source rock and a reservoir sandstone calls for some explanation, since they are typically produced by deposition in different environments. Sometimes a sand bar may have been formed by local current conditions in the marine gulf in which organic source material was being deposited. Similarly, barrier sand beaches, of lenticular cross-section, and many different kinds of sandhooks and spits, are the result of local current action. They are usually flanked to landward by lagoonal deposits, and grade outwards into marine silts and clays. Scouring of the sea floor by currents heavily-laden with sediment can result in the deposition of long, narrow belts of coarse sand. Any of these sand deposits, when subsequently buried by impervious material as a result of subsidence, could become a local oil or gas reservoir. Also, since the coarseness of the sediments deposited in the sea generally varies with the depth of water and distance from the shore, vertical movements of the sea floor could cause a change in the coarseness of the sediments deposited at any point, thus bringing about the juxtaposition of a fine-grained source or cap rock and a coarse-grained reservoir rock.

Many examples of sand lens fields are known all over the world, and the "shoestring" fields of Kansas and Oklahoma are typical examples of accumulation in sinuous sand bodies.

Other Siliceous Rocks

Conglomerates and grits, formed of larger component particles than sandstones, make excellent reservoir rocks. An example is the coarse, conglomeratic Lower Mississippian Big Injun bed of Ohio-West Virginia.

Greywackes are rocks with important amounts of each size-range from mud to sand, and in each of the composition groups—quartz, felspar, sheet silicates and rock fragments. Non-felspathic or "low-rank" greywackes (particle size 0·0625–0·25mm) provide excellent reservoirs in Trinidad and some parts of the USA—notably Pennsylvania, the Gulf Coast and the Los Angeles Basin, where they are thought to be deposits characteristic of submarine turbidity currents.

Arkoses with a high proportion of felspar with the quartz grains are important reservoir rocks in many oilfields of Western Oklahoma and the Texas Panhandle. They are derived from the disintegration of local acid igneous rocks, and are therefore sometimes called "granite washes".

Siltstones are composed of relatively fine particles of silt (1/256 to $\frac{1}{16}$mm diameter) and are consequently generally too low in primary permeability to act as satisfactory reservoir rocks. However, fracturing can improve the reservoir properties to such an extent that large commercial oil accumulations sometimes occur in siltstone reservoirs. The most notable example is the Permian Spraberry formation in West Texas, which has a productive area of about 1,000 sq miles.

Fractured Miocene cherts are the reservoir rocks in most of the Santa Maria and San Joaquin Valley fields in California. The chert consists of chalcedony, with a proportion of quartz grains, and has been formed either as a chemical precipitate or as an erosional end-product. Detrital cherts are oil reservoirs in some Kansas fields, where they are locally named "chat", while novaculite, a massive siliceous chert, produces oil in Arkansas and some other parts of USA. It is thought to have formed as an agglomeration of fine-grained, organically precipitated, silica particles below the local wave base. A rock largely made up of the tests of diatoms, microscopic silica-secreting plants, provides reservoirs of Pliocene age in California.

Carbonate Reservoir Rocks

Carbonate reservoir rocks are particularly important in the Middle East area, where many of the world's largest oilfields occur. The rocks vary from nearly pure calcitic limestones ($CaCO_3$) to mixtures of calcite with different proportions of dolomite–$CaMg(CO_3)_2$. (A dolomite rock is usually defined as a limestone containing at least 50% of the mineral dolomite.) Limestones are nearly always of biochemical origin, being formed either through the agency of lime-secreting organisms and the accumulation of "bioclastic" shell material, or as a consequence of the precipitation of calcium carbonate from solution; in both cases there is often admixture of the calcium carbonate with other sedimentary materials, to give a wide range of "impure" limestones containing varying proportions of clay, iron oxides, etc. Normally, massive marine limestones do not contain many macro-fossils; they are largely made up of the remains of algae, with biochemically precipitated calcite. Where no organic remains at all are visible, the limestone would appear originally to have been chemically precipitated as a "lime mud".

Limestones tend to be lower in porosity and permeability than sandstones, and particular units can be quite irregular in the distribution of these properties. Fractures, when present, commonly affect the reservoir properties of limestones more than is the case for sandstones. The size, shape and distribution of the pores in limestones range from fairly uniform to extremely heterogeneous, even when the rock is of a single type; this contrasts with the common fair degree of uniformity which exists in a homogeneous sand body. In limestones, the porosity is largely inter-particle, but intra-particle and other types of porosity may be important. Post-depositional changes can affect the original porosity markedly; porosity may be created, obliterated, or extensively modified by solution, cementation, recrystallisation or dolo-mitisation[21].

A number of different sets of terms are in use for describing and classifying limestones. Some, such as "reef limestone" and oolite are to a large extent self-explanatory; others, including certain terms concerned with grain sizes and structure, are not. In classifying limestones according to their grain structure, a *mudstone* is defined as being mud-supported, with less than 10% of grains, whereas a *wackestone,* while also being mud-supported, has more than 10% of grains. On the other hand, a *packstone* is grain-supported, but also contains particles of clay and fine silt size (mud), while a *grainstone* is grain-supported and lacks mud. In a *boundstone* there is evidence that the original larger fragments became bound together at the time of deposition; some of the openings are too large to be ordinary interstices, and may be floored by fine sediment[3].

Dolomite can be primary or secondary in origin, and it must be recognised that only when dolomitisation is a secondary process and takes place on a rigid framework can the increases in porosity calculated for the calcite-dolomite transformation be claimed to have been effective.

Limestones may be formed in place (i.e. autochthonous) by the accumula-tion of organic débris, e.g. bioherms, or by chemical precipitation, e.g. chalks, or they can be allochthonous—formed from transported material, rather like sandstones and shales, with the terms *calcarenites* and *calcilulites* being used, according to the dominant grain diameters.* The distances of material transportation will, however, be much less than for sandstones and shales. Oolitic limestones belong to the calcarenite group, the oolites having been formed by a process which involves precipitation of calcite in shallow, agitated water. Oil reservoirs are known in calcilutites as well as calcarenites, but whereas solution voids may be essential in the former, the latter may still function in the presence of some void infilling with sparite (comparatively coarse calcite cement). However, for the finer-pored rocks, a tall oil column is essential for water-free oil production to be possible.

Calcarenites form many important reservoir rocks, notably in the Saudi Arabian "Arab Zone D" fields. Finer-grained limestones provide reservoirs in the Tampico-Tuxpan area of Mexico, and the Asmari fields of Iran and Iraq.

Chalks are mainly fine-grained friable limestones derived from the tests of floating organisms. They can act as reservoir rocks when adequate porosity and permeability have been induced by fracturing; however, they can, under other circumstances, act as cap rocks. Normally the porosity is high (about 50%) while the permeability is low. Coarse varieties of chalk of Thamama age produce oil at *Zakum,* the offshore Abu Dhabi field. Finer varieties have been filled with oil when a sufficiently tall oil column has provided the necessary buoyancy (e.g. *Gach Saran* and *Agha Jari,* Iran), although natural or artificial fracturing (coupled with acidization) is necessary for the pro-duction of oil at acceptable rates. On the other hand, a number of Middle East oilfields are actually sealed by fine-grained chalky limestones.

*Calcarenites have average grain sizes of $\frac{1}{16}$-2mm; calcilulite grains have diameters of less than $\frac{1}{16}$mm.

Dolomite is rarely a primary deposit; it is usually the result of replacing calcium by magnesium in a pre-existing limestone by the action of magnesium-rich brines or meteoric waters. Primary dolomites have been produced in the laboratory by the evaporation of sea-water at low temperatures (near 0 °C), after the precipitation of calcium carbonate, gypsum and halite. Very fine-grained "micro-dolomites" have been found forming in coastal salt flats (e.g. the "sabkha" areas which exist today around the coast of Qatar and the Trucial States). They are generally very porous but poorly permeable, and are often found to be interbedded with anhydrite.

The occurrence of dolomite in the matrix of the Middle Devonian Onondaga Limestone in New York State in the form of very fine grains suggests that it may be detrital—possibly wind-transported from supratidal sediments of that age. Hence, the dolomitisation process may include the deposition of fine-grained dolomite (associated with detrital quartz and calcite), with subsequent diagenetic overgrowth on such nuclei. Access of magnesium-bearing water is obviously needed for converting calcite to dolomite, so a measure of permeability and a source of such water are essential[1,16].

Very large dolomitic limestone reservoirs include the Ordovician Trenton limestone of the *Lima-Indiana* field, the Permian reservoir at *Yates*, Texas, and the Cretaceous Tamabra reservoir of *Poza Rica*, Mexico.

Secondary porosity and permeability in limestones can be produced by the solvent action of meteoric waters containing dissolved atmospheric carbon dioxide. Uplift and exposure of limestones can afford an opportunity for such action, and it is believed to have produced the good reservoir properties found in some limestone oilfields. The effects may be limited to the top few tens of feet in some limestones; in other instances, however, thicknesses of several hundred feet of rock may be affected. Calcite is much more soluble than dolomite, so a mixed carbonate rock subjected to atmospheric weathering and solution processes may develop "cavernous" secondary porosity.

Carbon dioxide produced bacterially or otherwise from the subsurface breakdown of organic matter may also be instrumental in dissolving parts of limestones. In such cases, surface exposure may not be involved.

Many prolific oilfields occur in rocks with high fracture permeability, the unfractured rock having relatively poor oil- or gas-producing characteristics. Indeed, within a single reservoir there can be marked changes in the productivity associated with the non-uniform distribution of the fractures. An example is the significance of fracture porosity in the important Asmari limestone reservoir rocks of Iran.

The fractures arise from the stresses involved in folding movements. Their occurrence and frequency may vary with the degree of fold curvature. In addition, fractures can be associated with faulting and with change in the vertical load on the rocks.

REFERENCES

1. J. E. Adams and M. L. Rhodes. "Dolomitization by seepage refluxion". *Bull. Amer. Assoc. Petrol. Geol.*, **44** (12), 1912–1920 (1960).
2. E. G. Baker. "A geochemical evaluation of petroleum migration and accumulation". pp. 299–329, in: "Fundamental Aspects of Petroleum Geochemistry" (eds. B. Nagy and U. Colombo), Elsevier Publishing Co., Amsterdam (1967).

3. P. W. Choquette and L. C. Pray. "Geological nomenclature and classification of porosity in sedimentary carbonates". *Bull. Amer. Assoc. Petrol. Geol.,* **54** (2), 207–250 (1970).

4. R. J. Cordell. "Depths of origin and primary migration; a review and critique". *Bull. Amer. Assoc. Petrol. Geol.,* **56** (10), 2029–2067 (1972).

5. R. J. Cordell. "Colloidal soap as proposed primary migration mechanism". *Bull. Amer. Assoc. Petrol. Geol.,* **57** (9), 1618–1643 (1973).

5a. H. V. Dunnington. "Aspects of diagenesis and shape change in stylolitic limestone reservoirs". *Proc. 7th World Petroleum Congress, Mexico,* **2**, 339–352 (1967).

6. W. C. Gussow. "Differential entrapment of oil and gas; a fundamental principle". *Bull. Amer. Assoc. Petrol. Geol.,* **38** (5), 816–853 (1954).

7. B. Hitchon. "Origin of oil; geological and geochemical constraints". pp. 30–66, in "Origin and Refining of Petroleum". Advances in Chemistry Series 103, *Amer. Chem. Soc.,* Washington (1971).

8. G. D. Hobson. "Compaction and some oilfield features". *J. Inst. Petrol.,* **29** (230), 37–54 (1943).

9. G. D. Hobson. "Some Fundamentals of Petroleum Geology". Oxford University Press, London (1954).

10. G. D. Hobson. "Problems associated with the migration of oil in 'solution'." *J. Inst. Petrol.,* **47** (449), 170–173 (1961).

11. G. D. Hobson. "Factors affecting oil and gas accumulations". *J. Inst. Petrol.,* **48**, 165–168 (1962).

12. G. D. Hobson. "Petroleum accumulation". *J. Inst. Petrol.,* **59** (567), 139–141 (1973).

13. G. W. Hodgson, B. Hitchon and K. Taguchi. "The water and hydrocarbon cycles in the formation of oil accumulations". pp. 217–242, in: "Recent Researches in the Fields of Hydrosphere, Atmosphere and Nuclear Geochemistry". Maruzen, Tokyo (1964).

14. M. K. Hubbert. "Application of hydrodynamics to oil exploration". *Proc. 7th World Petroleum Congress, Mexico,* Vol. 1, 59–75 (1967).

14a. A. L. Kidwell and J. M. Hunt. "Migration of oil in recent sediments of Pedernales, Venezuela". In: *Habitat of Oil* (Ed. L. G. Weeks), Amer. Assoc. Petrol. Geol., Tulsa, pp. 790–817 (1958).

15. A. I. Levorsen. "Time of oil and gas accumulation". *Bull. Amer. Assoc. Petrol. Geol.,* **29** (8), 1189–1194 (1945).

16. R. C. Lindholm. "Detrital dolomite in Onondaga Limestone (Middle Devonian) of New York: its implications to the 'Dolomite Question'." pp. 216–223, in: "Carbonate Rocks, I. Classification—Dolomite—Dolomitization". AAPG Reprint Series No. 4 (1972).

17. C. McAuliffe. "Solubility in water of paraffin, cycloparaffin, olefin, acetylene, cyclo-olefin and aromatic hydrocarbons". Paper presented to Division of Petroleum Chemistry. *Amer. Chem. Soc.,* Chicago meeting, 1964.

18. E. Peake and G. W. Hodgson. "Alkanes in aqueous systems. I. Exploratory investigations of the accommodation of C_{20}–C_{33} n-alkanes in distilled water and occurrence in natural water systems". *Amer. Oil. Chemists Soc. J.,* **43** (3), 215–222 (1966).

19. E. A. Perry and J. Hower. "Late-stage dehydration of deeply buried pelitic sediments". *Bull. Amer. Assoc. Petrol. Geol.,* **56** (10), 2013–2021 (1972).

20. M. C. Powers. "Fluid-release mechanisms in compacting marine mud-rocks and their importance in oil exploration". *Bull. Amer. Assoc. Petrol. Geol.,* **51** (7), 1240–1254 (1967).

21. J. L. Stout. "Pore geometry as related to carbonate stratigraphic traps". *Bull. Amer. Assoc. Petrol. Geol.,* **48** (3), 329–337 (1964).

22. P. D. Trask. "Proportion of organic matter converted into oil in Santa Fe Springs field, California". *Bull. Amer. Assoc. Petrol. Geol.,* **20** (3), 245–247 (1936).

CHAPTER 4

The Accumulation of Petroleum

The Nature of Traps

THE traps in which oil and gas accumulate take a number of fundamentally different forms, as regards their mode of origin and detailed geometry. However, so far as their *shapes* are concerned, and omitting the case of hydrodynamic trapping, all can be considered as variants of a single basic form in which the oil or gas is trapped against a boundary between reservoir rock and non-reservoir rock which is *convex upwards*. The convexity may vary from a gentle curve to a sharp angle, but the result must always be that the trap is *closed* in both vertical and horizontal planes by the local confining factors; hydrodynamic traps appear to be exceptional in this respect.

Having recognised that the essential characteristic of a trap is this convex upward boundary between the reservoir rock and the sealing rock—cap rock, cap rock-seat seal combination, or these combined with a lateral rock seal provided by faulting—then the most important factor in exploration is the appreciation of the possible ways in which these forms and associations can arise. For "anticlinal" or "dome-type" traps, a rock seat-seal becomes involved only when the amount of hydrocarbons accumulated is such as to provide a hydrocarbon column greater than the thickness of the reservoir rock at the crest. Otherwise, the bottom seal is everywhere provided by water-bearing reservoir rock. Should a reversal of dip not be a feature of a trap, seat-seal rock is automatically involved in retaining the hydrocarbons in all cases except those of hydrodynamic trapping, and lateral seal provided by faulting when the hydrocarbon column does not exceed the height of sealing rock at the fault contact.

Domed forms, i.e. traps dependent on dip reversal, may arise in different ways: from folds caused by lateral compression of the beds; by differential compaction or "draping" over buried basement hills or limestone "reefs"; by non-uniform vertical reduction in thickness owing to solution, the gypsum-anhydrite transformation, or the formation of stylolites; by irregular uplift and settling associated with salt or other intrusions, or block faulting; and as a result of the formation of "roll-over" structures associated with so-called "growth" faults[8, 34].

In addition to lateral trapping by faults, trapping against a salt or other intrusion is also effectively a fault trap. A porous and permeable horizon can be terminated locally not only by faulting, but also by an unconformity or by a lateral change in the nature of the sediment deposited or in its later modification, or by a change in the degree of cementation, including the formation of clogging petroleum residues in the pore spaces. In the case of an unconformity, many examples are known of accumulations in beds overlying and abutting against an unconformity, as well as instances of an erosion plane cutting across a reservoir horizon or even developing the porosity and permeability in a reservoir rock beneath an unconformity[21].

A number of classifications of oilfield traps have been published, ranging from some which are extremely elaborate to others which are simple and effectively cover only the main types of trap referred to above. In particular, attempts have been made to differentiate between those traps thought to have been produced mainly by structural influences and others which are due to so-called "stratigraphic" factors. A further grouping of "combination" traps has then had to be created, to cater for the multitudinous accumulations which exhibit both "structural" and "stratigraphic" trapping features. There seems little virtue in a complex classification, and simple combinations of common geological terms can be adequate for any general description. This is equally true when the trap involves more than one of the basic features or when there is need to refer to some special association[5, 21, 32].

The following descriptions of "anticlines", faults, salt domes, reefs, unconformities, etc., indicate important features with which one or other of the fundamentally different types of traps can be associated (Table 29). It is obvious, for example, that a salt dome can have associated fault traps, wedge-out traps of various kinds, traps due to dip reversal, and even a trap with a lenticular form (cap rock).

TABLE 29

TRAPS AND THEIR ASSOCIATIONS

Form	Association
Anticlinal (dip reversal)	Folds
	Salt domes and other intrusions
	Uplifted blocks (horsts) ⎫
	Buried hills ⎬ compaction,
	Reefs ⎪ draping
	Sand lenses ⎭
	Growth faults—"roll-over"
	Rock removal or redistribution—solution, stylolites
Fault	Faulting
	Salt domes and other intrusions
	Rock removal or redistribution—solution, stylolites
Up-dip wedge-outs or termination ("stratigraphic")	Depositional changes
	Cementation/diagenesis
	Asphalt clogging
	Unconformities
Lens ("stratigraphic")	Sand lenses
	Salt dome cap rock
	Carbonate reefs

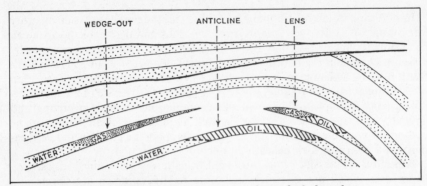

Fig. 18. Cross-section showing several types of traps for hydrocarbons.

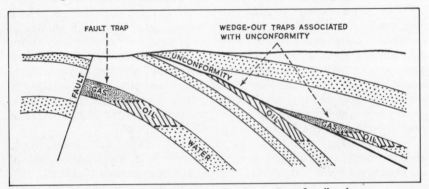

Fig. 19. Cross-section showing fault and wedge-out traps for oil and gas.

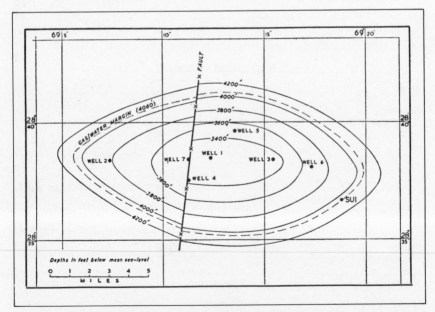

Fig. 20. Example of a faulted but symmetrical anticline at Sui, Pakistan. The structure contours are drawn on the upper surface of the main Eocene reservoir limestone formation.

STRUCTURAL TRAPS

Structural traps account for by far the largest group of hydrocarbon accumulations known at the present time, primarily because methods have been devised for locating them in the subsurface with some degree of accuracy. The other less easily detectable ("subtle") types of trap have up to the present usually been discovered more or less by accident during the course of drilling to discover or evaluate structural accumulations, or by careful studies of well and/or seismic data, combined with the ability to deduce that such traps are likely to occur in certain regions. Almost inevitably, structural traps are the first targets sought, considered and possibly tested in all previously unexplored areas.

1. ANTICLINES

Where there is a reversal of dip in a permeable bed along which secondary migration is proceeding, hydrocarbons will tend to accumulate at the crest of the resultant arch, and then spread down its flanks. Upward convexity (anticlinal arching) in a permeable bed therefore provides a means of concentrating and separating the relatively small proportions of hydrocarbons contained in migrating fluid currents.

Because of the comparative ease with which they can be detected by geological or geophysical techniques, the search for oil has always been very largely a search for anticlines. Hence, the dominant trapping factor in most of the world's known oil and gas accumulations is their anticlinal form (Table 30).

Anticlinal traps vary greatly in their shapes; they include folds the flanks of which dip at less than one degree, domal folds whose lengths are about the same as their widths, and many examples of symmetrical folds which are oval in plan with dips approximately equal on either side of their axial planes. There are also asymmetrical anticlines which have their axial planes dipping towards one flank, which in extreme cases may form "recumbent" folds, with one flank overturned. When the longitudinal axis of an anticline is inclined, the anticline is said to *plunge* in the direction in which the axis becomes lower.

The *closure* of an anticline is the vertical distance between its highest point and the plane of the lowest closed structural contour—the "spilling plane" which is the level at which hydrocarbons could escape if the trap were filled to its maximum capacity. The "*closed volume*" is the volume of the structure between its highest point and the "spilling plane".

When anticlines contain hydrocarbons they may be filled down to the "spilling plane", or on the other hand they may contain considerably smaller volumes of oil or gas than they could accommodate. A useful indication of the hydrocarbon content of a structure is given by the height of the *oil and gas column*—i.e. the vertical distance between the oil/water contact plane and the highest point at which oil or gas is found in the trap.

When the thickness of the reservoir rock exceeds the height of closure of the arched structure, the seat seal is provided throughout by water-bearing reservoir rock: this gives the maximum storage capacity for a single reservoir rock. However, for a structure filled with hydrocarbons to the spilling plane, a series of superimposed reservoir rocks (each with its own cap rock and spilling plane) aggregating in thickness more than the height of closure are potentially able to hold more oil or gas than a single reservoir rock equalling this total thickness in a structure of identical form.

When surface mapping has located an anticline, the problem arises of the relationship between the structure at the surface and the structure in the subsurface, even in the absence of an intervening major unconformity. Thus, for a very gentle fold, whether symmetrical or not, wedging of the group of beds between the surface marker and the reservoir rock can give a considerable difference in the plan positions of the crests of the surface and subsurface arches, provided that the reservoir rock is deep. On the other hand, with low dips, failure to hit the subsurface crest by an exploratory well will not mean a large drop below the crest of the subsurface reservoir. It is also possible with wedging for the subsurface arched beds to be associated with a surface feature which is not arched, but merely appears as a steepening of the dip, namely, a "terrace". The inverse situation is also possible.

An anticlinal feature superimposed on a regional monocline, its axis lying in the direction of the regional dip, will lack closure and produce an open anticlinal "nose" when its axis dips throughout in the direction of the dip of the monocline. An effective trap can then result only if the "open" side is closed by some additional structural or stratigraphic feature, e.g. a fault or a permeability barrier.

TABLE 30

TRAPPING FEATURES

Percentages of fields or pools

		USA[1]	Major fields in USA[2]	Non-Communist World majors[3]	Giant oil- and gasfields[4]
Type of trap	Anticline	65·4 ⎫	77·8	58·2	89·1*
	Fault	5·2 ⎬		7·8	
	Unconformity	0·7 ⎭		6·0	(29·3)†
	Reef and other		10·8		
	stratigraphic	10·2 ⎰		16·1	9·4
	Combination	18·5	11·4	11·9	
Lithology of	Sandstone	66·8		61·7	55·1
reservoir rock	Carbonate	31·9		32·0	41·9
	Fractured shale, igneous and metamorphic rocks	1·3		6·3	3·0

[1] Based on information in *J. Petrol. Tech.*, 1950.

[2] Fields expected to yield over 100×10^6 brl of oil or 1×10^{12} cu ft of gas (259 oilfields, 47 gasfields). From data assembled by M. T. Halbouty, *Bull. Amer. Assoc. Petrol. Geol.*, **52** (7), 1115–1151 (1968).

[3] From compilation by G. M. Knebel and G. Rodriguez-Eraso, *Bull. Amer. Assoc. Petrol. Geol.*, **40** (4), 547–561 (1956).

[4] Fields expected to yield over 500×10^6 brl of oil or $3·5 \times 10^{12}$ cu ft of gas (187 oilfields, 79 gasfields). Based on information in "Geology of Giant Petroleum Fields", Memoir 14, Ed. M. T. Halbouty, *Amer. Assoc. Petrol. Geol.*, Tulsa, Oklahoma (1970).

* This should be viewed as an indication of the association. For fields with more than one pool or accumulation some may be trapped by other features.

† Fields with important production related to an unconformity.

Should the flank dips of a fold be steep and the structure decidedly asymmetrical, the plan positions of the fold at the surface and at depth may differ considerably. The displacement will depend on the degree of asymmetry, the depth of the reservoir horizon, and the thickness of "incompetent" beds lying between the surface and the deep structure, as well as on any depositional wedging in the group of beds above the reservoir rock. Other things being equal, the shift may be expected to be greater when the covering beds are mainly competent rather than incompetent. However, there are cases, as in Iran, where extreme mobility of the cover rocks causes them to have complexities not shown by the underlying competent reservoir rock, with the result that their form does not provide a good guide to the subsurface structure.

In the case of asymmetrical structures with strong dips, a well which is drilled substantially off the crest and on the steep flank of the reservoir formation may be structurally below the level of its crest. When an exploratory well misses the crestal region of the reservoir and directly enters water, this is not necessarily a serious failure, so long as the volume of hydrocarbons which could be present in the untested crestal region is not of commercial magnitude.

The stronger anticlinal structures are formed by lateral pressure or folding movements. The very gentle structures can be formed not only in this way, but by other mechanisms: differential compaction or "draping" over buried topography or carbonate reefs; the effect of salt or other intrusions; the uplift resulting over a buried horst; and by the "roll-over" associated with growth faulting. Non-uniform rock removal or redistribution by solution or stylolite formation, respectively, may also cause the arching of overlying beds.

Whether a particular anticline contains hydrocarbons depends on a number of factors, which are sometimes obscure. An endeavour has been made later in this Chapter to explain the disappointing failures which sometimes occur. Nevertheless, the anticline discovered in a reasonably thick sedimentary sequence is still in general the trap most likely to contain petroleum.

Since most of the world's oilfields are anticlinal structures, only a few notable examples call for special mention. Thus, the *Burgan* field in Kuwait and the *Romaschkino* accumulation in the Ural-Volga petroleum province of the USSR are examples of relatively gentle domal structures containing enormous concentrations of oil. The *Kirkuk* field in Iraq is one of the longest oil-producing anticlines known, with a productive area extending in three culminations over a length of nearly 30 miles. In Mexico, the "Golden Lane" fields occur as local "highs" along a huge, undulating limestone arch some 51 miles long. Regional folds or arches several hundred miles long are known in the USA and USSR, but in general the hydrocarbon accumulations are limited to relatively small productive areas on the crestal areas of such major folds. Most oil-producing anticlines are only a few miles long and wide, although they vary in their size, dip, pitch and closure between wide limits.

In general, strong folds are usually associated with extensive faulting and likely fracture of the reservoir, although some accumulations are known in overthrust structures—e.g. *Boryslaw* in the West Ukraine and *McKittrick* in California. At the other extreme, some anticlines produced by epeirogeny in "shelf" areas can have extremely gentle flank dips as for examples *Bab* and *Bu Hasa* in Abu Dhabi. Some anticlines are filled with hydrocarbons to the limits of their closures, while others contain considerably smaller volumes

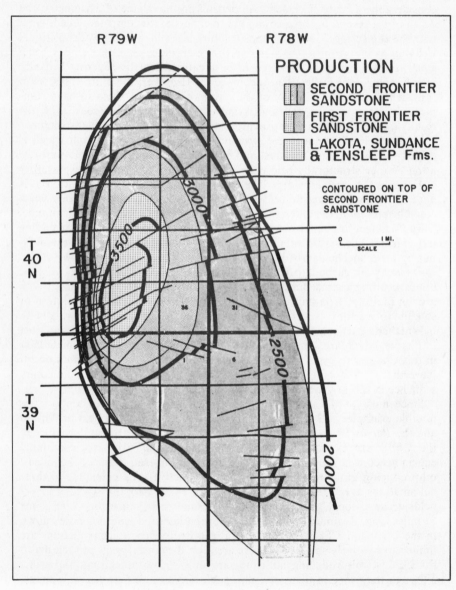

Fig. 21. Structure contours of the Salt Creek oilfield, Natrona County, Wyoming.
(J. A. Barlow and J. D. Haun, "Regional Stratigraphy of Frontier Formation" in "Geology
of Giant Petroleum Fields", Amer. Assoc. Petrol. Geol., Tulsa, 1970.)

of oil or gas than they could accommodate. The latter have been termed "starved anticlines"; the *Swanson River* field of Alaska is an example.

Examples of anticlinal trapping

(a) The *Salt Creek* oilfield, on the southwest flank of the Powder River Basin in Wyoming is an asymmetrical dome, elongated approximately north-south (Fig. 21); a considerable number of faults cut the crest and flanks, and the closure is 1,500ft. The overall productive area is nine miles from north to south and four miles from east to west. The producing area of the First Frontier sandstone (Cretaceous) is 2,500 acres, whereas that of the Second Frontier sandstone is over 22,000 acres. Fissured shales above and below the Frontier produce, while there is also deep production from the Lakota (Cretaceous), Sundance (Jurassic) and Tensleep (Pennsylvanian) formations. The main reservoir, the Second Frontier sandstone, is an offshore bar with a maximum thickness exceeding 100ft, a width of about 10 miles from east to west, and a north-south extent of more than 60 miles. There are marked lateral and vertical changes in grain size in the bar sandstone; bedding is indistinct in some places; elsewhere thin-, thick- and cross-bedding occur.

The anticline was formed in latest Cretaceous or early Tertiary time, and it has been suggested that the Second Frontier sandstone acted as a "stratigraphic" trap before the oil moved into the present structurally-determined position. To the south, the contiguous Teapot Dome had a gas cap in this sand, whereas the Salt Creek dome, with its crest 1,300ft higher, had no gas cap. This situation has been explained as an example of differential entrapment, involving migration from south to north along the anticlinal trend.

The Second Frontier sandstone outcrops south of Casper and again northwest of Salt Creek, indicating a potentiometric surface with a mean gradient of 4–6 psi/mile. Early publications showed a drop of 130ft in the oil/water contact from south to north. However, there is some doubt about this interpretation, and because the faults may affect the water flow pattern in the sandstone a simple relationship between the mean hydraulic gradient and the reported drop in level of the fluid contact is not to be expected.

(b) *Masjid-i-Sulaiman* is an asymmetrical anticline of Asmari limestone, in the foothills area of Iran, with the rocks overlying the Asmari reservoir rock having a structurally more complex form than the limestone (Fig. 22). The latter has dips of 20–30° to the northeast, and a steep southwest flank which is almost vertical at one place. The oilfield is 18 miles long and four miles wide. It had no initial gas cap, and its oil column was 2,200ft tall[19].

In this general area, the Asmari limestone consists of packstone and wackestone. The ratio of micrite to grain development differs in broad belts, as does the diagenesis involving dolomitisation and anhydritisation. These belts appear to be related to the overall basin shape. An arbitrary division of the limestone into pay and non-productive rock takes note of porosity, permeability and water saturation. The non-productive section has porosity under 5%, and permeability less (often much less) than 1 mD. Pay averages 9–14% porosity and 10 mD permeability, rising as high as 20 mD in some parts. On this basis "pay" ranges between 25% and 75% of the total Asmari thickness.

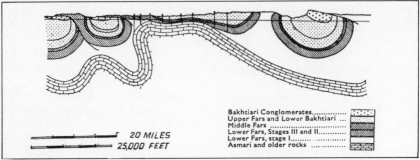

Bakhtiari Conglomerates...............
Upper Fars and Lower Bakhtiari ...
Middle Fars ....
Lower Fars, Stages III and II...........
Lower Fars, stage I.......................
Asmari and older rocks ....

20 MILES
25,000 FEET

Fig. 22. Cross-section of the Masjid-i-Sulaiman oilfield, Iran (after G. Lees).

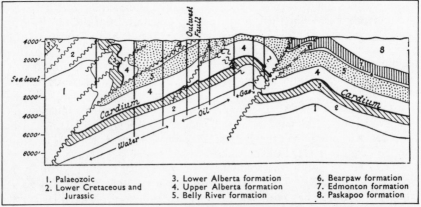

1. Palaeozoic   3. Lower Alberta formation   6. Bearpaw formation
2. Lower Cretaceous and   4. Upper Alberta formation   7. Edmonton formation
   Jurassic   5. Belly River formation   8. Paskapoo formation

Fig. 23. Cross-section of the Southern Turner Valley oilfield, Alberta (after G. Hume).

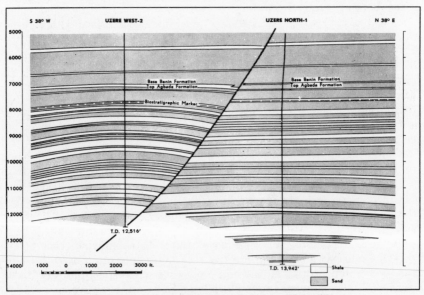

Fig. 24. Typical "growth" fault in the Uzere oilfield, Nigeria.
(K. Short and A. Stauble, "Outline of Geology of Niger Delta", *Bull. Amer. Assoc. Petrol.
Geol.*, p. 777, May 1967.)

A very compact nodular anhydrite with a few thin interbedded shales constitutes the Cap Rock over the Asmari, and this is the lowest part of the Gach Saran, a plastic group, 2,000–4,000ft thick, made up of alternations of anhydrite, salt, grey marls and a few thin limestones.

The indicated amplitude of some of the folds in the Iranian oilfield belt is as much as 30,000ft. The folds, running northwest-southeast, have critical closure for the "highs" at their southeast ends, because of the regional tilt down to the northwest. On large structures, the closure is between 4,000ft and 7,000ft. Generally, such structures are filled with hydrocarbons down to the spill point. *Gach Saran* has a 6,000-ft hydrocarbon column. Gas caps range from very large to small, while nearby fields can have undersaturated oil. Usually, the oil in any accumulation is very uniform in composition, because a very good fracture system in the limestone has allowed mixing by means of convection currents. There is rapid pressure adjustment over considerable distances during production, via the fractures.

Some 2,000–3,000ft of shale and marl separate the Asmari from the underlying Bangestan limestone, yet because of fractures across this interval the two formations have accumulations with identical oil/water levels, essentially the same composition and type of oil, as well as sympathetic fluid movements. Where there are no fractures, the Asmari and Bangestan behave as independent reservoirs, with different oils, and no fluid interchange or linked pressure behaviour during production.

A regional pressure gradient exists in the water in the Asmari, involving a high pressure (above hydrostatic) near the mountains and a lower pressure, approximately hydrostatic, near the Gulf. The water flow has given differences of as much as 500ft for the oil/water contacts on the two flanks of some structures.

(c) The *Turner Valley* oilfield in Alberta is an example of an oil and gas accumulation in an *overthrust fold structure*. There is oil in Upper Mesozoic sandstones, but principally in a Palaeozoic limestone, known as the Madison or Rundle limestone. This limestone has been crumpled and thrust eastwards along a west-dipping major thrust, and to the west of this structure is yet another overthrust (Highwood or Outwest overthrust sheet)[20].

The producing zones of the Palaeozoic limestone are dolomites separated by dense, hard zones with considerable chert. Porosities of 1–20% have been measured in the reservoir rock. There are marked variations in the porosity, permeability and degree of fracturing of the limestone, and also in the amounts of secondary calcite in the producing zones.

The heaviest oil is in the younger rocks. The Palaeozoic limestone reservoirs have tall gas columns (over 1,000ft), with considerable oil columns (about 2,000ft) down dip on the west flank (Fig. 23). For a number of years it was believed that the Palaeozoic limestone contained only wet gas.

On the basis of observations made during production it has been argued that the reservoir horizons in the Palaeozoic limestone are isolated from the limestone in the Alberta syncline to the east, and from the limestone in the overthrust Highwood structure to the west. However, in view of the limestone having non-reservoir zones as well as reservoir zones the significant point must be isolation of the latter, not necessarily isolation of the limestone as a whole.

(d) A number of "roll-over" anticlines are known at depth in the *Niger delta* area. These are usually about 4–5 miles long and 2–3 miles wide, and in some cases are cut by minor "growth faults" (p. 95) like the main one, and by antithetic compensatory faults. There is evidence of deep-seated anticlines, possibly associated with a shale bulge, occurring behind and beneath the main growth faults (Fig. 24).

Most of the oil accumulations in the Niger delta area are in such "roll-over" anticlines in Agbada sandstones. The fields are normally multi-reservoir, with few of the reservoirs full to spill point. Some of the sandstones are barren. The extent of the accumulations may or may not be limited by any subsidiary growth faults or antithetic faults. Limitation of this kind is more evident on the larger "roll-over" anticlines, because the greater extent of their crestal area tends to provide a less efficient focus for migration[34].

The reservoir rocks are typically channel and barrier sandstone bodies, developed during the delta's long history. Porosities up to 40% are known, and permeabilities as high as 1–2 darcys. The low-sulphur oils are paraffinic, and usually have wax contents of 1–10%. In a single multi-reservoir field it is not uncommon for the oils in the different reservoirs to differ considerably in properties, although there is a tendency for lighter oils and more gas to occur with increase in depth. The interbedded shales of the Agbada and the shales of the upper part of the underlying Akata close to the reservoir sandstones are considered to be likely source rocks. The ages of the producing horizons range from the Eocene probably into the Pliocene.

2. FAULT TRAPS[15, 16]

A fault is a rock fracture which results in the relative displacement of the strata on the two sides of the break. Faults are generally the result of shearing forces related to earth movements, and may vary in size from very small local dislocations to major breaks more than 100 miles long and involving many miles of displacement.

Fault planes, which are often complex fracture zones rather than simple surfaces, have a general dip and strike; since they produce displacement in the beds they cut, it is necessary to define the relationship between the resultant blocks on either side. The "throw" of the fault is the vertical component of the relative displacement of the blocks, the "heave" is the horizontal component of the displacement normal to the strike of the fault, whereas the "strike slip" is the horizontal component of the relative displacement parallel to the strike of the fault. The total relative displacement on the fault plane is referred to as the "slip". However, fault terminology is complex and has not always been applied consistently, even though the basic concepts are simple.

Faults are usually classified by reference to the relationship between the dip and strike of the fault plane and the dip and strike of the strata it intersects. Thus, "strike faults" parallel the local strike, "dip faults" parallel the local dip, and "diagonal" or "oblique" faults cut across the local strike. There are a number of well-defined geometrical relationships between groups of faults, resulting from their origin; they may, for example be "radial", "peripheral", "en échelon" or "antithetic", depending upon their mutual relationship in either the horizontal or vertical planes.

"Tear-faults" or "wrench faults" (also called "transcurrent faults") result from a generally horizontal shearing movement. "Step-faults" are groups of roughly parallel normal faults, and the affected beds all drop downwards in the same direction. Parallel thrust faults which dip in the same general direction can produce a number of repetitions of the same formations termed "imbrications".

A structural depression bordered by roughly parallel faults with their downthrow inwards is termed a "graben"; while the faults concerned here are usually normal, they may also be reverse. "Half-grabens" result from faults with their downthrown sides in the same direction. A structurally elevated block bounded by roughly parallel faults downthrown outwards is termed a "horst". When grabens and horsts take the form of large-scale regional features, they produce the characteristic fault-block and block-mountain structures called "basin and range", i.e. linear mountains and parallel valleys (as in Nevada and the adjoining States, where the valleys or "basins" are in fact filled with sediments from the adjacent mountains).

A long, narrow graben which has sunk as a result of crustal tension while being covered with sediments forms a "rift valley", if it is of regional dimensions and is bordered by normal faults.

Faults may lie entirely in the subsurface or they may reach the surface and affect the local topography. If the exposed fault zone contains resistant material which weathers more slowly than the surrounding rocks, it may be marked by a distinct ridge. Outcropping fault traces are generally most easily detected by means of air photographs through the recognition of any topographical feature (ridge or depression) associated with the fault zone, changes in soils or vegetation, and any offset formation boundaries. Strike faults raise special problems in detection and, when the rocks exposed on the two sides differ in resistance to weathering and erosion, may give a fault scarp which simulates a scarp formed in the erosion of a normal geological sequence of rocks. Such a fault scarp, depending on circumstances, may either face towards the downthrown block or towards the upthrown block.

The beds on either side of a major fault will be displaced in the subsurface sometimes by hundreds or even by thousands of feet. This may have been deduced from the prior seismic survey, or may only be discovered for the first time in the course of drilling, when correlations show discrepancies between the forecast and actual depths at which a particular "marker" bed is found. Alternatively, beds may be unexpectedly repeated, or apparently thickened, thinned or even completely eliminated as a result of faulting. In the latter cases, care must be taken to differentiate between the effect of faulting and the presence of unconformities, which may perhaps only be discovered later when further drilling data become available.

Elimination resulting from faulting gives a linear band of anomalies, and can appear at different stratigraphic levels in a series of wells which penetrate the fault; elimination caused by an unconformity has effects which will be noted over a wide area, and there will be some consistency in the group of stratigraphic levels involved.

Confirmation of a subsurface fault may be obtained by seismic detailing and also by the recovery of "slickensided" cores—i.e. rock samples with grooves, scratches or friction-polished surfaces.

From the point of view of petroleum geology, the most important basis of classification of faults is the relationship between the directions of movement of the blocks of strata intersected by the fault. If the block towards which the fault plane dips has moved *downwards* relative to the other block, resulting in a local tensional lengthening of the crust, the fault is called "normal"; but if it has moved relatively *upwards,* the fault is termed "reverse". A shallow-angle reverse fault, where the average dip of the fault plane is less than 45° is called a "thrust fault". Major thrust faults are known along which large blocks of rock appear to have travelled for considerable distances, often of several miles. Such "allochthonous" travelling masses move outwards along "overthrusts" from an inner "hinterland" towards a stable, autochthonous "foreland". In such cases, the faulted area has apparently been shortened as a result of lateral compression.

Relatively few oilfields are associated with reverse and thrust faults, pro-bably because the intense tectonic activity in such areas, involving strong distortion and extensive adjustments, increases the chances of oil and gas escape from any potential traps. The accumulations, when they exist, are generally found in the overturned folds which have been carried forward along low-angle fault planes. Examples are the *Boryslaw* field in the West Ukraine (previously Poland) (Fig. 25), the *Turner Valley* field of Canada (mentioned above), and the *Talar Akar* field of Sumatra.

To some degree, there is a general relationship between faulting and folding. Thus, while some folds are not faulted at all, it is more common for faults to develop in regions of maximum folding stress—i.e. the zones of sharpest curvature at the culminations of anticlines. These faults may be sufficiently important to have allowed the partial or complete escape of hydrocarbons if the fracturing occurred during the later phases of folding after initial accumu-lation had taken place. On the other hand, faults can provide physical barriers to oil movement by setting an impermeable bed against a reservoir formation. The trapping capacity of a fault is in fact directly related to the differences in displacement pressures between the reservoir and boundary rocks which it brings into contact. This sealing capacity of a fault may be related to any impervious "gouge" material present between the fault walls. "Gouge" is the characteristic finely-ground, clayey material which results, often with angular tectonic breccias, from the shearing of sedimentary rocks in fault planes. Cementation by calcite- and quartz-bearing fluids is also com-mon in fault planes along which such liquids have permeated. Heavy oils, waxes and asphalts are also sometimes found in fault planes, and oil and gas "shows" or seepages frequently occur at fault outcrops[35].

The relationships between faults and petroleum accumulations are therefore complex and are largely dependent on the local structural circumstances and timing in relation to oil and gas migration. Although surface mapping or seismic reflection surveys may show a faulted marker bed on a monocline to have a form which indicates that the rocks could provide a trap, this deduc-tion is not conclusive, because there must also be appropriate juxtaposition of reservoir rock against sealing rock at the fault plane at a structurally-high point of the reservoir rock. When the seal is not material in the fault plane or zone, sealing must depend on the detailed distribution of reservoir rock and non-reservoir rock in relation to the fault displacement, at a locally structurally-high point of the reservoir rock. Reservoir beds dipping away from a fault on its two sides raise different considerations with respect to the possible existence of a trap.

In general, faults produce oil accumulations when in combination with other structural and stratigraphic factors; thus, the intersection of a fault plane with an anticlinal "nose" can form a sealed trap, provided that an impermeable surface is brought into contact with the reservoir formation, while fissuring and fracturing associated with faulting may induce secondary porosity and permeability in compact rocks and hence favour oil accumulation. Faults also often break up reservoirs into non-communicating sections and are thus important features in oilfield development. They may provide lateral as well as up-dip seals for accumulations.

Faults which cut across a number of formations may in some cases have provided the channels whereby oil and/or gas has passed from an initial accumulation to a subsequent one, as is believed to have been the case in the oilfields of Northern Iraq. Interchange of oil and water must occur in these cases, with the oil moving upwards and the water downwards.

Deep-seated "basement" faults are thought to underlie many apparently anticlinal fields in the Middle East and the USSR. The faults associated with salt domes have also had a considerable influence on the accumulation of oil in many areas, particularly on the Gulf Coast of the USA, acting sometimes as seals and sometimes as channels of migration.

Where accumulations are directly attributed to faulting, they are nearly always on the upthrown side of the fault plane, provided that the downthrown block is in the up-dip direction, since in these cases oil and gas will tend to flow towards the fault plane. In addition, for throws less than the thickness of the cap rock immediately above the reservoir rock, the latter will be opposite the reservoir rock across the fault plane. However, this is not the only favourable circumstance.

Important examples of oil accumulations on the upthrown side of major faults are the Balcones-Mexia-Luling fault fields in the USA. Downthrown side traps are very exceptional, since in general the hydrocarbons would tend to escape up-dip around the end of a strike fault, when the upthrown side was up-dip.

A type of fault which is particularly important on the US Gulf Coast is the "growth" fault, defined as a normal fault which has a substantial increase in throw with depth and across which, from the upthrown to the downthrown block, there is a great thickening of correlative section.

Many other terms have been used to describe faults of this nature—"syndepositional faults", "flexure faults", "progressive faults", etc. They have been found, apart from the Gulf Coast, in modern deltaic areas, such as the Niger and Ganges-Brahmaputra deltas, as well as in ancient sedimentary basins (e.g. the East Venezuela and Vienna basins).

Growth faults usually form arcuate patterns and when apparently regional have often been shown to be made up of a series of curved segments. They are usually (but not always) downthrown and concave towards the present coast, and anticlines commonly form on the downthrown side. Throws of 3,000ft or

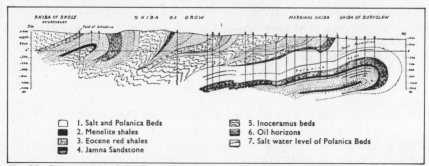

Fig. 25. Cross-section of the Boryslaw oilfield, West Ukraine (after K. Tolwinski). Length of section is about six miles.

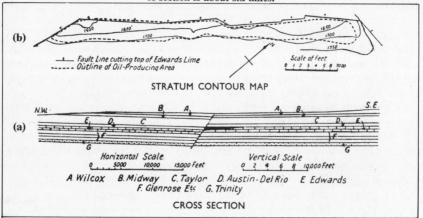

Fig. 26. Cross-section (a) and stratum contour map (b) of the Luling oilfield, Texas (after E. W. Brucks).

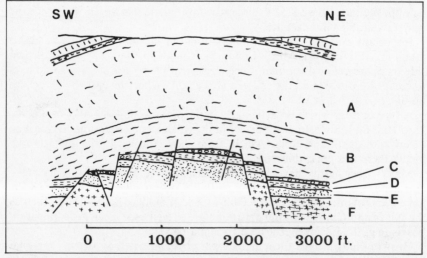

Fig. 27. Cross-section of the Hurghada oilfield, Gulf of Suez, Egypt (after P. van der Ploeg). A: Miocene shale and gypsum series. B: Miocene Globigerina marls. C: Miocene conglomerate. D: Nubian series—Garnet Zone. E: Nubian Series—Staurolite Zone. F: Pre-Cambrian.

more are common, and the dip of major growth faults ranges from 30°to 70°, the faults becoming less steep at depth. Although the causes of growth faulting are not fully understood, it seems likely that one cause is the greatly increased overburden pressure created by the influx of large quantities of sediment.

Normal faults generally affect the strata which they cut in such a way that beds close to the fault are tilted up towards the break on the downthrown side, and tilted down towards the break on the upthrown side. This fault "drag" may be caused by friction on the broken surfaces during the fault movement, or it may represent a flexure on the rocks which preceded the actual fracture. The dip associated with drag is likely to be at right-angles to the fault plane, and therefore dip-meter measurements on the strata just above or below the fracture in a well may make it possible, in particular when the normal formation dips are small, to deduce the trend of the fault. In the absence of such a measurement, a single well may show only the existence of a fault and the thickness of the beds eliminated.

However, for growth faults, with the fault plane becoming less steep at depth, the beds on the downthrown side must have tendency to be made to dip towards the fault for some distance, leading to a "roll-over" effect. Where the normal dip would otherwise be away from the fault on the downthrown block the combination creates the anticlinal feature which can trap oil moving up dip. "Roll-over" anticlines in Nigeria with oil accumulations have already been described on p. 92. Other oil accumulations of this kind occur in the Vienna basin, and there is an important series of similar accumulations south of the Vicksburg fault zone in southwest Texas, e.g. the *Seeligson* and *Agua Dulce* oilfields. If the general dip on the downthrown block is towards the fault the "roll-over" effect will cause steepening, but there will be no anticline.

Examples of Fault Trapping

(a) The *Luling* field of Caldwell and Guadalupe counties, Texas, produces from the Edwards limestone (Cretaceous). The general dip of the strata, 1°–2° to the southeast, is modified in direction at the northeast end of the field in particular (Fig. 26a). A northeast–southwest fault, downthrown to the northwest, provides the northwest boundary of the accumulation. Its throw is 450–500ft, and the fault plane dips at 48°–65°. The cap rock above the Edwards limestone is the hard, grey, impure Georgetown limestone, while the beds across the fault opposite the reservoir zone belong to the Taylor formation which consists of calcareous shales and argillaceous limestones[4].

The stratum contour map (Fig. 26b) draws attention to the fact that for fault trapping there must be either suitable structural warping against a strike fault, or else the fault itself must have a bent trace to provide lateral closure. In both cases escape of hydrocarbons round the ends of the fault is prevented.

(b) The *Hurghada* oilfield in Egypt is associated with a faulted uplift of the Mesozoic and Palaeozoic Nubian series lying on Pre-Cambrian igneous and metamorphic rocks. These old rocks are strongly faulted, whereas the overlying Miocene beds show arching and only minor faulting, as a result of renewed movements along faults in the pre-Miocene rocks (Fig. 27)[29].

The bulk of the oil has been produced from the Nubian series (Devonian to Cretaceous in age). However, oil has also been obtained from the dolomites of the Miocene Lagoonal Series, as well as from the basal beds of the Miocene.

SALT DOME STRUCTURES[24]

The natural evaporites are deposits apparently formed by the evaporation of brines to dryness in arid conditions; they are often found interbedded with carbonate rocks and red and green shales in cyclical sequences. The principal evaporites are halite (sodium chloride) and gypsum or its dehydrated form, anhydrite. Evaporite beds of almost every age are known, although Silurian, Permian and Triassic times seem to have been most favourable for their formation.

Notable examples of ancient evaporites are the Silurian Saline series which underlie about 10,000 sq miles of the eastern United States, and are often as much as 250ft thick, and the Permian Castile anhydrite which has a maximum thickness of some 1,500ft over an area about 200 miles across in Texas and New Mexico.

In some parts of the world, buried evaporite beds, lying perhaps thousands of feet below the present surface, have generated plugs or "domes", which have moved upwards through the overlying beds in "diapiric" or piercing, plastic flow. Investigations have shown that as the salt masses move upwards they typically raise and rupture the immediately overlying strata, which then settle down to fill the peripheral spaces left by the salt. The pillar-like masses are often several miles in diameter. They may penetrate considerable thicknesses of sediments—including potential reservoir beds—and may even form surface hills or (as in Iran at the Kuh-i-Namak, near Bushire) mountains with a salt core above the general ground level. Where the salt, instead of moving in plug-like form through the overlying sediments, develops semi-horizontal bulges or buried "pillows", i.e. salt masses with convex upper surfaces, the overlying beds may show arching and anticlinal flexuring due to the vertically upward pressure that was applied.

Salt domes have long been known in the USA, in the Gulf Coast area and elsewhere, in Mexico, in Germany, Transylvania, Romania, the Ural-Emba region of the USSR, in North and West Africa, the Red Sea area, in Iran, and in other parts of the world. More recently, they have been found in the North Sea. Deeper offshore seismic profiling has shown that salt diapirs also exist off the margins of Labrador, Newfoundland, Morocco and Portugal. They have been found under the lower continental rise and abyssal plain off northwest Africa, indicating that the salt basins of the continental margin may have extended seaward during Late Triassic and early Jurassic times. Some of these salt masses have pierced through overlying sediments and produced submarine topographic highs with marked relief. These results imply the existence off the continental margins of the North Atlantic of a deep-sea salt layer. It has been suggested that the formation of this layer is related to the initial rifting phases of the development of the Atlantic Ocean (p. 241). Similar diapirs have also been discovered by the deep-sea drilling operations in the western Mediterranean.

Salt domes are associated with oil accumulations in several parts of the world, most notably in Germany, the US Gulf States, the Emba-Volga region of the USSR, Romania, West Africa and Mexico. (It should be pointed out, however, that even in the halokinetic* regions such as the US Gulf Coast and Germany, only a relatively small proportion of salt domes present have been found to have associated oil, while in other salt areas, such as the Dead Sea region, Sind, and Transylvania, no oil at all has yet been found associated with salt uplifts).

The relationship of salt domes to oil accumulations is essentially structural:

(i) The salt may seal the upper broken end of a reservoir bed along which oil is migrating and thus give rise to an accumulation, as an example of trapping in a sealed monocline.

(ii) The intrusion of the salt may produce arching in overlying sediments, with the resultant formation of an anticline in which oil may subsequently accumulate.

(iii) The beds on the flanks of the penetrating salt mass will be disturbed and faulted, and the faults may serve as channels of oil migration or as barriers. In addition to radial and peripheral faulting, a complex series of faults—almost a trellis-work—can be formed in the beds above a salt mass, as is the case, for example, in the *Reitbrook* oilfield in Germany (Fig. 28).

(iv) Reservoir rocks may wedge out or be bevelled by unconformities in the flank areas, in this way forming traps.

(v) Porous and permeable salt-dome cap rock can serve as a reservoir rock.

Because of the many different kinds of trap which can be associated with salt intrusions, they constitute one of the most difficult geological features to test exhaustively for possible associated oil and gas accumulations.

Generally speaking, the oil-producing zones on the US Gulf Coast are reservoir sands lying on the flanks and over the tops of the salt domes—Miocene "super-cap" sands (e.g. *Goose Creek*), Oligocene flank sands (e.g. *Spindle-.top*), Miocene-Eocene lateral sands (e.g. *West Columbia*, Fig. 29), and Eocene and even Cretaceous re-worked sands and limestones.

A classification of the salt domes of the US Gulf Coast was originally made on the basis that they were either "piercement" or "non-piercement", i.e. "deep-seated". This was quite arbitrary, since "deep-seated" only meant too deep to be reached locally by the drill, and this level has obviously varied with circumstances and become progressively deeper with time. A more practical classification has been to call "deep-seated" any domal structure the top of which lies below 10,000ft.

It was originally believed that the close association of salt masses and oil accumulations found in several parts of the world was due to a common volcanic origin, the salt having been injected into the overlying sediments in much the same way as igneous veins are injected from subcrustal magma. This idea was reinforced by the frequent occurrence of sulphur deposits close to the salt. The *Spindletop* salt dome oilfield was successfully drilled in 1901 in the belief that the salt and oil both were of volcanic origin; it was

---

* *Salt basins, in which the evaporites characteristically increase in thickness towards the centre, have been classified as "halotectonic" when these bedded deposits have been subsequently deformed by lateral tectonic forces, or "halokinetic" where vertical and autonomous movements of the salt have occurred.*

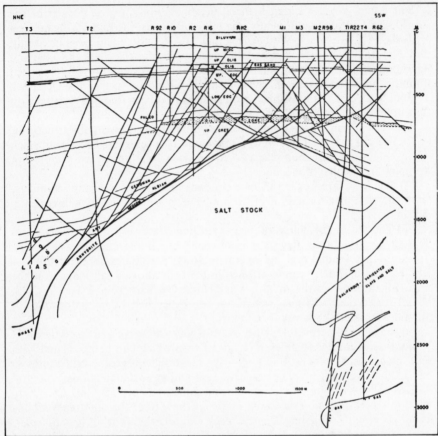

Fig. 28. Cross-section of the Reitbrook oilfield, West Germany (after F. Reeves).

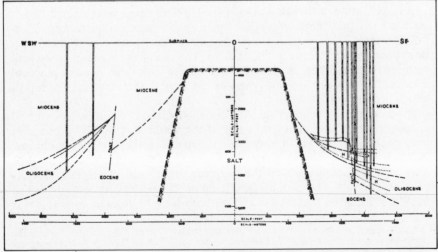

Fig. 29. The West Columbia salt dome, US Gulf Coast, which produces oil from lateral sands (after Carlton).

the first of many important Gulf Coast oil discoveries related to intrusive salt masses. However, excavations in the salt masses revealed evidence of a stratigraphical mode of deposition, contortions, as well as occasional algal and other organic remains, showing that the salt had a sedimentary origin.

There have been two general theories about the mode of origin of salt domes, which, while agreeing on the necessity for some kind of pressure to inject the salt into the overlying sediments, have disagreed over the source of this pressure. The evidence differs in the two classical areas of salt dome occurrence, Germany and the US Gulf Coast.

In the German area, and also in Romania, the salt intrusions are elongated in plan view parallel to known structural trends, whereas in the Gulf Coast area of USA they are more nearly circular in plan. The main German salt basin is of Upper Permian Zechstein age, and salt is seen *in situ* in the Stassfurt deposits. Drilling in the Hanover area has shown that the salt in its upward motion seems to have taken advantage of two regional structural trends (Rhenish and Hercynian). Lateral stresses were considered to be the mechanism responsible for the vertical movement of plastic salt in this area. It was envisaged as an extreme case of the disharmonic folding, by lateral pressure, of an incompetent group interbedded between competent rocks, with the former first thickening in the core of an anticline and ultimately breakthrough to give a diapiric structure. The further upwards the salt travels, the more like a volcanic neck in shape does it become, often developing with a marked overhang. In fact, in many German fields drilling has passed through the lip or edge of an overhanging salt mass to enter a reservoir bed below. However, even in Germany, doubts have since been expressed about a link between salt domes and the conditions which lead to folding.

A large number of salt domes have been found in the United States, principally in the Gulf Coast area of Louisiana and Texas, many of which are associated with hydrocarbon accumulations. Some of these are exposed at the surface, and others are buried at various depths. More than 400 salt domes have in fact been explored by drilling in the Gulf Coast salt basin, which contains five areas of maximum salt deposition ("depocentres") and extends as far as southeastern Vera Cruz and western Tabasco in Mexico.

There are two groups of domes in this region: an interior group, lying in an arc round the Gulf of Mexico, sometimes as much as 100 miles north of the coast; and an outer or coastal group. Nearly all the interior domes are cylindrical plugs of salt at least a mile high, with diameters within 3,000ft of the surface of $\frac{1}{2}$–$1\frac{1}{2}$ miles. The surface evidence for underlying salt domes includes saline prairies or "salt licks", circular drainage systems round central hills, unusual soils with calcite crystals, pieces of deep-seated rocks mixed haphazardly at the surface, and gravity anomalies which are usually negative.

Deep-water drilling evidence from the Sigsbee Deep on the outer (Yucatan) Continental Shelf, has shown that salt domes extend far out into the outer Continental Shelf areas of the Gulf of Mexico, some, at least, being associated with hydrocarbon accumulations. It has in fact been suggested that buried Mesozoic salt is present across the entire Gulf of Mexico basin, in which case the present Gulf represents an ancient ocean basin which has been a site of deposition, at least since Mesozoic time[1].

Opinion is divided upon the age of the salt in the US Gulf Coast area;

Comanchean (Cretaceous), Triassic and Permian ages have been suggested. It is certainly older than Lower Cretaceous, since it is found intruded into beds of this age. It was first thought that the Gulf Coast salt domes occurred along fault lines or at the junction of fault planes; but whereas there is the appearance of association with known structural trends in Germany, on the Gulf Coast there is, in fact, no clear relation to regional faulting. The salt is usually found to have penetrated great thicknesses of Tertiary beds which show no sign of lateral compression, but rather give evidence of considerable down-sagging.

It is believed that in this area *vertical pressure* ("geostatic load") was the principal cause of the motion of the salt, the result of the accumulation of sedimentary overburden, which, when it produced certain conditions, caused the underlying salt to flow laterally and thence forced it upwards through the overlying beds in characteristic diapiric shapes.

Evaporite deposits are variable in their make-up, but sodium chloride is one of the potentially most mobile of the components, and can also be dominant as regards bulk. Its mobility undoubtedly increases (i.e. its apparent viscosity decreases) with rise in temperature, and therefore on progressive burial its "viscosity" diminishes. Differences in vertical load on the salt bed will give a tendency for salt to flow laterally towards points of locally lower load, causing initial upward arching of the overlying strata at these points[12].

Lateral tension develops in these strata, as in well fracture treatments using fluid pressure, and fractures are created and/or opened in the strata, into which the salt moves. The low density of the salt compared with that of other deeply-buried rocks enhances the pressure differential as the salt rises; the fractures are propagated and the rocks are forced aside by the moving salt. Other things being equal, it appears that the rate of flow during intrusion will be a function of the cube of the thickness of the salt bed. There is evidence of the earlier and better development of salt domes in the central part of a salt basin, a feature which is consistent with the greater depth of burial (higher temperature) and often greater salt thickness in that area.

Intrusion of salt by this means and the associated adjustments in the surrounding strata offer a reasonable explanation of the striking features associated with the Gulf Coast salt domes, such as up-turned beds, overhang, down-faulted blocks and rim synclines.

To illustrate some aspects of the process, a series of experiments may be cited in which two liquids of different densities and viscosities were used to show the effects of differential loading and buoyancy when the lighter and less viscous fluid originally lay below the denser one. Thus, a thick layer of concentrated corn syrup was placed at the bottom of a glass box, and a layer of crude oil introduced above it. When the box was completely filled, it was covered with a tight rubber diaphragm and carefully inverted. The system was then in unstable equilibrium, with the denser syrup layer resting on top of the less dense oil and the rubber diaphragm at the bottom. It was found that if the diaphragm was at this stage pressed slightly upwards at one point, the equilibrium was disturbed and an upwelling of the lower layer took place, which developed more and more rapidly until a "dome" of oil protruded upwards into the syrup layer, with a trough surrounding its periphery. It has been suggested that this process approximately illustrates the growth of salt domes as a result of the upsetting of local equilibrium between the underlying salt layer and the overlying rock, as a consequence of differences in the weight of overburden. (In the experiment the "overburden" could flow but not fracture). It is also possible that a limit is put on the growth of the dome by the fact that the surrounding troughs ("rim synclines") will have become so deep that very little fresh salt can be fed in from the original salt bed[25].

It is interesting to consider some other reasons why a salt mass, once having started to flow from the parent salt horizon should eventually cease this growth. The following are possibilities:

(a) The mother-bed is limited in volume, and after a certain amount of salt has been squeezed out of it to form a dome, the supply of salt is exhausted, and hence the dome ceases to grow.

(b) If the growth is related only to differences of density between the salt and the surrounding rock, under certain circumstances and at some point in its upward progress, equilibrium may be attained for the two rock columns, and therefore no further motion will occur.

(c) Possible resistance to continued fracture, uplift or displacement of the overlying rocks could finally bring the upward motion to a stop.

It has been inferred that the plastic flow of salt usually occurs only after the deep burial of an evaporite bed—perhaps to 25,000ft or more. At depths such as these, temperatures in excess of 200°C are probable, and it has been shown in the laboratory that at this temperature salt becomes decidedly more plastic and mobile. It has therefore been suggested that a marked rise in temperature is necessary before salt will flow; however, there is also ample evidence that salt flows at surface temperatures, so temperature increase is not the critical factor.

The Gulf Coast salt stocks, plugs or domes are remarkably pure, presumably because of the derivation of this salt from a single deep-seated layer. On the other hand, the salt plug necks in Iran are fllled with impure salt and blocks of sediment, probably because of the derivation of the salt from a number of beds which are interbedded with normal sediments.

It is relevant to note that diapirism occurs with other sedimentary materials besides salt—mud, clay, shale and peat. Clay diapirs are known in Trinidad (the *Forest* field) and the "mudlumps" of the Mississippi delta area are also diapiric. Many of the Baku oil-bearing anticlines have diapiric clay cores.

Salt may have an additional structural significance in that, being highly soluble, it can be removed after formation and burial in a sedimentary sequence by the action of waters which reach it subsequently, due perhaps to faulting or erosion. The result may be the formation of *subsidence structures* as a result of underlying salt removal. Some of the oil accumulations in the Williston Basin of North America are believed to have been trapped in structures resulting from the removal by solution of underlying Palaeozoic salt beds[23,28].

Cap Rocks

In many US salt-dome oilfields, a rock known as "cap rock" lies on top of the salt mass. This is typically made up of a series of zones dominated, from the base upwards, by anhydrite, gypsum, and then "limestone", often with pockets of sulphur. The rock can have cavities and show evidence of brecciation. On some salt domes this cap rock acts as a reservoir rock for oil and gas.

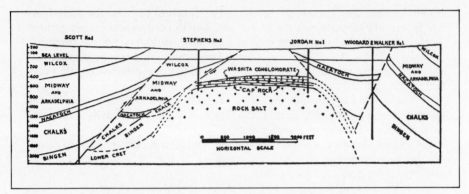

Fig. 30. The Vacherie salt dome, Louisiana, which has about 130 ft. of conglomerate above the cap rock. The conglomerate is of Cretaceous age and appears to have been carried upwards by the rising salt mass (after W. Spooner).

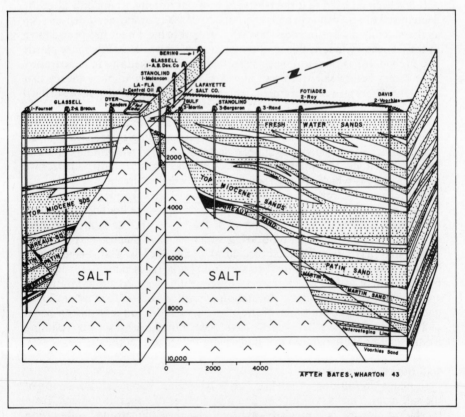

Fig. 31. Block diagram of Anse la Butte salt dome (view toward northwest), St. Martin Parish, Louisiana (after Bates and Wharton, 1943).
(F. B. Rees, "Methods of mapping Stratigraphic Traps" in "Stratigraphic Oil and Gasfields", *Amer. Assoc. Petrol. Geol.* 1972, p. 180.)

Measurements of the $^{13}C/^{12}C$ ratio for the "limestone" (calcium carbonate) have given values more characteristic of the carbon in organic matter than of the carbonate in sedimentary limestone.

Several suggestions have been made with regard to the mode of origin of salt-dome cap rock:

(i) The rock may be part of a bed which immediately overlay the salt bed from which the salt dome was developed, and has been carried upwards by the rising salt. This might be viewed as a mechanism capable of explaining "limestone" cap rocks if the carbon isotope ratio was normal; but if anhydrite had originated as gypsum from the same desiccating sea that produced the salt, it could be expected to have formed first, and hence should underlie rather than overlie the salt. Conversion of gypsum to anhydrite (a reversible process) is a consequence of increase in temperature and pressure on deep burial, and the presence of these sulphates in the cap rock could be explained by assuming that the gypsum was deposited in a later sea, after the main salt layers had been laid down and buried.

(ii) The anhydrite of the cap rock may, in fact, be residual, resulting from solution of the far more soluble sodium chloride from the top part of the salt intrusion. The formation of a thick deposit of salt must call for the addition of more sea-water to the basin in which the deposition of evaporites is taking place. Although the evaporation of a fixed volume of sea-water would lead to calcium sulphate crystallising first, some would also crystallise during the phase in which sodium chloride was the main precipitate. Hence, sulphate-free sodium chloride cannot be expected, and when new additions of sea-water take place there must be some mixing of solid sodium chloride and calcium sulphate on a granular and perhaps also on a layer basis. Consequently, the low-solubility materials are to be expected to occur in the main salt mass.

(iii) The possibility of relatively late chemical reactions being involved in the formation of the cap rock cannot be excluded. Meteoric water carrying calcium in solution might react with more soluble sulphates in the salt mass to deposit calcium sulphate; or water carrying sulphates leached from the salt mass might come into contact with limestone. In both cases, the gypsum or anhydrite would be secondary, and would overlie the salt mass.

A hypothesis which links several features in cap rocks is interaction between sulphates and organic matter, probably oil or gas, with the formation of calcium carbonate and sulphur. Hydrogen sulphide might also be formed. The general zonation, the association of sulphur and calcium carbonate, and the carbon isotope ratio of the latter would all be compatible with such a mechanism.

The *Vacherie* dome in the interior salt dome province of the US Gulf Coast (Fig. 30) is remarkable in that there is a thickness of about 130ft of conglomerate lying above the cap rock; this conglomerate is identifiable as being of Cretaceous age, and appears to have been carried upwards by the rising salt mass.

Attempts have been made to link the presence or absence of cap rock with the degree of purity of the intruded salt. However, other factors could be involved, such as depth to the top of the salt, the characteristics of the rocks at that level, the general topographical and geological setting of the salt intrusion.

Salt was met as shallow as 160ft on the *Anse la Butte* salt dome, St. Martin

Parish, Louisiana. No clear evidence of salt-dome cap rock was found. This salt dome has flanking wedge-shaped blocks of sediments, separated by at least five major radial faults, with displacements of 150 to 900ft. These faults are groups of parallel, en échelon breaks, in zones several hundred feet wide, not single fault planes. There are other major faults, as well as numerous minor faults. Unrecognised tangential and concentric faults may exist. The formation dips are up to 75°.[3, 31]

Gas seeps above the salt dome directed attention to this oilfield. The shallowest recorded oil production came from about 400ft; the deepest is over 9,000ft. There are minor post-Miocene (Plio-Pleistocene) oil sands. In the Miocene the Fleming is productive, but the main production is from the Upper Catahoula, while the Lower Catahoula and the Chickasawhay also have oil sands.

The oil accumulations are controlled to a considerable extent by the radial faults, certain blocks being highly productive and others barren. It has been suggested that the original oil accumulations were in the Chickasawhay sandstones, and that the oil has escaped in part or entirely to the prolific shallower sandstones, via the numerous fault zones. Fig. 31 shows trapping against the salt and against faults; it also indicates the presence of wedge-outs of the flanking sandstones. Production has been obtained from such wedge-outs.

REEF CARBONATES[26]

Carbonate thickenings in the form of "mounds", "build-ups" or "reefs" form extremely prolific petroleum traps in several parts of the world. "Bioherm" is the term now commonly used for this type of thick, organic carbonate deposit when it covers a limited area and occurs as a steep-sided mass (Fig. 32), while "biostrome" describes a somewhat similar deposit of relatively uniform thickness laid down over a wider area[2].

Fossil organic reefs have been found by drilling in many areas; they range in age from Pre-Cambrian to Tertiary, but have their principal developments in Palaeozoic and Mesozoic rocks. The largest examples can be very large indeed—extending laterally for several hundred miles and with a thickness of hundreds of feet. Thus, for example, the Capitan reef complex in the Permian Guadalupe series of West Texas and southeastern New Mexico is more than 400 miles long, and has a maximum thickness of 1,200ft.

If, as seems probable, hydrocarbons are locally generated from the decomposition of the organisms whose skeletons went to make up reefs, then it might be expected that the most porous parts of reef masses would subsequently contain hydrocarbons, and that these would be confined by the changes of permeability at the interfaces between the reefs and the surrounding less permeable deposits.

Large oil accumulations associated with organic reefs have been found in several parts of the world, notably in Mexico, the southwestern United States, Alberta, the West Ural area of the USSR, the Middle East and Libya. An oil production rate of some 260,000 barrels/day, recorded from a single well in the Golden Lane of Mexico, exemplifies the enormous potential productivity of these reef reservoirs, the result of the volume of extremely porous and permeable rock they may contain. Thus, a thickness of 959ft. of oil "pay" was found in one 1,200ft-thick section of the *Intisar A* reef in Libya. This oval bioherm (one of several discovered in a small area) is about $2\frac{1}{2}$ by 3 miles in its greatest dimensions, but it contained at least 2 billion barrels of oil in place before production began in 1968.

Although the term "coral reef" is widely used, many other organisms are involved in reef construction, in particular the essential strength of ancient reefs was markedly dependent on these other reef-building organisms with comparable habits.

Corals need sea-water at a temperature of at least 20°C, and thrive best at 25°–30°C. Moving water is necessary to provide their oxygen and food requirements. Corals proper do not live at depths greater than about 150ft, because of their dependence on sunlight, and they cannot survive exposure out of water for more than an hour or two. Where the surf is strong, corals will not normally grow at water depths less than 6–8ft. The water must not be very silty, because burial by silt prevents their development.

A coral reef consists of coral skeletons, coral débris and sand produced by boring organisms and wave action, together with algae, foraminifera, polyzoa, molluscs, echinoderms, etc. Encrusting coralline algae grow on the coral surfaces and in crevices, especially where wave action is strongest and dissolved gas most abundant, cementing the corals and giving greater strength and compactness. The most rapid growth of the reef is always on the seaward side, because of the better food supply, and on that side in particular a steep bank of coral and other débris is produced, with an outer slope which may extend to water depths of 1,000ft or more. The talus slope, of which the surface angle of slope may be as much as 40°, has fragments provided by wave action and boring organisms, the latter being the most destructive. The borers produce fines in making holes, whilst the material between the holes crumbles to give coarser fragments. Waves move the débris, the fines being carried far from the reef, whereas the coarser débris forms the talus and undergoes little sorting in its limited transport. Bedding is commonly good, because of variations in erosive activity. Borers are active in the talus, as well as in the reef proper.

"Fringing reefs" extend out from the shores of islands or the mainland; "barrier reefs" are separated from the land by a channel or lagoon, which may be many miles wide. "Atolls" ("pinnacle reefs") take the form of a ring of reef with a central lagoon. Low-lying islands may occur on atolls and barrier reefs; they are rarely co-extensive with the reefs. These islands are built of coral sand and boulders formed by the waves, deposited on the coral flat, and built above low-tide level by the waves; wind can move sand on the incipient island. Foraminiferal tests are other commonly important components of coral sand.

Reefs may be from one mile up to 90 miles long, with the top surface a few hundred to several thousand feet wide. They grow up to about low-tide level; on the flat top the water is warm and low in dissolved gases, and the top may even be exposed at low tide. The sharp changes in temperature in this area lead to solution and precipitation of calcium carbonate, filling-in and compacting the inner part of the reef. In the quiet lagoon waters, some coral and algal growth adds to the accumulation of calcium carbonate.

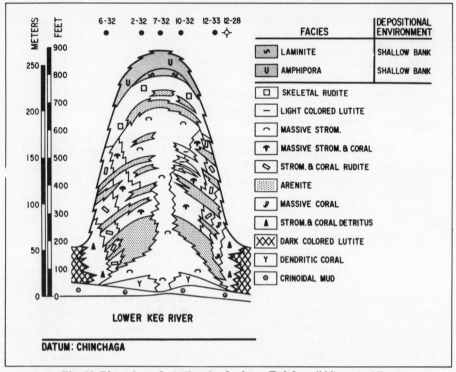

Fig. 32. Pinnacle reef, shallow-bank phase, Rainbow "A" pool, Alberta.
(D. L. Barss *et al.,* "Geology of Middle Devonian Reefs" in "Geology of Giant Petroleum Fields", Amer. Assoc. Petrol. Geol., Tulsa, 1970, pp. 148.)

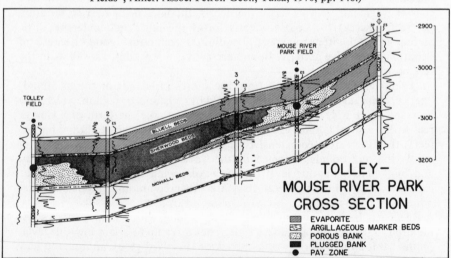

Fig. 33. Southwest–northeast cross-section from Tolley to Mouse River Park fields. Structural datum: top of upper Mission Canyon Formation. Contour interval = 100ft Elevation datum: sea-level.
(S. Harris *et al.,* "Relation of Mission Canyon Formation to Oil Production", *Bull. Amer. Assoc. Petrol. Geol.,* **50** (10), 2275, Oct. 1966.)

Reef development is clearly linked with the local sea level, and changes in sea level influence the form of the developed reef. In southern Florida, reef growth is estimated to be 1·5 to 3·0ft/year. Slow sinking of the sea floor encourages the upward growth of the reef, whereas a stable sea level encourages outward growth, as does a slowly rising sea floor. Rapid sinking could stop the growth of corals. Hence, the form of the reef and some features of its internal structure depend, among other things, on the behaviour of the sea level.

The basic components of a reef tend to give the mass a high initial porosity, but this is modified later by processes of solution, cementation, precipitation and replacement, commonly resulting in irregularities in the distribution of porosity and permeability, as well as a considerable range in the sizes of the individual open spaces. In many examples of present-day and recently emergent reefs, inter-skeletal, intra-skeletal and inter-granular voids exceed 50% of the bulk volume of the rock.

"Draping" of younger beds over reefs, as an original feature or as a consequence of differential compaction, gives an arched form to these beds, and this feature may be used under favourable circumstances to locate ancient reefs by seismic reflection surveys. When these overlying beds include reservoir rocks, they can obviously provide structural traps of anticlinal form.

Unlike the siliceous and clayey rocks, the components of the carbonates have, in general, travelled comparatively short distances as solids between the areas of their formation and deposition. Indeed, in certain types of "deposit" some components have not moved at all. The relatively high solubility and low strength of carbonate fragments, as compared with silica particles, militate against the carbonate solids as a whole surviving long-distance travel. However, certain geographical and climatic conditions favour more or less "on-the-spot" formation of solid carbonate rocks and hence the fragments or grains are able to give rise to some deposits which are similar in form to those common as siliceous sediments (sands and sandstones). On the other hand, the coral reef, with the significant *in-situ* development of solid material, has no counterpart in form among the siliceous rocks. Furthermore, some carbonate banks incorporate components which may be viewed as abortive attempts at reef building. Banks are formed from the débris of organisms not able to build, under the conditions of growth, a rigid three-dimensional frame-work. In spite of this, there may be effectively *in-situ* formation of banks by organisms, involving local growth, binding and baffling action, quite apart from banks resulting from transport and accumulation.

A special feature of areas of carbonate sedimentation is that the climatic conditions, in conjunction with certain local geographies, can cause the crystallisation of comparatively highly soluble calcium sulphate and sodium chloride within the granular carbonate sediments, thereby filling the pores with solids to give "plugged" carbonate deposits. Such crystallised solids may also be present, inter-layered with the carbonates.

SAND BODIES[30]

All sedimentary rocks are of limited extent, and are, therefore, in principle lenticular. Nevertheless, the term *lens* in the macro sense is commonly applied in oilfield work only to bodies of reservoir rock which are smaller than the structural "wavelength" in a given area. For sandstones which extend over a number of structures, the description "sheet sandstone" has been used. However, ribbons of sandstone, some labelled "shoe-strings", may stretch over several "highs". A salt-dome cap rock or a carbonate reef is a lenticular body of rock (commonly porous and permeable), and to that extent is a lens, but a lens in a structurally-high position. Because of this special position, the likely sites of lenses of this type can often be pin-pointed by seismic means ahead of drilling, whereas other lenses are not usually susceptible to such direct discovery of their existence and position. Nevertheless, there are cases of the preferential deposition of sands over bottom "highs", and consequently the possibility of these sites being recognisable in advance of drilling.

A study of 7,241 sandstone reservoirs in the United States has indicated that 56% of the petroleum accumulations were in structural traps, 10% in "stratigraphic" traps and 34% in "combination" traps. The "stratigraphic" group itself was divided into 23% bar deposits, 19% beach deposits, 19% near-shore marine deposits, 19% dependent on a regional facies change and 4% associated with unconformities[6].

Discussion of the nature and origin of sandstone bodies has led to the recognition of a number of types: alluvial fans and cones, bars, beaches, deltas, dunes, estuary sands, channel (river) sands, sub-aqueous bars, wave-built terrace sands, turbidity current and shelf-edge sands.

There have been arguments about the ways in which some sandstone bodies are formed, although the associations may be clear. Furthermore, observations on modern deposits need to take note of certain events in the geologically-recent past which may cause some processes to be relatively more active in areas accessible to study than has generally been the case. In addition, the ultimate consideration is *what is preserved,* not merely how the deposits are laid down.

Carbonates may show lithological trapping features comparable with those of sandstones. For example, in north-central North Dakota, the Mission Canyon formation shows, from southwest to northeast, a change from basin to shelf depositional conditions, and then to lagoonal deposits. The dark fine-grained deposits of the basin are commonly argillaceous, indicative of normal to slightly reducing conditions, with fossils rare. Clastic limestone, light in colour and fine- to coarse-grained, was deposited on the shelf in a medium energy environment. It consists primarily of fragments of pre-existing limestones and organisms, especially bryozoans, crinoids and corals, together with thin beds of oolitic and algal limestones. Nearer shore bank deposits of light, fine to very coarse-grained limestones, largely oolites and pisolites, as well as algal material, were laid down in a high-energy environment in shallow water, possibly being sub-aerial in part. This facies provides the commonest reservoir rock in the Upper Mission Canyon formation. Generally, the porosity is good, but there can be very fine pore-filling clastic carbonate. In addition, there are bank deposits plugged by the deposition of calcium carbonate or sulphate, or by dolomite. The low-energy zone lagoonal deposits consist of grey and white anhydrite and evaporitic dolomite, thin beds of oolitic limestone, shale, salt and dolomitic limestone. The various zones shifted over time, so that interfingering is common. Fig. 33 shows various stratigraphic traps in the Mission Canyon formation, with up-dip seals of plugged bank deposits and evaporites[14].

Dunes and Sheets[10]

Winds may sweep sand from beaches and other sources to form dunes whose spread may be wide, and whose upper parts at least will undergo reworking should a sea advance across them later. The rates of formation of deposits developed at or near sea-level relative to changes in sea-level will determine, with local topography, whether the sea margin moves seawards or significantly landwards, i.e. there is regression (offlap) or transgression (onlap), respectively. Each of these processes may develop a sheet sand, which is admittedly composite and strictly diachronous, when the lateral shift of the shore line is extensive.

Bars[17, 33]

None of the processes leading to the formation of sand deposits are simple; all involve the availability of sand-size grains, the vigour and kind of fluid motion, the interplay of river and tide-induced currents, and local topographical conditions, such as water depth. Experiments suggest that longshore bars develop at the point of wave break; that in very shallow water an emergent bar is commonly formed, and then the formation of a lagoon shorewards of a barrier becomes a possibility, leading to limitation of beach growth. However, not all are agreed that the development of a lagoon takes place shorewards of an offshore barrier[18, 22].

Beneath the Mississippi "bird's foot" delta, elongated lenticular sand bodies underlie the 15–20-mile long major distributaries. These bar-finger sands branch, have maximum widths of about 5 miles, and reach thicknesses of over 250ft. They tend to widen downstream. Typically, each finger has a central zone of clean sand with minor amounts of silt and clay. There is a relatively thin upper transition zone with more silt and clay, grading upwards into natural levées and delta-plain deposits; a lithologically similar and comparatively thick lower transition zone grades downwards and laterally into delta-front deposits. Bar-finger sands are believed to originate as bar deposits at distributary mouths, their maximum widths being similar to the widths of the present bars. The lenticular form and considerable maximum thickness of the bar-finger sands are considered to be the results of the accumulating sand compacting and perhaps displacing highly-water-charged delta-platform clayey silts[9].

The Booch sandstone (Lower Pennsylvanian) of the Greater Seminole district of eastern Oklahoma is considered to be an example of bar-finger deposits in which oil has accumulated.

In the Rhone delta, the ratio of sand deposited in distributary-mouth bars and the amount removed by coastal currents is smaller than for the Mississippi delta, and sand initially deposited in the Rhone delta bars is subsequently transported laterally to be re-deposited as barrier complexes parallel to the coast. The relatively slow coastal advance has caused these barriers to merge into a 10-m thick coastal sheet sand of wide lateral extent[27].

The Saber bar of Logan and Weld Counties, Colorado, effectively an elongated lens, was apparently deposited on a relatively smooth, gently

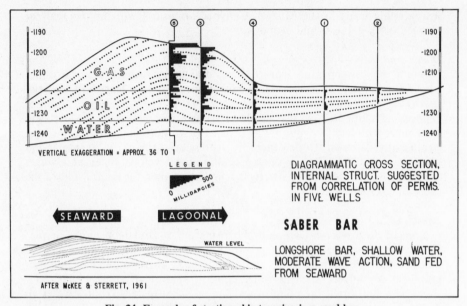

Fig. 34. Example of stratigraphic trapping in a sand bar.
(E. G. Griffith "Geology of Sabre Bar", *Bull. Amer. Assoc. Petrol. Geol.,* **50** (10), 2117, Oct. 1966.)

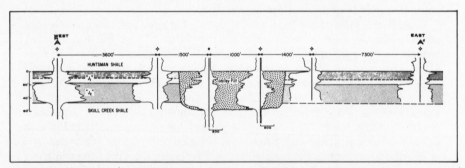

Fig. 35. Example of stratigraphic trapping in valley-fill, Nebraska (cross-section).
(J. Harms, "Stratigraphic Traps in a Valley Fill", *Bull. Amer. Assoc. Petrol. Geol.,* **50** (10), 2134, Oct. 1966.)

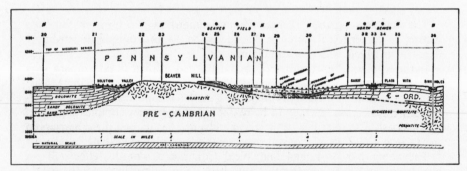

Fig. 36. Cross-section of the Kraft-Prusa oilfield, Oklahoma (after R. F. Walters).

sloping surface (Fig. 34). It has been traced for about 10 miles, and at its southern end it is thin and loses its bar-like character. The width is nowhere defined by wells, but it is most likely less than 1·5 miles, and generally less than about 1 mile. The maximum drilled thickness is 45ft, although it could exceed 50ft along the crest. The thickness and width change together, giving a series of pod-shaped sand bodies, connected by bar sandstone of smaller width and thickness. On the lagoon (coastal) side of this offshore bar the top of the sandstone drops at first by about 1ft/100ft for some 700–800ft horizon-tally, and then it steepens to 4ft/100ft for 300–400ft, before becoming nearly flat, which is the structural attitude of the lagoon floor. The seaward slope of the top of the sandstone is some 3ft/100ft[11].

The sand is believed to have been obtained from the seaward side of the bar, possibly by a combination of longshore and rip currents. The bar is one of several situated north of a postulated delta of Upper Cretaceous age. Per-meability measurements suggest that the bar has a broadly layered structure; the permeability generally increases towards the upper surface of the sand-stone and towards the seaward side. Lateral changes are gradual, whereas vertical changes in the bar sandstone are abrupt, indicating differences in the water energy levels at the times of deposition of the various layers. The bar is underlain by the Huntsman shale and overlain by the Graneros shale, both of Cretaceous age.

Channels

Streams erode channels sub-aerially, and currents can carve out somewhat similar features sub-aqueously. Flowing water may then deposit coarse sediments in these depressions, giving rise to sand bodies bounded laterally and sharply by older sediments. The deposits laid down by rivers may become wider and branch in a downstream direction as broadly indicated by cross-bedding. Oil and gas have been found in what are believed to be fluvial sand-stones.

The Niger delta is tide-and wave-dominated. Fluviatile sands appear to be deposited first at river and estuary mouths as bars, only to be picked up later by wind- and tide-induced currents, transported parallel to the coast, and then deposited in considerable amounts as tidal-channel-fill complexes running normal to the coast[27].

Valley fill can produce a stratigraphic trap. An example of this kind pro-vides a series of oil accumulations in Western Nebraska (Fig. 35). The fill is made up mainly of porous and permeable sandstone; it is about 50ft thick, some 1,500ft wide, and has been traced in a north-south direction for approxi-mately 20 miles. The oil accumulations are located where the fill crosses the axes of very gentle northwest-plunging anticlines.

The Skull Creek dark marine shale is overlain by the J interval, a sandstone and siltstone unit, deposited in a predominantly marine environment, and 38–77ft thick. The lower part, the J_2 member, is 20–40ft thick, consisting of very fine sandstone with a clay matrix. Its top shows evidence of having supported plant growth for a time before the deposition of the J_1 member, of which the lower part is shale and siltstone (5–10ft thick), and the upper part has sandstone as well as some siltstone and shale. Emergence then occurred and a meandering stream cut a valley in the J interval deposits; in places it even cut slightly into the Skull Creek shale. The stream also deposited the valley-fill sandstone before the area was again submerged, deposition of the Huntsman shale taking place[13].

The valley fill has coarse cross-laminae, with dips generally over 20°. Cross-lamination sets are 3 inches to 2ft thick, and bounded by nearly horizontal surfaces.

Lateral escape of oil from the valley-fill sandstone is prevented by the J interval discontinuous sandstones with a low oil-entry pressure, separated by sandstones or shales with a higher oil-entry pressure. The oil columns are 40–50ft thick. The oil/water contacts are horizontal at Lane and Reimers, but inclined at North Faro and Faro because of lithological variations. Oil staining is common in structurally-low valley-fill sandstone bodies. Unstained core plugs have higher capillary pressures than stained plugs.

Turbidites

When there are multiple repetitions of sandy and pebbly deposits *interbedded* with shales containing foraminifera which indicate sedimentation in depths of, for example, 2,000–8,000ft, it is not possible to accept a shallow-water origin for the sands. The implication is that the coarse deposits are turbidites—sediments carried from their original site of deposition suspended in a density current. However, turbidites are not confined to deep water. Poor sorting is to be expected in turbidite sands, for they have not been winnowed, and hence they have lower porosities than shallow-water sands. Graded bedding is a feature of such deposits, and there can be convolute bedding, current bedding and shale inclusions, in addition to shallow-water faunas. There is a sharp contact with the underlying bed, whereas the top grades imperceptibly into normal pelagic deposits. Furthermore, the top is commonly planar and laminated[36].

It seems likely that the late Miocene Stevens formation of the San Joaquin Valley (California) is a turbidite sequence, and that the Upper Miocene and Lower Pliocene of the Los Angeles Basin include turbidites. The former has a maximum thickness of more than 3,000ft, of which two-thirds is sand, and it occurs in an area 30 miles by 50 miles. In both areas, large amounts of oil have been produced from the apparently turbidite sands.

UNCONFORMITIES

An unconformity is the result of a phase of uplift and erosion of areas which have previously been sites of sedimentation. Weathering and erosion of the rocks give rise to sculpturing, and under suitable conditions there can be changes in the more permeable rocks in particular, in addition to their erosion. Percolating waters, by solvent action, may remove carbonates, increasing the permeability and porosity of limestones and of carbonate-cemented sandstones. Carbonates may also suffer other changes. However, there can be instances of the near-surface zones of rocks being rendered less permeable and porous, even destroying instead of improving reservoir characteristics. The removal of parts of evaporite sequences by solution leads to the settling of overlying beds, with the creation of fractures and at times brecciation. Subsequent burial and sealing of those unconformity-exposed rocks which have suitable porosity and permeability can provide traps for oil and gas.

In places, during the period of exposure, residual deposits may be formed, and other deposits can be laid down on the land surface, e.g. screes, dunes, etc.

When the area is next invaded by the sea various new deposits are formed, the earlier ones abutting against the topographical features developed during the period of exposure. Should a reservoir-rock type deposit wedge out against a non-reservoir type of rock beneath the unconformity, the combination provides a possible trap.

Eventually, the old land surface may be completely buried by new deposits, and high points of the old topography may have smaller thicknesses of such deposits than low areas of that topography. When the areas with the greater thicknesses of new deposits include greater thicknesses of compactible beds beds than are present in the areas with thinner new deposits, differential compaction will lead to arching of some of the beds which may have been essentially horizontal at the time that they were laid down. In this way, traps may be formed in rocks overlying buried hills.

The *Kraft-Prusa* area of central Kansas, from which Fig. 36 is taken, shows some of the features associated with buried hills and unconformities which can provide traps. Thus, there are structural traps (not without some stratigraphic influences in this case) in Pennsylvanian limestones well above the unconformity which underlies the Pennsylvanian, and on the section shown the top of the Missouri series is indicative of their form. At the bottom of the Pennsylvanian there are stratigraphic traps in sandstones in the basal marine conglomerates. On the pre-Pennsylvanian unconformity are residual sands from the Cambro-Ordovician rocks which provide other traps. At this unconformity, the underlying Cambro-Ordovician Arbuckle dolomites constitute traps. Lastly, the Pre-Cambrian quartzite, in the form of a buried hill cloaked by Cambro-Ordovician sandstones, sandy dolomites and dolomites, is yet another reservoir[37].

HYDRODYNAMIC INFLUENCES[7]

In Chapter 3 (p. 73), the causes of inclined fluid contacts were discussed. One cause is flowing water, and when the flowing pressure gradient, expressed in terms of change in the head of water of density ρ_w, is dh/dl, the slope of the oil/water contact is: $dz/dl = \dfrac{dh\,(\rho_w)}{dl\,(\rho_w - \rho_o)}$, where ρ_o is the density of the oil. Thus, the slope of the fluid contact depends on the difference in density of the two fluids in contact, and for a fixed value of the flowing pressure gradient the slope will be less for a gas/water contact than for an oil/water contact.

When oil and gas are accumulated in an anticline in a reservoir rock of uniform properties, in the absence of water flow along the reservoir rock the gas/oil and oil/water contacts will be horizontal (Fig. 37). Low rates of flow of the water will distort the oil mass, giving an inclined oil/water contact, but the gas/oil contact will remain horizontal. At a higher flowing pressure gradient in the water the oil will be displaced more, with steepening of the oil/water contact, and ultimately the formation of a gas/water contact which

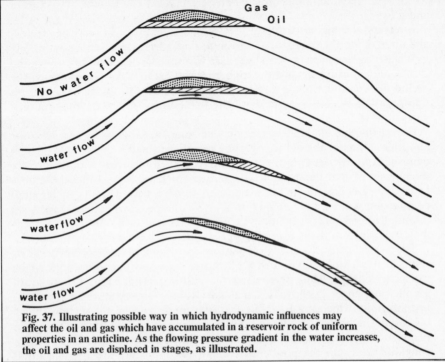

Fig. 37. Illustrating possible way in which hydrodynamic influences may affect the oil and gas which have accumulated in a reservoir rock of uniform properties in an anticline. As the flowing pressure gradient in the water increases, the oil and gas are displaced in stages, as illustrated.

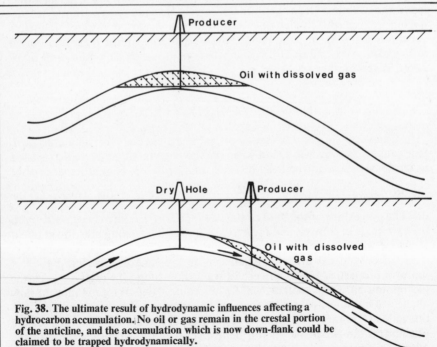

Fig. 38. The ultimate result of hydrodynamic influences affecting a hydrocarbon accumulation. No oil or gas remain in the crestal portion of the anticline, and the accumulation which is now down-flank could be claimed to be trapped hydrodynamically.

will be less steep, although where the oil and gas are still contiguous their contact will be horizontal. Further increase in the flowing pressure gradient will separate the oil mass from the gas mass, and both will have steeper fluid contacts than for the previously described conditions. The anticline must have a suitable profile for the oil to remain in it, since the cap-rock/reservoir-rock boundary must locally be steeper than the hydrocarbon/water boundary for the hydrocarbons to be retained.

Circumstances can exist under which no hydrocarbons will remain in the crestal part of the anticline (Fig. 38), with the consequence that an exploratory well in that part will find only water (an apparently barren structure), whereas a well some distance down the flank towards which water is moving, will find hydrocarbons—the hydrocarbons could be claimed to be trapped hydrodynamically. Displacement of all of the hydrocarbons from the crestal area is obviously more likely for an undersaturated accumulation than for one with a free gas cap.

Down-dip water flow on a monocline with a low-permeability zone up dip of a high-permeability sector will involve a higher flowing pressure gradient in the former zone than in the deeper sector. The oil/water contact will be steeper in the low-permeability rock than in the high-permeability rock, causing the fluid contact to curve upwards to meet the cap-rock/reservoir-rock boundary. This, too, would constitute hydrodynamic trapping. (It is assumed that there is no flow across the reservoir rock boundaries and that the rock has a constant thickness. A comparable condition would arise if the cross-section of the reservoir rock available up dip for water flow was smaller than that down dip, the rock having uniform permeability, since the former would have the greater flowing pressure gradient).

Clearly, with flowing water trapping is possible in the absence of structural closure. In the second case, where a stratigraphic change is involved, it is assumed that this in itself is not adequate to trap the oil in the absence of down-dip water flow.

Fully built-up bottom-hole pressures (adjusted to a datum level in water) for wells in specific reservoir horizons have been used to determine the directions of water flow in those horizons, and hence the directions in which hydrocarbon accumulations in them would be displaced in structural traps.

Numerous examples of inclined fluid contacts have been reported, and undoubtedly many of them are valid and represent a condition which existed at the time when the field was discovered. On the other hand, in the case of some of the older fields especially, the observations available are not adequate to demonstrate conclusively the real explanation of the condition. Reference is made to inclined fluid contacts at *Masjid-i-Sulaiman* and *Salt Creek* on p.89–91.

An oft-quoted example of inclination arising from production elsewhere from the same reservoir horizon is the *Cairo* field in Arkansas. *Cairo* lies two miles northeast of the much larger *Schuler* field, which also produces from the Smackover limestone and was put on production eleven years earlier. The large extraction of fluids from *Schuler* is believed to have created significant pressure gradients in the surrounding area. As a result, wells in the southwestern part of the *Cairo* field produce from levels below what seems to have been the original oil/water contact, whereas cores from the northeastern part show evidences of oil at points somewhat above that level, but produce only water when tested over the same intervals.

The *Hugoton* gasfield on the Oklahoma-Texas border appears to be dependent in part at least on hydrodynamic trapping. The productive Lower Permian limestones and dolomites are on a monocline with dips of 15–25ft/ mile to the east. There is no structural closure, but the limestones and dolomites are replaced laterally by red shales and silts some 20 miles west of the field. The westernmost wells have gas/water contacts at +500ft, wells in the central part at +250ft, and those in the east at +120ft relative to sea level. The initial pressure in the gas is reported to have been 435 psi throughout the area, pointing to continuity of the reservoir horizons. Eastward flow of water with a diminishing pressure gradient would be compatible with the observations.

PRESERVATION OF ACCUMULATIONS

The initial retention of oil or gas in a reservoir rock would seem to depend on the presence of a cap rock, and with one exception the continued existence of any hydrocarbon accumulation is dependent on the cap rock functioning as the only top seal. The development of fractures in the cap rock or its removal by erosion impairs or destroys the seal.

The present shape of the top of the reservoir rock and hence of its cap rock, as well as its depth of burial, depend on the geological history of the area, and these factors may well have changed since the hydrocarbon accumulation was first formed. Any changes may range from slight to very strong, and it is evident that the older the accumulation, the greater may have been the opportunities for changes to take place. When subjected to folding movements, the behaviour of the rocks depends on whether they are "competent" or "incompetent" (i.e. rigid or plastic). Both varieties of rock may be able to develop slight curvature without the formation of joints, whereas greater degrees of deformation lead to the appearance of fractures, which will first appear in the competent rocks. However, it should be noted that the extent of competency or incompetency is a function of the temperature and pressure, i.e. the depth of burial. The ideal cap rock should not only have a high capillary pressure; it should also behave plastically, for then it will flow rather than fracture when it and the adjacent rocks are folded, and even when folded quite strongly.

The sequence—salt, gypsum/anhydrite, clay, shale—is one of diminishing plasticity, yet all these beds are more plastic than the compact or very fine-pored limestones and other rocks which sometimes act as cap rocks. A thick cap rock of the plastic type is more likely to remain a seal than a thin one, when the rocks are strongly folded.

Clearly, the tectonic setting is important with regard to the chances of preservation of hydrocarbon accumulations; the timing of tectonic events and the extent of uplift and erosion are also important. Orogenic activity and/or igneous intrusion reduce the chances of survival of oil and gas accumulations. Geologically "quiet" areas, subjected perhaps only to mild epeirogeny, may retain accumulations intact, even though they are of great age. Moreover, the smaller thicknesses of rocks which may be laid down in some nonorogenic areas can mean less extensive maturation of the hydrocarbons, i.e. they are less likely to have been changed into the terminal products, methane and graphite, while the reservoir rocks themselves may have suffered less impairment of their desirable properties.

The Iranian oilfields provide examples of the fracturing of the Asmari reservoir rock, while the overlying Fars series, with its evaporite beds, constitutes an excellent cap rock. The even more strongly disturbed overthrust Oligocene sandstone of the *Boryslaw* oilfield in the Western Ukraine region has retained its hydrocarbons to a considerable extent because of the cover of plastic Miocene clays. Search is currently being carried out for oil and gas accumulations in overthrust structures in the US Appalachians, on the basis that Lower Palaeozoic shales may have been able to protect local accumulations of oil and gas, in spite of the intense structural disturbance.

The ingress of meteoric waters will change the composition of the water in reservoir horizons, and in one way or another will affect accumulations of oil and gas in the path of water flow. The accumulations will be distorted by the flow; they may even be displaced partially or wholly from a trap; and when no other more effective trap exists to capture the displaced hydrocarbons, they will be dissipated. Alternatively, when the intruding water makes contact with the oil there may be preferential removal of certain components in solution by partition effects (this may happen at an even earlier stage), or there will be bacterial or chemical reactions leading to the formation of tarry substances in the basal part of the oil accumulation. A tar mat may thus be created which "freezes" the oil/water contact.

Erosion, by reducing the thickness of the rocks over an accumulation, may lead to the formation of joints or to the opening-up of incipient joints in rocks which, under a greater rock load, were previously an effective seal for underlying accumulations of hydrocarbons. Upward escape of hydrocarbons may then become possible, again causing gradual dissipation of an accumulation.

BARREN AND GAS-ONLY TRAPS

Cases have been reported in which apparently properly sealed traps do not contain gas or oil. It has been assumed that the exploratory well (or wells) has penetrated the reservoir rock in a crestal position, for otherwise the well could be too low structurally to find any hydrocarbons which were present. (A single well on some indicated subsurface features of necessity leaves doubts on certain points). The further assumption is that there is no water flow which has moved the oil or gas from a crestal position.

Truly barren, yet properly sealed traps may be explained in more ways than one:

(a) There were no oil- or gas-generating rocks in the appropriate part of the geological succession in the area.

(b) The trap, involving dip reversal or faulting, was formed after oil and gas migration had taken place. Both faults and folds develop slowly, and the present form may post-date the principal phase of migration. Lenses and some kinds of wedge-out of reservoir rocks are potential traps from the time immediately after their deposition and the formation of a seal at the top surface.

(c) The trapping feature did not lie in an oil or gas migration path at any time, or was shielded by another structure in the migration path.

(d) Water flow in conjunction with the height of closure (on the "downstream" side), the amount of hydrocarbons and their properties, has given a pressure gradient large enough to carry the previously accumulated hydrocarbons out of the trap. Flow rates, reservoir permeability, geometry and setting are clearly of importance. This possibility would apply only to a trap dependent on the arching of an extensive reservoir rock.

Basically, there are three kinds of explanation for traps which contain only gas:

(a) The type of organic matter deposited in the area and the conditions favoured the generation of gas only, and not oil. Certainly, there appear to be gas provinces.

(b) The geological conditions were such that maturation has led to previously accumulated oil, and "oil" in source rocks, being transformed to gas. This would seem to be a matter of rock temperatures of the order of 180°C having been attained by deep burial.

In the above two cases, all the traps in the given geological setting would be gas-bearing only.

(c) A phenomenon which is effectively displacement can operate in three ways to leave a series of traps in an area with gas only, gas and oil, or water in a single extensive reservoir rock on a regional monocline. This has been described in detail in Chapter 3, the three mechanisms being:

(i) differential entrapment; (ii) tilting which reduced the "up-dip" closure of a trap, so that any oil in it escaped to another area; and (iii) reduction in reservoir pressure as a result of erosion of the cover rocks, causing gas to come out of solution in the oil, while free gas expanded; when the new volume of the gas/oil system exceeded the volume of the hydrocarbon storage space in the trap, oil would be the first material to be forced out of the trap.

(iv) Another mechanism which belongs to the displacement category is water flow in the reservoir rock with a pressure gradient suitable for carrying oil only from an accumulation with a free gas cap, away underneath the spilling plane.

Lake Kivu in the former Belgian Congo has water highly charged with dissolved methane at depth. On bringing this water to the surface, and therefore to a lower pressure, much of the gas is released. A comparable mechanism could operate in a reservoir rock, with gas initially provided by a gas-only source. The latter can arise in more than one way: association with organic matter which generated gas alone, as under (a); or flowing water underneath an oil accumulation with its inevitable dissolved gas could pick up gas by partition with the gas/oil solution, and on moving to a place where the saturation pressure of the gas/water solution was exceeded, some of the gas would be set free.

REFERENCES

1. J. W. Antoine. "Geology and hydrocarbon potential, deep Gulf of Mexico". *Bull. Amer. Assoc. Petrol. Geol.,* **54** (5), 834 (1970).
2. D. L. Barss, A. P. Copland and W. D. Ritchie. "Geology of Middle Devonian reefs, Rainbow area, Alberta, Canada". pp. 19–49, in: "Geology of Giant Petroleum Fields" (ed. M. T. Halbouty), *Amer. Assoc. Petrol. Geol.,* Tulsa (1970).
3. F. W. Bates and J. B. Wharton. "Anse la Butte dome, St. Martin Parish, Louisiana". *Bull. Amer. Assoc. Petrol. Geol.,* **27** (8), 1123–1156 (1943).
4. E. W. Brucks. "Luling oilfield, Caldwell and Guadalupe counties, Texas". pp. 256–281, in: "Structure of Typical American Oil Fields", Vol. I, *Amer. Assoc. Petrol. Geol.,* Tulsa (1929).
5. F. G. Clapp. "The occurrence of petroleum". pp. 44–73, in: "A Handbook of the Petroleum Industry" (ed. D. T. Day), Vol. I, John Wiley and Sons Inc., New York (1922).
6. B. F. Curtis *et al.* "Characteristics of sandstone reservoirs in United States". pp. 208–219, in: "Geometry of Sandstone Bodies" (eds. J. A. Peterson and J. C. Osmond), *Amer. Assoc. Petrol. Geol.,* Tulsa (1961).
7. P. A. Dickey and J. M. Hunt. "Geochemical and hydrologic methods of prospecting for stratigraphic traps". pp. 136–167, in: "Stratigraphic oil and gasfields—classification, exploration methods and case histories" (ed. R. E. King), AAPG Memoir 16, *Amer. Assoc. Petrol. Geol.,* Tulsa (1972).
8. H. V. Dunnington. "Aspects of diagenesis and shape change in stylolitic limestone reservoirs". *Proc. 7th World Petroleum Congress, Mexico,* Vol. 2, 339–352 (1967).
9. H. N. Fisk. "Bar-finger sands of Mississippi delta". pp. 29–52, in: "Geometry of Sandstone Bodies" (eds. J. A. Peterson and J. C. Osmond), *Amer. Assoc. Petrol. Geol.,* Tulsa (1961).
10. K. W. Glennie. "Desert Sedimentary Environments". Elsevier Publishing Co., Amsterdam (1970).
11. E. G. Griffith. "Geology of Saber bar, Logan and Weld counties, Colorado". *Bull. Amer. Assoc. Petrol. Geol.,* **50** (10), 2112–2118 (1966).
12. W. C. Gussow. "Salt diapirism: importance of temperature, and energy source of emplacement". pp. 16–52, in: "Diapirism and Diapirs" (eds. J. Braunstein and G. D. O'Brien), *Amer. Assoc. Petrol. Geol.,* Tulsa (1968).
13. J. C. Harms. "Stratigraphic traps in a valley fill, Western Nebraska". *Bull. Amer. Assoc. Petrol. Geol.,* **50** (10), 2119–2149 (1966).
14. S. H. Harris, C. B. Land and J. H. McKeever. "Relation of Mission Canyon stratigraphy to oil production in north-central North Dakota". *Bull Amer. Assoc. Petrol. Geol.,* **50** (10), 2269–2276 (1966).
15. G. D. Hobson. "Some Fundamentals of Petroleum Geology". Oxford University Press, London (1954).
16. G. D. Hobson. "Faulting and oil accumulation". *J. Inst. Petrol.,* **42** (385), 23–26 (1956).
17. C. T. Hollenshead and R. L. Pritchard. "Geometry of producing Mesaverde sandstones, San Juan Basin". pp. 98–118, in: "Geometry of Sandstone Bodies" (eds. J. A. Peterson and J. C. Osmond), *Amer. Assoc. Petrol. Geol.,* Tulsa (1961).
18. J. H. Hoyt. "Chenier versus barrier, genetic and stratigraphic distinction". *Bull. Amer. Assoc. Petrol. Geol.,* **53** (2), 299–306 (1969).
19. C. E. Hull and H. R. Warman. "Asmari oilfields of Iran". pp. 428–437, in: "Geology of Giant Petroleum Fields (ed. M. T. Halbouty), Amer. Assoc. Petrol. Geol., Tulsa (1970).
20. W. D. C. Mackenzie. "Palaeozoic limestone of Turner Valley, Alberta, Canada". *Bull. Amer. Assoc. Petrol Geol.,* **24** (9), 1620–1640 (1940).
21. R. Martin. "Palaeogeomorphology and its application to exploration for oil and gas (with examples from Western Canada)". *Bull. Amer. Assoc. Petrol. Geol.,* **50** (10), 2277–2311 (1966).
22. E. D. McKee and T. S. Sterrett. "Laboratory experiments on form and structure of longshore bars and beaches". pp. 13–28, in: "Geometry and Sandstone Bodies" (eds. J. A. Peterson and J. C. Osmond). *Amer. Assoc. Petrol. Geol.,* Tulsa (1961).
23. G. de Mille, J. R. Shouldice and H. W. Nelson. "Collapse structures related to evaporites of Prairie Formation, Saskatchewan". *Geol. Soc. Amer. Bull.,* **75**, 307–316 (1964).
24. G. E. Murray. "Salt structures of Gulf of Mexico Basin—a review". pp. 99–121, in: "Diapirism and Diapirs" (eds. J. Braunstein and G. D. O'Brien), Amer. Assoc. Petrol. Geol., Tulsa (1968).
25. L. L. Nettleton. "Fluid mechanics of salt domes". *Bull. Amer. Assoc. Petrol. Geol.,* **18** (9), 1175–1204 (1934).
26. N. D. Newell *et al.* "The Permian reef complex of the Guadalupe Mountains region, Texas and New Mexico". W. H. Freeman, San Francisco (1953).

27. E. Oomkens. "Lithofacies relations in the late Quaternary Niger delta complex". *Sedimentology*, **21**, 195–222 (1974).
28. J. M. Parker. "Salt solution and subsidence structures, Wyoming, North Dakota and Montana". *Bull. Amer. Assoc. Petrol. Geol.*, **51** (10), 1929–1247, (1967).
29. P. van der Ploeg. Egypt. pp. 151–157, in: "The Science of Petroleum" (ed. V. C. Illing), Vol. 6, Pt. 1, Oxford University Press, London (1953).
30. W. A. Pryor. "Reservoir inhomogeneities in some recent sand bodies". *Soc. Petrol. Eng. J.*, 229–245 (June 1972).
31. F. B. Rees. "Methods of mapping and illustrating stratigraphic traps". pp. 168–221, in: "Stratigraphic oil and gasfields—classification, exploration methods and case histories" (ed. R. E. King), AAPG Memoir 16, Amer. Assoc. Petrol. Geol., Tulsa (1972).
32. G. Rittenhouse. "Stratigraphic-trap classification". pp. 14–28, in: "Stratigraphic oil and gasfields—classification, exploration methods and case histories" (ed. R. E. King). AAPG Memoir 16, Amer. Assoc. Petrol Geol., Tulsa (1972).
33. F. P. Shepard. "Submarine Geology". 3rd Ed. Harper & Row, New York (1973).
34. K. C. Short and A. J. Stauble. "Outline of geology of the Niger delta". *Bull. Amer. Assoc. Petrol. Geol.*, **51** (5), 761–779 (1967).
35. D. A. Smith. "Theoretical considerations of sealing and non-sealing faults". *Bull. Amer. Assoc. Petrol. Geol.*, **50** (2), 363–374 (1966).
36. H. H. Sullwold. "Turbidites in oil exploration". pp. 63–81, in: "Geometry of Sandstone Bodies" (eds. J. A. Peterson and J. C. Osmond), Amer. Assoc. Petrol. Geol., (1961).
37. R. F. Walters. "Buried Pre-Cambrian hills in northeastern Barton County, central Kansas". *Bull. Amer. Assoc. Petrol. Geol.*, **30** (5), 660–710 (1946).

CHAPTER 5

Surface Exploration for Petroleum

PETROLEUM AT THE SURFACE

PETROLEUM occurs at or near the surface of the ground in many parts of the world, in the form of liquid or solid hydrocarbons or as impregnations of hydrocarbons in sedimentary rocks of many types. Some of the major examples are themselves of commercial significance, and there are also innumerable minor seepages, occurrences or "shows" of gas, oil and solid hydrocarbons, which are evidence of the ubiquity of the generation processes involved. There are probably also many "micro-seepages" of oil and gas which are only detectable by sophisticated instrumental techniques.

The basic condition for the presence of a "show" is that oil or gas must have been generated and accumulated somewhere in the region at some time in the past, and that a proportion of these accumulated hydrocarbons should have escaped upwards from the original trap and eventually reached the surface of the ground. This may have resulted from the fracturing of the reservoir by earth movements after the time of accumulation; it may also have been the consequence of erosion which has exposed part of the reservoir rock at the surface. Clearly, if the process has been active on a large scale or over a long period of time, it may lead to the extensive escape of oil and gas to the surface, with the consequent serious depletion of the hydrocarbon contents of the reservoir.

Liquid petroleum exposed to the effects of atmospheric oxidation and bacterial attack becomes *inspissated*—i.e. the lighter components evaporate and the remainder polymerize and oxidize to form a variety of viscous, tarry, semi-solid or solid residual products, whose nature depends upon their geological history and the chemistry of the crude oil from which they were derived.

In the following description of the main varieties of surface petroleum occurrences, it should be borne in mind that there is seldom a clear-cut division between the different hydrocarbon phases involved. Thus, where "wet" gas is present there is usually also some liquid oil; conversely, where liquid oil occurs there is nearly always some gas. Where an asphaltic oil is entirely residual it may grade from a heavy liquid into a sticky "tar", and

123

various semi-solid or solid asphalts. Similarly, a paraffinic crude oil may give rise to residual deposits of paraffin wax. Any classification of surface occurrences can therefore be only very general.

In the early history of exploration for petroleum, wells were often drilled in the immediate neighbourhood of oil and gas seepages, since these were correctly considered as providing evidence that petroleum had accumulated in the vicinity. In areas such as Pennsylvania, where there were many seepages derived from shallow subsurface pools, this technique was quite successful. However, although a seepage is usually an indication of an accumulation not far away, the oil may have reached the surface by lateral migration along a fault or bedding plane, so that a well drilled near the "show" can still be far removed both horizontally and vertically from the underlying accumulation. Furthermore, the existence of a seepage must indicate the loss of a portion of the subsurface oil and gas; the larger the seepage and the longer it has been active, the less "live" oil will remain and the greater will be the chance that its quality will have been affected by the action of meteoric waters and bacteria. There are, nevertheless, many large deposits of heavy, viscous oils known in various parts of the world which are today potentially of commercial significance[22].

In general, however, the occurrence of asphaltic solids and heavy liquid hydrocarbons may indicate serious reservoir depletion, although it is sometimes still possible to discover liquid crude oil trapped behind a semi-solid bituminous seal by drilling down-dip from the surface outcrop. A "show" of light, relatively unaltered—i.e. "live"—oil, associated with high-pressure gas, is the most optimistic indication.

On the other hand, it must also be borne in mind that the *absence* of surface "shows" by no means condemns a sedimentary area as regards potential subsurface accumulations. Many very large oilfields and gasfields have been discovered for which there has been no surface hydrocarbon evidence at all.

(i) Surface Gas

Natural gas migrating along joints, faults and bedding planes in the strata overlying a subsurface hydrocarbon accumulation can sometimes reach the surface after travelling distances of several thousand feet vertically as well as horizontally. Hence, gas escaping anywhere from the Earth's surface is an important clue to possible subsurface accumulations. The periodic evolution of bubbles of gas produced by vegetation decaying in a stagnant pool can be deceptive, but evanescent "shows" of this type can easily be differentiated from deep-seated natural gas seepages, which will continue to evolve gas at a steady rate and at the same point, and will also generally be accompanied by a film of oil on the water surface, not to be confused with the more common films of iron oxides. (Gas samples are always collected for laboratory analysis if the presence of paraffin gases other than surface methane is suspected.)

In some well-known cases, the ignition of a natural gas seepage by lightning or friction has resulted in a continuing pale flame which may burn for centuries, as at Surakhany (Baku) and Kirkuk. Increasing the reservoir pressure by compression and reinjection of the gas as a conservation measure may sometimes result in an unwelcome increase of the extent of these "eternal fires", so that in such cases the gas is clearly escaping from the main reservoir and there must be fairly direct connection between the reservoir at depth and the surface of the ground. However, gas seepages of this nature, although they appear to be almost continuous, are probably in fact only intermittent; otherwise the large volumes of oil and gas still remaining in the subsurface reservoirs would have largely been dissipated.

Where natural gas burns at the surface, the surrounding rocks generally show signs of scorching, and if the gas contains appreciable quantities of hydrogen sulphide, the vegetation in the area is likely to be poisoned and stunted. The pungent odour of sulphur dioxide is produced by oxidation, and there may also be widespread deposition of sulphur particles. In some areas of the southern United States where there are many active gas seepages, the occurrence in the surface soil of brown "paraffin dirt" is often noted; it comprises agglomerations of yeasts and fungi produced by micro-organisms which feed on the escaping hydrocarbons. In Iran, a characteristic type of soil called "gach-i-turush" ("sour earth") is found in the oilfield regions, where large areas of limestone have been altered to gypsum by interaction with escaping hydrogen sulphide gas. (It is possible that this is a reversible process, so that interaction of gypsum with hydrocarbons under different conditions may produce secondary calcium carbonate, as in the cap rock zones of salt domes on the US Gulf Coast.)

Altered, highly acidic soils occur in the Plio-Pleistocene Tulare beds in the San Joaquin Valley of California and are thought to be the result of the interaction of ascending natural gases containing hydrogen sulphide with calcium-iron compounds in the sediments. Clusters of pyrite concretions occur in this area along the migration passages of rising gases. The basic iron sulphate jarosite $[K Fe_3(SO_4)]_2(OH)_6$ is also typically found.

The phenomena of "sedimentary vulcanism" are characteristic of gas shows in some parts of the world, where semi-explosive upwellings of gas, oil and loose sediments, can build up surface mounds, some of which may be of remarkable dimensions, called "mud volcanoes"[7, 17].

When high-pressure gas moves upwards through soft, water-bearing sediments, it can carry considerable volumes of clay with it in the form of a liquid mud, which, emerging at the surface, is spattered around the vent until a mound is built up with a crater at the summit. This "volcano" tends to palpitate as further escaping gases and fluids disturb the loose mass of sediments. If the proportion of oil to mud and water is high, the production of "tar cones" may result, as at Mene Grande in Venezuela. Groups and rows of "mud volcanoes", occurring above plug-like mud intrusions, on the fractured crests of anticlines, or along faults, are found in Burma, Trinidad and California. Eruptions may be spasmodic or continuous, according to the degree of build-up of gas pressure and the nature of the obstacles to be overcome in the passage of the gas to the surface. Apart from some similarity in the gas escape mechanism, there is, of course, no connection with magmatic volcanic activity. Like true volcanoes, however, "mud volcanoes" can grow to considerable proportions and they also resemble salt domes in the piercement structures they produce.

The "mud volcanoes" of Trinidad, associated with the unstable crestal region of the Southern Range, have been classified as of two types: a "fluid" variety, mainly muddy salt-water springs with much gas; and a "sticky" variety, periodically erupting viscous mud and large boulders which can be identified as coming from the Middle Miocene strata from depths as great as 10,000ft.[11, 19]

Submarine sedimentary vulcanism can result in the creation of new islands. The development of islands of this nature is preceded by the under-sea release of gas, which can be a very violent process. Once cratering occurs, a rapidly increasing free passage is created for the escape of ejected material, and in Trinidad about a million tons were explosively ejected in one day (August 1st, 1964) to form a new island[8].

A number of similar islands appeared in the Arabian Sea, west of Karachi, at the time of the Makran earthquake in November, 1945, probably produced by a similar undersea release of gas pressure.

In the Baku area of the USSR, as many as 160 "mud volcanoes" have been listed which are either currently active or only recently extinct, and many of the islands off the coast of the Apscheron peninsula have been formed by sedimentary vulcanism[5, 15]. The periodicity of the Caspian "mud volcanoes" has been shown to be related to the gravitational influence of the Moon.

In Burma, a basin (or "paung") filled with agitated soft mud is often produced rather than the more common mound or cone. Factors which determine whether a mound or mud-filled basin will be formed include the viscosity of the mud-sand mixture, the rate at which it is ejected, the local topography, and also the climate.

In Central Australia, the huge crater-like landmark of Gosses Bluff, originally believed to have been caused either by igneous extrusion or meteoric impact, is now thought to be the expression of an enormous "mud volcano" with a disturbed roughly circular surface area some seven miles across.

Sand dykes are often associated with "mud volcanoes", and may vary in thickness from a few inches to more than 15ft. The sand is forced by escaping hydrocarbon gases along fissures and cracks in the overlying strata. *Mud dykes* also occur, some of them extending in linear flows across as much as a mile of sediments. They probably act as channels to feed the "mud volcano" vents, throwing up into them rock fragments of every age that the mud flow has penetrated. Fossil mud dykes are also sometimes found in the course of drilling. Mud dykes occur particularly along fracture lines, providing evidence of the ability of natural gas to migrate for considerable lateral as well as vertical distances where there are channels along which it can travel, after the overlying sediments have become compacted[10].

Although "shows" at the surface can be satisfactorily explained by gas movements of this nature along "macro-fissures" in the sediments overlying a subsurface accumulation, there is also a possibility that much smaller escapes of gas can periodically occur by movement along hardly-discernible "micro-fissures"[4]. The fact that various types of instrumental geochemical survey have been developed and applied at the surface with at least partial success seems to confirm this belief. There is also some evidence of an increase in hydrocarbon gases dissolved in offshore waters overlying shallow oil accumulations (e.g. some of those off the coast of California), and it is possible that the monitoring of the gas content of sea water in selected areas by marine "sniffer" instruments may become a valuable future exploratory technique (p. 172).

For the petroleum geologist, the discovery of a "show" of natural gas is a source of encouragement. It indicates that an accumulation of "live" oil may well be in the neighbourhood—although, of course, it must be remembered that there are also many "non-associated" gasfields in various parts of the world which produce gas but little or no oil[20].

However, although favourable from the regional point of view, the presence of escaping high-pressure gas may complicate local drilling, since obvious mechanical difficulties are likely to be encountered, as has been the case in areas such as Assam. In Trinidad also, many early wells drilled near gas "shows" and "mud volcanoes" were unsuccessful, because of the distance the gas had migrated from the original reservoirs, which meant that such wells were often drilled off-structure.

Since most exploration operations have been carried out with the objective of discovering oil rather than gas, the obvious course of action after a gas seepage has been discovered on an anticlinal structure is to drill down-flank in the hope of avoiding the gas cap, and of finding the underlying oil, if any is present.

(ii) Surface Oil[14]

Liquid oil "shows" occur in the form of pools or seepages, but more commonly the oil impregnates the surface rocks. Presumably, the crude oil has escaped in such cases along faults and fissures from underlying accumulations, and is therefore often accompanied by varying proportions of associated natural gas and salt water. Oil seepages usually take the form of heavy, viscous liquids, since exposure to the atmosphere and bacterial attack tend to oxidize and polymerize crude oils, particularly the asphaltic varieties. Major heavy oil seepages of this type are found in many of the world's oilfield areas, notably in the USA[3], Iraq, Venezuela and Colombia.

Seepages of light, fluid oil are also known, although they are infrequent; examples from Borneo and Eastern Venezuela are said to be light enough to be directly usable as fuels in internal combustion engines. Such oils are generally the result of "wet" gas condensation.

Escaping crude oil cannot travel far from its point of efflux, since the oil is very soon precipitated in streams or rivers, particularly those containing suspended clays. In countries with tropical rainfall, the intensity of weathering is very great, and all traces of oil are usually washed from the surface rocks, unless the seepage is of a viscous, asphaltic oil, when some evidence may be retained.

During exploration, care must be taken not to confuse with petroleum seepages hydrocarbon oils of different origin and chemical constitution, derived for example from the contact metamorphism of coals by igneous intrusions—as in the Karroo beds of South Africa, or in some Scottish coalfields. Certain types of resin are also difficult to distinguish from hydrocarbons in field conditions. If, however, they are readily soluble in alcohol, this generally indicates their vegetable nature. The laboratory determination of acid value and iodine value will be additional distinguishing tests. In any case, true surface oil "shows" are seldom confined to a solitary and doubtful occurrence.

Surface seepages were the only sources of petroleum for many centuries, and up until quite recent times shallow, hand-dug pits provided local supplies of oil in some parts of the world, such as Burma and China. At a few places—Pechelbronn in Alsace, Grosni in the USSR and Wietze in Germany—commercial volumes of oil were obtained by drainage into galleries driven into shallow, oil-saturated sandstone beds outcropping at the surface.

Reports of oil seepages in marine waters are relatively rare, perhaps because any crude oil released in this way will tend to be rapidly polymerized and precipitated as a result of contact with clay particles held in suspension in the water. In offshore areas of California, for example, where sea-floor seepages would be likely to occur due to the relatively thin "cover" overlying many subsurface oil accumulations, there is often difficulty in determining whether the semi-solid asphaltic material washed onto the beaches is in fact the result of crude oil seepage or of contamination from tanker washings. Recent investigations seem to favour crude oil seepages, which coagulate into a tar-like "slick" when the sea is calm[12]. A similar effect has been known for many years on the southern beaches of Australia.

(iii) Oil Impregnations

When crude oil has entered or had originally accumulated in a porous bed which is now exposed at or near the surface, an oil-impregnated rock results, the nature of which will depend on the lithology of the rock, the chemical composition of the oil and the degree of oil alteration that has taken place[21]. Oil-impregnated rocks are generally tough and weather-resistant, although some relatively uncemented rocks may become quite plastic if saturated with crude oil. Fluctuations of the water table may flush the oil from the surface zones, although rocks which have contained heavy asphaltic crudes sometimes retain traces of these—particularly along joint planes and fissures.

Impregnated sandstones[18] are illogically but traditionally classified as "tar sands" and are described separately on pp. 131–3. Impregnated clays are usually black in colour, while oily marly clays have a blue appearance. Impregnated limestones often show dark blotches and spots, but the presence of oil may be missed if the sample is not crushed and extracted with a suitable solvent such as carbon tetrachloride or chloroform.

Limestones impregnated with heavy oils or asphalts are widely used for building purposes (roofing and flooring) and also as road surfaces. The principal commercial deposits in Europe are in France (Gard and Seyssel), Switzerland (Neuchatel*), Germany (Verwohle) and Sicily (Ragusa).

(iv) Vein Bitumens

Solid bitumens† occur in various forms, usually as near-vertical veins penetrating surface beds[9]. They are characterised by their solubility in liquids such as chloroform and carbon tetrachloride, and are as a result distinguish-

---

*In Dr. Johnson's dictionary (1755), "Asphaltum" is described as "a bituminous stone found near the ancient Babylon, and lately in the province of Neufchatel; which mixed with other matters, makes an excellent cement, incorruptible by air, and impenetrable by water; supposed to be the mortar so much celebrated among the ancients, with which the walls of Babylon were laid."

†The terms "bitumen" and "asphalt" are unfortunately largely interchangeable in normal usage.

able from kerogenous materials such as oil shales (p. 287) from which oily liquids can only be obtained by distillation. They appear to have been formed as a result of the passage of asphaltic crude oils up zones of rock permeability, or along faults, fissures and bedding planes between the subsurface reservoir and the surface. Deposits and impregnations could be expected to form in local cavities or wherever the beds along the route to the surface were sufficiently porous. There often seems to have been a selective retention of the migrating material, the asphalts being retained while the lighter components were lost.

Asphaltic material may sometimes have been precipitated from crude oils when large quantities of natural gas have been dissolved in the oils as a result of increasing pressure. This process may also account for the particles of asphaltic matter found in some oil reservoir rocks.

Veins of so-called "natural" or "native" bitumens have been found (e.g. in Turkey)[13] which are locally as long as 3500m, as wide as 80m, and which cut through several thousand feet of strata, although such large-scale examples are relatively rare. The bitumens can be recovered by shallow mining operations when the economic return is attractive.

Since the chemical composition of crude oil is variable, a similar variety can be expected in the constitution of its residual products, but many of the apparent variations between bitumens may be comparatively local; vertical exploration can often reveal homogeneity which is not apparent on the surface. The varieties of bitumens probably represent time-stages in the solidification of different oils, although there is also some evidence of subsequent changes brought about by pressure metamorphism in the course of time.

For purposes of classification, solid bitumens have been subdivided into "fusible bitumens" (asphaltites), and "infusible bitumens" (asphaltic pyrobitumens). This differentiation is dependent upon the degree of fusibility, the specific gravity and the amount of free carbon present. Several other methods of classification have been suggested, depending principally on such factors as the solubility and sulphur content. The following are the principal groups; there are many other varieties, some of which have picturesque names*:

(a) *Uintaite* (named after the Uinta Mountains, Utah, USA), also called *Gilsonite,* is one of the purest and most valuable of "native" bitumens, found in eastern Utah and western Colorado. It is recovered by mining from vertical, semi-parallel veins varying in thickness from a fraction of an inch to several feet, which traverse the local Tertiary sediments.

(b) *Manjak* or *Glance-Pitch* is typically found in Barbados in veins and fissures connected with faults, bedding planes and joints in Tertiary strata.

(c) *Grahamite* contains a much higher content of mineral matter than the other fusible bitumens, and is generally found in thicker veins. It was originally mined in West Virginia, where it occurs in vertical fissures in Carboniferous sandstones and shales, but is now also found in other parts of the USA, Cuba and Mexico.

As examples: Tabbyite, Piauzite, Nigrite, Courtzilite, Libillite, Chapapote.

(d) *Elaterite* was originally discovered in lead mines in Derbyshire in the Carboniferous Limestone. It is comparatively soft and elastic, although classed as an asphaltic pyrobitumen. Detailed analysis has shown that it is really an amorphous branched polyethylene, with a sol fraction and a gel fraction with a network structure.

(e) *Wurtzilite* occurs in thin, nearly vertical veins traversing Tertiary limestones in the Uinta river basin, Utah. It resembles elaterite in its elasticity and decomposes upon heating. (The different types of solid bitumen found in the Uinta basin seem to represent variations in the local depositional environments, i.e. a derivation from source materials of progressively younger ages.)

(f) *Albertite* is found in veins intersecting the Lower Carboniferous shales of New Brunswick, Canada. The type occurrence is a vertical vein 17ft thick, from which extend numerous small, lateral veins.

(g) *Impsonite,* which occurs typically in the Impson Valley, Oklahoma, appears to be an end-product in the process of bitumen formation, having a very high proportion of fixed carbon and being infusible. It also occurs in fissures and veins in Oklahoma, Arkansas and Nevada in sandstones and shales of Carboniferous age.

TABLE 31

CHARACTERISTICS OF SOME "NATURAL" BITUMENS

Type	Streak	Sp. gr. at 77° F	Fusibility	Fixed Carbon %
Uintaite (Gilsonite) .	Brown	1·05–1·10	250°–350° F	10–20
Manjak (Glance Pitch)	Brown-Black	1·10–1·15	250°–350° F	20–30
Grahamite . .	Black	1·15–1·20	350°–600° F	30–55
Elaterite . . .	Light Brown	0·90–1·05	Decomposes	2–5
Wurtzilite . . .	Light Brown	1·05–1·07	Decomposes	5–25
Albertite . . .	Brown-Black	1·07–1·10	Fuses on de-composition only	25–50
Impsonite . .	Black	1·10–1·25	Infusible, de-composes	50–85

(v) Mineral Waxes

A number of waxy solids are found associated with seepages and impregnations of paraffinic (as opposed to asphaltic) crude oils. *Ozokerite,* the best-known native petroleum wax, derives its name from the Greek ("odoriferous wax"), and has been found most typically at Boryslaw (formerly in Poland, now in the USSR) and in some localities in the USA.

At Boryslaw, ozokerite is mined from fissures and veins in the Miocene rocks overlying an overthrust "nappe" (p. 94). The wax seems to have been derived from the paraffinic oil which has percolated upwards along fractures in the cap rock of the reservoir. Where oil occurs in the galleries of the Boryslaw ozokerite mines, it is free of wax, and it is notable that the paraffins present in the oil differ from those in the ozokerite.

Ozokerite is in fact composed of higher members of the C_nH_{2n+2} and C_nH_{2n} series, with n ranging from 22 to about 29. It usually contains a certain amount of liquid oil diffused through the wax, whose presence may lower the melting point of the solid from about 180°F to about 140°F. The colour of ozokerite depends on the proportion of impurities present, and ranges from transparent straw to dark brown. It has a conchoidal fracture and a typical waxy lustre.

Several other hydrocarbon mineral waxes are known which differ slightly from ozokerite in their chemical constitution, texture and melting point. *Zietrisikite* is found at Zietrisika in Moldavia; *Neft-Gil* occurs on the island of Cheleken in the Caspian Sea; and *Baikalite* is found near Lake Baikal in Siberia.

These petroleum waxes must not be confused with vegetable waxes derived from lignitic material or curiosities such as Bute Island wax, which is thought to come from the local coelenterates, or Mumiyo, a waxy substance of obscure origin apparently produced by certain birds in Asia and Antarctica.

(vi) "Tar Sands"[1, 16]

Sandstones impregnated with heavy residual petroleum and occurring at or near the surface are traditionally termed "tar sands", although "tar" is more correctly the term for an end-product of the destructive distillation of organic matter. A more apposite title would be "heavy oil sands", since there is no evident distinction between such "tar sands" and subsurface reservoir sandstones containing heavy crude oils. Thus, in general, the oil in "tar sands" seems to have been derived from a migrated crude, generally asphaltic in nature, which has lost its lighter constituents in the course of migration or as a result of subsequent bacterial action. The sandstone bodies themselves, however, are in many cases of estuarine or brackish water rather than marine facies. They are of many different geological ages.

The largest "tar sand" occurrence in the world is in northern Alberta, near Fort McMurray on the Athabaska River—where there is an area of some 13,000 sq miles of sandstones impregnated with heavy oils. The sandstones here are of Early Cretaceous age, of both marine and non-marine facies, and rest unconformably on the tops of eroded Devonian beds.

From east to west they form three distinct stratigraphic units—Wabiskaw-McMurray, Bluesky-Gething and Grand Rapids. The "tar sands" are partially exposed along the river banks in the Mildred Lake area, but most of the deposits, which average about 150ft in thickness, are covered by an overburden of glacial drift or other surface formations which are often several hundred feet thick[2].

The heavy oil forms a film around each grain, so that when it is removed the remaining sands are completely uncemented. Presumably, these sands are the remains of deltas deposited by rivers flowing northwestwards from the pre-Cambrian "Shield" into an elongated lagoon or lake, part of which may have been formed in a depression resulting from extensive evaporite solution. It has been variously suggested that the oil (a) has been formed *in situ* from deltaic deposits in the Lower Cretaceous; (b) has migrated laterally from marine sediments to the south and southwest; (c) has been formed in the

overlying Clearwater shales; (d) is a residual deposit; or (e) has entered the
Cretaceous sands from the underlying Palaeozoic beds. Generation during the
Lower Cretaceous is favoured from the evidence of isotope studies.

The Athabaska deposits form part of a belt of Lower Cretaceous heavy,
viscous oil accumulations extending from the Peace River area of Alberta
to Lloydminster in Saskatchewan. It is probable that all these oils, which are
relatively young, are connected in their origin. One hypothesis is that the
hydrocarbons in this area moved out of the deep parts of the basin held in
micellar or colloidal solution in the compaction waters and were then
"precipitated" in existing structural or "stratigraphic" traps as a result of a
change of salinity of the formation water.

The relationship between oil-in-place, and recoverable "synthetic" (or "non-
conventional") oil for the Wabiskaw-McMurray sand deposit, which is by far the
most important of these groups, is shown in Table 32, from which it will be
seen that the Athabaska deposits may contain as much as 267 B brl of
extractable "non-conventional" oil, a volume equivalent to nearly half the
total proved "conventional" crude reserves of the whole world in 1971.

TABLE 32

ATHABASKA "TAR SANDS"

Saturation (oil % by wt.)	Oil in place B brl	Recoverable "synthetic" crude oil (B brl)
Rich sands (>10%)	440·8	188·3
Intermediate sands (5–10%)	145·7	61·9
Lean sands (2–5%)	39·4	16·7
Total >2%	625·9	266·9

Source: P. Phizackerly & L. Scott[16].

Since nearly 75% of the Athabaska reserves lie under more than 250ft of
overburden, open-cast mining of the sands is never likely to be economically
attractive, in view of the very large amount of inert material that has to be
handled per barrel of oil obtainable. (It must be remembered that a cubic yard
of sand must be mined for each barrel of oil obtained even *after* the overburden
has been removed.) Various *in situ* processes are under development, however,
which are more promising. They include a technique whereby a mixture of
steam and sodium hydroxide is injected into the sands through specially
drilled shallow wells. The raw oil obtained in this way is then mixed with
lighter oil to make it fluid enough to be pumped away for full refining treatment.

The first modern plant designed to produce oil from Athabaska "tar sands"
was put on stream in 1967. Situated 260 miles north of Edmonton, it had an
initial output of 45,000 brl/day of oil (planned to increase to 65,000 brl/day),
which is produced by a basically simple process. In this, the tar is washed from
the sand grains by hot water, the alkalinity of the resulting mixture being

adjusted until the oil floats to the surface, to be collected and treated by delayed coking and hydrogenation refining operations. Most of the high (about 5%) sulphur content of the oil is removed in the process and can be recovered and sold. The end-product, which is termed "synthetic", but more correctly "non-conventional" crude oil, is usually of about 40° API gravity, with a low sulphur content. It yields on refining a higher-than-average proportion of light products. A number of other, larger, plants are now in the planning stages, the first of which may be in commercial operation by late 1977[6].

Another large "tar sand" deposit has been discovered in the Canadian Arctic on Melville Island, in Triassic sandstones. In the United States, the Bureau of Mines has estimated that 2·5–5·5 B brl of oil may eventually be recovered from various shallow "tar sand" deposits in Utah, Kentucky and California. In East Venezuela, an extensive "tar belt" covers some 9,000 sq miles in an arc about 375 miles long between the Orinoco delta and the central state of Guarico. In this area, there is an average thickness of up to 100ft of oil-impregnated sandstones of Oligocene age, which are occasionally exposed but generally covered with overburden. It is believed that this deposit contains at least 200 B brl of heavy oils, from which 10–50 B brl might one day be recoverable. The only other known "tar sand" deposits of the same order of magnitude as those of North America and Venezuela occur in the USSR— in the Fergana area of Soviet Central Asia, in the Volga-Urals area and in northeastern Siberia. Other smaller deposits are at Bemolanga in Malagasy, where there is an area of 150 sq miles of Triassic "tar sands" which contains about 1·75 B brl of heavy oil; in Albania (Selenizza), and in Romania (Derna).

(vii) Asphalt "Lakes"

The large, horizontal deposits of solid or semi-solid asphalts which occur in some parts of the world are sometimes referred to as "lakes"; they are believed to represent the last stages in the oxidation and polymerization of asphaltic crude oils which have been released in large quantities from the destruction of major subsurface hydrocarbon accumulations.

Many examples of such asphalt deposits are known, some of which have been worked commercially, as at Bermudez in East Venezuela, where asphaltic oil leaking surfacewards up a fault plane has formed a deposit at least 10ft deep. Other lake-like asphalt deposits occur in Iraq and in and around the Ouachita and Amarillo Mountains in the United States. The most famous example of all is the eponymous Trinidad "Pitch Lake", of which the earliest record comes from the diary of Sir Walter Raleigh, who visited Trinidad in 1595, and subsequently wrote in his "History of Guiana":

". . . I rowed to another part called by the naturals "Piche" and by the Spaniards "Tierra de Brea"; there is that abundance of stone pitch, that all the shippes of the world may be therewith loden from thence, and we made trial of it in trimming our shippes to be most excellent good, and melteth not with the sunne as the pitch of Norway, and therefore for shippes trading the South parts are very profitable . . ."

The Trinidad Pitch Lake nowadays covers an area of some 137 acres lying roughly on the axis of the La Brea anticline. Attempts to discover its depth by drilling have not found a definite bottom, and to the eye, the asphalt seems motionless and solid. There is only one small patch in the centre, known as the "Mother of the Lake" which is soft, the rest being hard enough to carry a light railway (which, however, must frequently be moved) for transporting the extracted asphalt to the refinery. Although apparently quiescent, the Lake is nevertheless in a complex state of motion, so that pits dug in its surface are gradually obliterated. The Lake thus gives the impression of being inexhaustible, although in fact a slight drop in its level has been noticeable in recent years.

There are several theories for its origin, but whatever the mechanism has been, the Pitch Lake must represent the last stages of depletion of a very large oil accumulation. Its almost constant (and unusual) composition is remarkable, containing as it does 40 % of bitumen, 31 % of silty clay and 29 % of water. One explanation of the Pitch Lake is that it is the combination of an oil seepage and a "mud volcano" on a tremendous scale. Thus, during late Tertiary times and after the local structure had been formed, part of the La Brea sands and clays was removed by weathering. At the same time, intense faulting resulted in the escape of the oil and gas from the underlying Morne l'Enfer oil reservoir.

A basin began to form along the main fault plane, which was gradually filled with a mixture of mud, sand and oil. The lighter fractions of the oil then evaporated, leaving an asphaltic residue. (This does not explain, however, the constant quantitative relation of the various constituents of the emulsion.) An alternative theory is that there was a slow, continuing seepage of oil taking place during the Pleistocene sedimentation, followed by folding to form a structural dome, the roof of which has since been eroded away. Evidence in favour of this theory is that the ground level around the Lake is gradually sinking, suggesting that the "pitch" is flowing in from the sides.

GEOLOGICAL SURVEYING

An important responsibility of the exploration geologist, where local conditions are suitable for such operations, is the preparation of a surface geological map from which a subsurface stratum contour map (p. 226) can subsequently be constructed to form the basis of the exploration drilling programme. If a reliable topographical map of the area being examined already exists (e.g. a Government or Ordnance Survey map) this can be used as a base map upon which geological information can be plotted as it is gathered. In undeveloped countries, it may be necessary for the field geologist to construct his own topographical base map, perhaps with the aid of a survey party, which will, however, usually only establish the primary triangulation; in such cases, the geologist must map both the detailed topography and the geology of the area more or less simultaneously, and for this purpose must be familiar with the appropriate techniques and instruments*.

The scales principally employed in geological mapping are: one inch to the mile (1 : 63,360), three inches to the mile (1 : 21,210), 2.53 inches to the

---

*Nowadays, aerial photography is commonly used as a means of assisting and accelerating the ground survey, as described on pp.143-8, and the appropriate prints of aerial photographs are often utilized as a basis for geological surveying in the field.

mile (1 : 25,000), six inches to the mile (1 : 10,560) and, in areas where great detail is required, 25·344 inches to the mile (1 : 2,500). Of these, the three inch to the mile scale is the most convenient, being large enough to show adequate detail without requiring excessive time to be spent on the survey.

The general system of height contouring used on British Ordnance Survey maps is: 50ft, 100ft then by 100ft contours up to 1,000ft, and thence every 250ft.* In field geological mapping, the height intervals selected will depend on the nature of the topography to be shown, but 20ft or 50ft contours are normal. The datum level is of course "mean sea level", but in some inland countries it may be preferable to use the mean level of the nearest large lake. In little-explored areas, an arbitrary or only approximately determined base height may have to be used as the local datum to which other height measurements are related.

Similarly, where no pre-existing topographical map is available, a local co-ordinate "grid" will have to be established. To do this, a base-line is selected and carefully measured. From this, a number of "first-order" triangulation points are determined with theodolites and marked on the ground as permanent stations. Upon the lines joining these points, which now provide subsidiary base-lines, a network of "second-order" triangulation points can then be built up by theodolite observations of stations selected on suitable prominences, distributed as evenly as possible over the area of the map. Cairns or marker posts are then erected at these stations. It is desirable for the geologist never to be out of sight of one (or better two) of these second-order markers, to which he may relate his observations. He can then set up his own system of tertiary triangulation points for subsequent detailing.

The Instruments

In place of the theodolite, which has for many years been the standard instrument used in primary and secondary triangulation operations, or whenever special accuracy was required, various electronic instruments have been developed whose use greatly facilitates basic surveying techniques[29, 31]. They include devices using micro-radio waves and laser beams. The *Tellurometer,* for example, operates on radar principles, sending out microwaves and recording the elapsed time for the signal to return. Its use greatly reduces the time spent on surveys, and makes possible rapid distance measurements even in areas of heavy forest. Tellurometer traverses can often be run through country in which normal triangulation would be physically or economically impossible. The latest models (e.g. CA 1000) have a simplified read-out method (calibrated directly digital) and an accuracy claimed to be better than ±15mm over a range of 30 km.

The change to the metric system currently being undertaken in Britain means that on new and revised Ordnance Survey maps on the 1/1,250 and 1/2,500 scales, heights will be shown in metres and areas of parcels of land in hectares as well as acres. The 6in map will be replaced by a 1/10,000 scale map with metric contours. The reader who wishes to gain an insight into the enormous amount of work that is involved in complete metric conversion is referred to the paper by Col. F. M. Sexton given to the Royal Geographical Society, London, on Feb. 19, 1968 and published in the Society's Journal,

The *Mekometer* (Fig. 39), due to the National Physical Laboratory, uses a light beam from a gas laser, which is visible up to three miles in daylight. The instrument is designed for use over distances of up to 10,000ft and to measure lengths to within two parts in a million. Over shorter distances—between 160ft and 200ft—the sensitivity of measurement is 0·004in. Portable gas lasers are also now used in survey operations to project a narrow beam of light which serves as a line of reference. The red spot of light projected onto a distant unattended target can also be used as a sighting point during triangulation operations.

However, in constructing surface geological maps, only a few relatively simple surveying instruments are generally necessary. These are as follows:

(1) Various types of *prismatic compass* can be used to define the position of rock outcrops by making a simple bearing and pacing traverse from known points, provided that a suitable topographical map is already available. They consist essentially of an aluminium compass ring graduated in degrees and half-degrees, a needle on an agate bearing, a sighting bar with an adjustable end, and a movable prism reader and back sight which enable the face of the compass to be read while a distant object is viewed through the sights. The instruments can be attached to collapsible tripods for detailed work.

The *Brunton Transit* has folding, open sights, a double level fitted on a clinometer with a vernier attachment, and a mirror-lined lid with a vertical slit which enables horizontal or vertical angles to be measured. By opening the lid so that it is flat with the case a long straight edge is available, and if this is laid on a rock surface the dip can be measured by moving the bubble arm till the bubble is horizontal and reading off the angle of tilt on a scale, accurate to one or two degrees.

The field geologist must know the length of his normal pace, and by pacing combined with hand compass sights it is possible to map considerable detail very rapidly. The best length of sight is about 600 to 1,200ft, and wherever possible closed traverses should be run so that the final closing error can be found and distributed.

(2) The portable *aneroid barometer* is particularly useful for measuring heights in reconnaissance surveying and in running rapid traverses. It is also useful as an aid in the determination of position on a topographical base map with a small contour interval. There are a number of different types of aneroid in use, but all consist essentially of an exhausted metal capsule with a diaphragm whose movement depends on the changes of external air pressure.

Aneroids reading to a maximum height of 5,000ft usually can be read in 1ft intervals; those reading to 10,000ft usually have 2ft calibrations in 50ft intervals.* Since aneroids depend for their action on variations in atmospheric pressure, they cannot produce reliable data in changeable weather unless a base barograph is used to make automatic records of pressure variations at, say, 10-minute intervals throughout the period of survey. Most aneroids are constructed in such a way that they are self-compensatory for temperature

In general, the aneroid must be used with a degree of caution, since the sense of accuracy conveyed by the scale calibrations is somewhat illusory.

variations. Where an uncompensated instrument has to be used, a table of corrections usually provides for correcting the observations for temperature change.

(3) The *plane-table* and *telescopic alidade,* which are employed for general mapping purposes, utilise tacheometric principles. They enable topographical and geological maps to be rapidly constructed, particularly in desert areas.

The *plane-table* outfit usually consists of a plane-table, tripod, alidade, a number of stadia rods, a surveyor's umbrella for sheltering the operator, and two or more flags for signalling to the rodmen. The plane table itself, which may vary in area from 18in by 18in to 30in by 24in, is supported on a collapsible tripod by a screw-and-socket joint controlled by two wing nuts, so that the board may be levelled and orientated without moving the tripod legs.

Telescopic alidades (Fig. 40) differ in details of design, but consist essentially of a telescope mounted on a vertical pillar which rests on a long metal base with a graduated edge. Attached to the base are usually a trough compass and a levelling bubble, and also a parallel rule which can be extended laterally from the main instrument base. The telescope of the alidade can swivel in a vertical arc and the angle through which it is moved can be controlled and measured by a gradienter screw, vertical arc and striding bubble level. The eyepiece is usually fitted with three fine horizontal stadia wires. The centre wire is on the axis of the instrument, while the two outer wires are placed such a distance apart that the distance between the points at which they appear to cut a stadia rod when this is observed through the telescope (i.e. the stadia intercept) multiplied by 100 gives the distance between the rod and the instrument. The eyepiece is also often fitted with a vertical wire for convenience of sighting.

The stadia rods used are the normal surveyors' collapsible rods, usually extensible to 10ft or so, sometimes with their markings modified for easier distant viewing.

The development of the "self-reducing level" (e.g. the *"S 700"* made by Cooke, Troughton and Simms and the Hilger & Watts *"Autoset 2"*), which incorporates an automatic self-compensating device, has greatly assisted rapid levelling operations. The speed of ground levelling may be virtually doubled as a result of the very rapid preparation of the instrument at each station which is now possible.

The Autoset level has an optical stabiliser which maintains a level line of sight even when the telescope is tilted, thus dispensing with the need for precise instrument levelling. In the stabiliser is one fixed prism, and two prisms on a mount supported on four metallic strips forming a cross-spring flexible pivot. When the Autoset is tilted, this mount swings like a pendulum and the rays of light are deflected so that points on a horizontal plane are always imaged on the cross-lines.

Theory of Tacheometry

The mathematical theory on which the use of the telescopic alidade is based assumes that the stadia rods are held vertically at each station, and that the instrument is itself perfectly level.

Fig. 39. The Kern ME 3000 Mekometer short-distance measuring device.

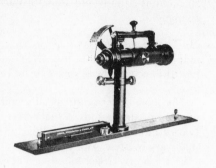

Fig. 40. Plane-table alidade with compass needle and Beaman arc.

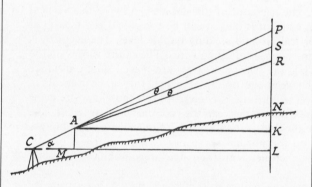

Fig. 41. Theory of Tacheometry.

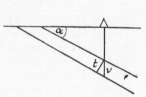

Fig. 42. Estimation of bed thickness from borehole measurement.

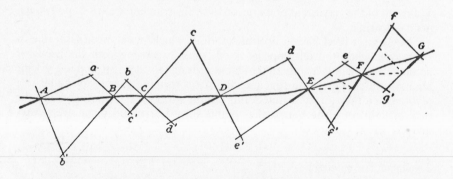

Fig. 43. Diagram illustrating determination of thickness of steeply inclined beds by construction method: the thicknesses of the beds are $\frac{1}{2}(Ab^1 + aB)$, $\frac{1}{2}(Bc^1 + bC)$, etc.

Consider Fig. 41 which represents the position when the alidade is used to determine the height of a point N which is higher than the alidade station. Let PR(h) be the rod intercept, A the anallactic point, C the centre of the instrument. Let AS be the central line of sight and PAS = SAR = θ; let the angle of tilt of the alidade be ACM = α. It can then be shown that, as a close approximation, for small angles, where K is a "stadia coefficient":

$$CL = Kh \cos^2 \alpha \text{ and } SL = Kh \sin \alpha \cos \alpha = Kh\tfrac{1}{2} \sin 2\alpha.$$

From these equations, the horizontal and vertical displacements of the rod relative to the plane-table can be deduced. Tacheometer Tables are available, showing all values of $\cos^2\alpha$ and $\sin \alpha \cos \alpha$ for angles 0° to 30°, and hence the calculation is simple, particularly as K will nearly always be 100.

As $Kh/2$ is used in the formula, and since by using only half-intercepts the range of the instrument is doubled, for rapidity of working it is best to use the intercepts between an outer and the central cross-hair.

The intercept between outer hair and central hair having been read off, this is then multiplied by 200 giving the value of Kh. The angle of tilt (α) is read on the vernier, the value of $\cos^2\alpha$ is read from the tables, and $Kh \cos^2\alpha$, the horizontal interval calculated. (For small values of α, Kh can be taken instead of $Kh \cos^2\alpha$ without appreciable error on the map.) The surveyor then scales off a line equivalent in length to this distance on the plane-table sheet, using the edge of the alidade as a ruler, and starting from the point representing his station. The corresponding value of $\sin 2\alpha$ is read off from the prepared tables. Upon multiplying the original $Kh/2$ by $\sin 2\alpha$ (a pocket calculator is used for this) the difference of height of the foot-mark on the rod from the table will be found. From this is subtracted the height of the foot-mark above the base of the rod, fixing the difference of height between the plane-table and the ground at the base of the rod. It is clear that if the rod is higher than the plane-table there must now be added to this distance the height of the plane-table above the ground at the station (usually about 4·8ft); or if the rod is below the plane-table this distance must be subtracted.

The above operations sound complex when recorded in detail, but in fact they can be performed very rapidly with a little practice. The whole process is greatly accelerated by using the "Beaman Arc," which is fitted on most present-day alidade telescopes. This device is for use only with a vertical rod, and is not precise enough for triangulation calculations, but adequate for normal geological detail, and its employment obviates the use of tacheometer or trigonometrical tables in the field.[37]

Field Techniques[31a]

The first step in plane-table mapping consists of setting up, levelling and orientating the plane-table at a convenient primary or secondary triangulation point, or on a suitable topographic elevation. The rod-carriers are then sent out to erect their rods at desired points, which may be geological contacts, marker beds, notable features, or items of topography such as stream bends or hilltops.

The first man, having reached the position selected for him, extends his rod and holds it vertically. The geologist rules a ray towards the rod on the plane-table, notes the intercept of the stadia hairs, makes a reading of the Beaman arc, and signals the rodman to proceed to the next point while he is computing the height and distance. Having marked off the position of the rod on the map, he writes against it its elevation and sketches in the general contour and topography in its neighbourhood. If several rodmen are at work, the geologist can rapidly record a great deal of information in this way, and in arid countries where the geology is exposed it is a simple matter to instruct a rodman to walk along a marker bed and flag him periodically to elevate his rod, so that a complete record of the outcrop and elevation changes of the bed can be recorded. Other features such as cliff exposures, etc. can be treated in more detail, and as much lithological information as possible recorded on the map for each survey point.

Having mapped in this way all the features visible from his first station, the geologist then recalls the rodmen and moves on to the next station. His choice of new station depends on the type of area he is working in, but in any case its position and elevation must first be fixed by reference to the initial point.

For purposes of the structural reconnaissance so frequently required in petroleum geology, it is generally unnecessary to produce a detailed outcrop map showing the local surface geology, but instead one or more suitable "marker" beds are sought which can be traced most of the way round any evident structures.

Marker beds need not have a particular stratigraphical significance in themselves, requiring only to have some persistent lithological distinction from the surrounding rocks. Flaggy, flinty or oolitic limestones often provide excellent markers, as do bands of ferruginous cherty sandstone. In tracing such a marker bed, the essence of the operation is speedy but accurate reconnaissance to permit the dip, pitch and closure of the structure to be assessed with a view to future detailing or geophysical surveys.

Distinctive stratigraphic junctions can be used and followed in the same way as marker beds, where local conditions permit. In general, however, a definite lithological horizon will be found most suitable, but care must obviously be taken to ensure that the same marker is always followed, and not another lithologically similar bed.

The field geologist must utilize all available evidence to locate beds that, through variations in dip or accumulations of surface alluvium, have become locally obscured. A sharp change of inclination on an otherwise even slope may mark the boundary of a harder bed, and lithological information may be inferred from local vegetation. Dip-slopes have low moisture and therefore little vegetation, while scarps may be too steep for any but the smallest plants. Limestones give grassy slopes on the tops of gently dipping beds, sandstones tend to favour the growth of scrub, and shales may be too compact to encourage the growth of any vegetation.

It is advisable to examine in detail the beds and banks of any streams encountered, since these often reveal useful rock exposures. For the same reason cliffs, excavations and railway cuttings require special attention. When the rock exposures are genuinely *in situ,* their dips and strikes are measured by clinometer, and recorded directly on the map, the point of the conventional dip arrow coinciding with the location of the exposure.

The true dip of an exposed bed cannot be directly measured from its trace on a rock face if the rock face is not normal to the strike. In this case, only an apparent dip can be measured, and the true dip must then be obtained by geometric or trigonometrical methods. The fundamental relationship between true and apparent dips is indicated by the following formula:

$$\tan \alpha = \tan \delta \cos \theta$$

where θ = angle between direction of true and apparent dips
δ = angle of true dip
α = angle of apparent dip

The hand-held clinometer cannot give an accuracy of dip measurement of more than about a degree, and where sufficiently long rock sections are available, the dip can be more accurately obtained by alidade measurements of heights at several points. Special attention must be given to good rock sections which may be of value in determining the vertical separation between different marker beds or between a marker bed and the potential oil reservoir horizon.

Examination of the overlying soil is also helpful in following hidden rock boundaries. Sandstones, for example, produce gritty soils, shales give heavy clays, and conglomerates result in pebbly earths. Small fragments of the parent rock can often be found if the soil is examined with a hand lens. Where the underlying beds are roughly horizontal, it is possible to map geological boundaries by these soil and fragment observations. Where, however, there is appreciable surface slope, errors may be made due to the tendency of soils to "creep" downhill. In such cases, it is best to take as genuine the topographically highest indication of a rock in the soil, and to treat with suspicion soils on hillsides.

In the course of his survey, the geologist will collect samples of rock for subsequent laboratory examination. These samples must be chosen to give as wide as possible a horizontal and vertical coverage of the beds examined. It is important that unweathered material should be collected and care must be taken in labelling and separating the specimens, and in marking their points of origin on the map, so that valuable stratigraphical determinations will not be confused or wasted.

Specimens which are selected for palaeontological examination should obviously contain as many macrofossils as possible; but even where no fossils are at once evident, a series of samples should still be collected at close intervals to permit microfossil determinations. About 1 lb weight of rock is usually adequate for each sample.

Where no rock exposures can otherwise be found it may sometimes be necessary to resort to digging pits or augering. Modern, easily-portable mechanical drills can take cores of about 1in diameter even from moderately hard rocks such as sandstones and can reach depths of as much as 150ft.

In offshore operations, divers are sometimes used to recover samples of submerged strata, and coring devices are also available which can be lowered onto the sea bed to take cores of the sub-bottom formations. Thus, the *Maricor* has developed the principle of diamond core drilling, using wire-line core barrels driven by compressed air. The machine operates by remote control from a vessel at the surface and is designed to recover a series of cores each 3 metres long from the sea bed in depths of up to 600 ft of water. Another device is the *Vibrocore,* which also operates on the sea floor, this time by means of a hydraulic motor. Cores of 2½in, 4in or 8in diameter, which may be 20 ft long can be recovered at the surface, and the intervention of divers is not necessary.

In desert or semi-desert areas, differential erosion of soft and hard beds may leave a "butte and mesa" type of countryside where the hills are capped with harder beds which reflect the general dip. Gently dipping structures when weathered in this fashion can be rapidly mapped by using successive hard layers as marker beds. Care must be taken to avoid spurious dips due to local surface effects, generally the result of weathering or solution. The absorption of water in surface rocks—notably clays and shales containing bentonite—causes swelling and buckling; anhydrite near the surface generally changes to gypsum by the addition of water, with an increase of volume of more than 60% and the consequent production of superficial local folds.

"Slumping" is a common effect wherever a fairly plastic formation outcrops in an area of sharp relief, often enhanced by water absorption in a shale over which sandstone or limestone beds will then move down the topographic dips. Soils and weathered rocks similarly tend to "creep" down slopes, giving rise to anomalous local dips, and carrying specimens of an outcropping bed far below the true level of exposure.

Solution effects are most common in areas of limestone and dolomite, with the production of characteristic "karst" topography, and it is important to distinguish between true dip and local tilting due to the solution of an underlying bed. Similar superficial structures may result whenever solution of an underlying salt or gypsum bed occurs. Structural features due to the removal of underlying salt on a large scale occur in some parts of the world, notably near the MacKinac Straits area of the North American Great Lakes, in West Texas and in Kansas.

In regions of "sedimentary vulcanism" (page 125) local dips are often related to the diapiric outflow of mud or salt; and where sandstone beds have been formed by deposition in rapidly fluctuating currents or in pre-existing channels, cross-bedding is common with resultant highly variable dips which must not be confused with true formation dips.

Topography can often be utilised as a guide to the underlying geology, except where there is a thick mantle of drift. Igneous rocks give a rugged, broken landscape, hard sandstones and limestones form steep escarpments, and shales and soft rocks erode rapidly to wide valleys. The drainage and arrangement of streams are other sources of evidence. Faults, or the contacts between porous and impermeable beds, may give rise to a line of springs or pools. Topography and the shape of the rock outcrops are inter-related. Thus, the outcrops of horizontal beds run parallel to the contours, but with dipping beds the shape of the outcrops will depend on the relative dips of ground and beds.

In measuring the thickness of beds in the field (an essential step in the preparation of a stratum contour map), the procedure is simple so long as the strata concerned are approximately flat. It is then only necessary to find an exposure of a suitable stratigraphic section and to measure the vertical distances between successive beds, using either a tape or the alidade.

When beds are steeply inclined, the observed dips often vary considerably. To estimate thicknesses in such cases, the following construction can be used with profiles that are approximately horizontal:

Measure horizontal distances at right angles to the strikes, and construct a normal profile to scale, upon which all the observed dips are plotted in their correct horizontal relations. Then extend the dip lines above and below their profile, and draw normals at the dip points. The thickness of each of the beds will then be approximately equal to half the sum of the normal line lengths intersected above and below the profile (Fig. 43).

When the thickness of a bed has to be estimated from borehole evidence, this can be done if its local dip is known from core samples or other evidence. If the borehole thickness is V, the true thickness t, and the angle of dip α. Then, from Fig. 42, the relationship is $V = t \sec \alpha$.

AIR PHOTOGRAPHY[27, 34, 35, 36] AND "REMOTE SENSING"[25]

Geological surveys are still extensively used in petroleum exploration, but their emphasis has shifted from the tactical—the discovery of anticlines— to the much more complex strategic requirement of unravelling the stratigraphic and tectonic history of the area under investigation and providing data upon which the overall probabilities of petroleum genesis, trapping and preservation can be assessed. Air photography is nowadays nearly always used to supplement the surface survey and can often indicate structural features and possible geological boundaries.

The first large-scale application of air photography in petroleum exploration was its use in 1935 to make a reconnaissance of an area covering about 40,000 sq miles in the western half of Dutch New Guinea (West Irian). The results obtained enabled topographical base maps and photogeological interpretations to be rapidly compiled, and a number of potential oil structures were located, several of which were eventually successfully drilled.

Since then, greatly stimulated by war-time technical advances, air photography has become an accepted exploration tool which is now widely used to supplement and accelerate ground geological surveys. While it cannot entirely replace the field party, an air survey can provide a useful overall record and in particular give valuable information about areas difficult of ground access. A preliminary air reconnaissance will reveal notable physical features and watersheds, enabling the most suitable regions for detailed investigation to be defined and possible routes and landing grounds to be selected. Air surveys can also be used for revising and amplifying details from previous maps, and for extending the boundaries of areas already mapped.

Some idea of the potentialities of air photography can be obtained from the fact that one vertical exposure with a modern "super-wide angle" high precision air survey camera (e.g. the Wild RC 9, Zeiss RMK 15, etc.) taken from a height of 27,000ft will cover over 200 square miles of country at 1/100,000 contact scale.

A ground survey is always necessary for detailed geological evaluation, but it will generally be greatly assisted by preliminary air photography, since, as will be seen, the air survey can provide a contoured topographic map from which valuable geological information can often be deduced. Thus, the topographical expression recorded will often be related to the underlying rock types, while soil colours, the growth of trees, and variations in the general vegetal zoning which are detectable in air photographs can often be used to differentiate the underlying strata.

Sedimentary rocks are characterised by stratification, which appears in aerial photography as banding. Arenaceous, argillaceous and calcareous sediments can be distinguished by their differing resistance to erosion; igneous outcrops often produce characteristic shapes. "Marker" beds can sometimes be picked out and their exposures traced for considerable distances across country. Air photographs also permit the thickness of exposed beds to be measured or calculated, and useful information can often be gathered about lithological features—as, for example, sink holes and solution depressions, which may reveal the presence of underlying soluble rocks such as salt or gypsum.

Air surveys will often reveal structural features, although thick vegetation and topsoil or detritus will tend to cloak structures from the air as they will from the ground survey. However, even in areas of dense jungle, the forest canopy often reflects the underlying morphology with surprising fidelity. Structural interpretation will generally be based on the form of outcrop, drainage patterns, topographical expression, and dip and strike data.

Where good exposures occur, the air photograph will directly show the presence of anticlines and domes in the surface rocks (Fig. 44), and even where exposures are poor and relief is low such structures may still be revealed by lines of broken bluffs, or by deflections in the regional drainage patterns. Strikes of individual beds can be measured from suitable photographs, and dips may be estimated from the outcrop shapes: dendritic drainage patterns are an expression of flattish beds, while parallel banding indicates steep dip.

Besides major structural features, secondary features such as faults, dykes, joints, foliation, etc., can be revealed from the air. Where a fault line is covered by soil, alluvium or detritus, local movements and drainage tend to alter the moisture content of the soil and thus to affect the type and growth of vegetation, thereby often revealing the underlying fault.

Several varieties of air photographs are used for topographical and geological work:

An assembly of rectified vertical photographs (i.e. photographs printed in such a way as to neutralize errors due to tilting), with the detail matched at the margins of the individual parts of the photographs used, is termed a *mosaic* and is of particular value in the initial surveying of unmapped areas. When the air photographs are brought to the same scale and fitted to ground control points, a *controlled mosaic* is then obtained. *Obliques* are useful for local views of profiles, cliffs, or sections, or to define geological boundaries, where a change of soil or rock, difficult to follow from the ground, becomes easily traceable from the air (Fig. 45). The maps eventually produced fall into three groups:

(1) Small-scale: 1/30,000 to 1/200,000—used principally by the mapping surveyor and cartographer.
(2) Medium-scale: 1/5,000 to 1/30,000—including most regional topographical and military maps and reconnaissances for oil or minerals.
(3) Large-scale: 1/500 to 1/5,000—general plans for major engineering projects.

Modern air survey cameras have lenses which may be "wide angle" (95°) or even "super-wide" angle (120°). These are designed for photography at scales of 1:50,000 and 1:100,000. Normal-angle (60°) and narrow-angle (40°) lenses are used for larger-scale work. Cameras may use either film or plates. Thus, for example, the Williamson Survey camera Eagle IX gives a picture 9in square and has a magazine of film for 200 semi-automatic exposures.

Survey photography is a specialised task, since exact courses must be flown and aircraft tilt avoided. For this purpose, some form of automatic gyro-pilot system is nowadays always used. The required area is covered by a series of overlapping "strip" photographs, i.e. the survey aircraft flies up and down a series of parallel tracks, taking vertical photographs at regular intervals until the whole area has been covered. For scales of the order of 1/25,000, it is usually difficult to fly accurately a strip longer than 20 miles unless intermediate ground control points are available. For scales of 1/5,000 or less, the maximum strip length used is somewhat shorter. Widths of strips are usually 5-6 miles, for an aircraft flying height of about 20,000ft.

It is customary to allow an overlap of 60% between adjoining photographs in the direction of flight, and about 25% lateral overlap between photographs in adjoining strips. Using modern equipment and techniques, it is not unusual to obtain 1,000 sq miles of vertical cover in a few hours of reasonably good weather.

When there is a lateral wind component, the aircraft will tend to drift out of its course; and if this is corrected by offsetting the aircraft it will "crab", so that the photographs would be taken at an angle. This effect may be corrected by traversing the camera in its mountings through an angle equal to the angle of lateral drift. The aircraft flight path is usually controlled nowadays by a radiolocation navigation system using ground stations (e.g. Decca, pp. 166-7), or audio-frequency phase comparison systems such as *Lorac, Rana, Raydist* or the Russian *Poisk*. The *Shoran* system depends on measurements of the travel times of radio waves. All the systems mentioned require special aircraft equipment. The *Doppler Navigator* device, on the other hand, is a more convenient method of aircraft position location, inasmuch as no auxiliary ground stations are required; it continuously monitors the aircraft velocity and the drift angle, with an accuracy of about ±0·5% for each measurement, and can feed this information into a computing device on the aircraft which activates an auto-pilot to keep the aircraft flying accurately along the desired ground track.

The "profile recorder" is a precise radar altimeter which when available is of great value in air surveying. The instrument uses radar signals to record the ground elevation along the flight path of the aircraft, and thus provides a method by which a network of height control lines can be rapidly established. More commonly, however, height determinations can be made from good air photographs by the process known as "photogrammetry", i.e. stereometric contouring, using successive pairs of vertical air photographs.

A ground party initially makes a number of selected control points clearly visible from the air, and measures the heights of these points by levelling from known triangulation points or by the use of aneroid barometers. Where the ground relief is less than 10% of the aircraft flying altitude and where a map without contours is required, graphical means can be used to extend

Fig. 44. An example of the interpretation of the aerial photograph of the pitching end of a steeply dipping anticline in sandstones and marls. The strata are overturned on one flank. Two lithological groups are visible, "A", in the core of the fold, is made up of soft, incompetent strata which have been partly extruded from below through the sandstone and marl group "B".

Fig. 45. An excellent example of the way in which geology is clarified by an oblique aerial photograph. This striking photo of an isoclinal syncline was made in the Asmari fold-belt system in Iran.

a chain of interlocking triangles between the control points on the photographs and thus to establish a true plane picture of the ground at the correct scale of the map which is to be drawn.

Where there is sharper relief, or where detailed height contouring is needed, a photogrammetric map is produced by the use of an elaborate instrument employing the principle of space vectors, whereby corresponding rays from a stereo-pair of photographs are made to intersect in space. Complex automatic plotting devices are used to eliminate errors due to tilt, inclined air-base, and unknown heights of aircraft. Such instruments, as for examples, the Zeiss *Stereoplanigraph* and *Stereotope,* the Wild *Autograph* and the Thompson-Watts *Stereoplotter,* can produce contoured maps with only very little ground control. This type of levelling does not supersede the theodolite where substantial vertical accuracy is required (e.g. for engineering projects), but is very useful for initial geological surveys. Contours can be drawn and the general form and shape of the ground clearly shown, even when much of it is covered by vegetation.

Stereoscopic pairs of photographs are most important for geological interpretation. They must overlap longitudinally by 50% to allow for subsequent stereoscopic examination; an extra 10% overlap is usual to allow a margin for tilt errors. They generally require the use of a radar altimeter in the aircraft for recording small differences of height between the individual stations.

Recent developments in air survey work include the increasing use of colour and "false colour" (Ektachrome I.R.) emulsions to assist in the differentiation of rocks and soils. Good quality colour photography replaces by an infinite number of colours and shades the limited grey tones of monochrome photographs and thus adds a new dimension of observation. The use of "false colour" film can then give further valuable information. Three layers of emulsion are sensitized to green, red and infra-red radiations, the combination of which produces a photograph quite "false" in normal colours but very valuable in locating drainage patterns, since the bright blue of the water lines contrasts vividly with the green, magenta or red of the land. Different types of vegetation are also sharply differentiated, enabling geobotanical boundaries to be located.

Laser scanning has also been applied to the interpretation of aerial photographs, and it is claimed that its use enables dominant features to be eliminated so that less developed ones can be discerned.

The increasing number of man-made satellites which have been orbiting the Earth since the late 1960's has led to the availability of *satellite photographs* which can be used to some degree to supplement conventional air photographic coverage of certain areas. Clearly, since most such photographs have been near-vertical colour by-products of *Gemini* and *Apollo* flights, their very small scales (of the order of 1:2,250,000) have necessarily limited their effective applications for geological mapping[30].

However, since mid-1972, the NASA Earth Resources Technology Satellite (ERTS-1) has been regularly scanning the Earth's surface from a near-polar 576-mile high orbit, using a multi-spectral scanner and relaying back to the surface images and radiation patterns which are converted into black and white and "false colour" photographic prints on a 1:1,000,000 scale. The results of these surveys have shown that they can be of particular value in delineating the gross structure and areal extent of sedimentary basins and revealing the patterns made by major faults and tectonic features[24, 28, 33].

Among the newly developed "remote sensing techniques", *infra-red* and *thermal infra-red photography* have been used to provide valuable geological information in certain circumstances. Using equipment sensitive to radiation in the 0·3 to 1·1 micron and 8 to 14 micron bands, with normal air photographs taken simultaneously for comparative purposes, fault lines have been traced over great distances. Pictures are taken before sunrise when the radiation that is picked up is most closely related to the thermal properties of the rocks or soil, rather than to the effects of the sun. Essentially, the warmer an object is, the lighter its image appears on infra-red sensitive film, and vice versa.

Rocks formed in different ways have different thermal properties, so that, for example, siliceous shales appear cooler than sandstones and conglomerates of the same age. Formation boundaries can therefore often be traced, even in hazy conditions or through clouds. Drainage patterns can be detected, as well as variations in types of forest trees and vegetation.

Ultra-violet scanning systems are being tested which may have some applications in geological mapping from the air; in particular, the response from calcareous rocks is stated to be characteristic.

Radar geology, i.e. the recording of radar images, can supplement photo-geology when light and weather conditions preclude normal photography, and also, to a limited extent, by providing data from beneath the soil cover. Side-looking airborne radar (SLAR) of relatively long wavelength, propagated from antennae mounted in the sides of a reconnaissance aircraft, is claimed to be capable of penetrating to bedrock through dense vegetation or several feet of soil, enabling changes of lithology to be detected[26, 38].

A number of other remote sensing techniques are currently being used in mineral exploration operations, some of which may prove to be of value in petroleum geology when fully developed[25, 32]. (The geophysical aspects of remote sensing operations are discussed in Chapter 6).

REFERENCES

PETROLEUM AT THE SURFACE

1. M. A. Carrigy, "The physical and chemical nature of a typical tar sand". *7th World Petrol. Congr., Mexico,* PD 13 (2), 31–50 (1967).
2. M. A. Carrigy, "Deltaic sedimentation in Athabasca tar sands". *Bull. Amer. Assoc. Petrol. Geol.,* **55** (8), 1155–1169, (1971).
3. P. A. Dickey and J. M. Hunt, "Geochemical and hydrogeologic methods of prospecting for stratigraphic traps". In "Stratigraphic Oil and Gasfields". Amer. Assoc. Petrol. Geol., Tulsa (1972).
4. T. J. Donovan, "Petroleum microseepage at Cement, Oklahoma: evidence and mechanism". *Bull. Amer. Assoc. Petrol. Geol.,* **58** (3), 429–446 (1974).
5. I. M. Goubkin and S. Fedorov, "Mud volcanoes of the Soviet Union and their connection with oil deposits". *18th Int. Geol. Congr., Moscow,* 4, 29–59 (1937).
6. J. O. Harvie, J. H. Nicholls and A. G. Winstock, "The outlook for Canadian Oil Sands Development". Paper to 71st Nat. Amer. Inst. Chem. Eng. Meeting, Dallas, Texas, Feb. 20–23, 1972.
7. H. D. Hedberg, "Relation of methane generation to undercompacted shales, shale diapirs, and mud volcanoes". *Bull. Amer. Assoc. Petrol. Geol.,* **58** (4), 661–673 (1974).
8. G. E. Higgins and J. B. Saunders, "Report on 1964 Chatham mud island, Erin Bay, Trinidad, WI". *Bull. Amer. Assoc. Petrol. Geol.,* **51** (1), 55–64 (1967).

9. G. D. Hobson, "Oil and gas accumulations and some allied deposits". In "Fundamental aspects of Petroleum Geochemistry", eds. B. Nagy and U. Colombo, Elsevier Publishing Co. (1967).
10. P. F. Kerr, I. M. Drew and D. S. Richardson, "Mud volcano clay, Trinidad, WI". *Bull. Amer. Assoc. Petrol. Geol.,* **54** (11), 2101–2110 (1970).
11. H. G. Kugler, "Contribution to the knowledge of sedimentary volcanism in Trinidad". *Journ. Inst. Pet. Technol.,* **19**, 743–759, (1943).
12. K. K. Landes, "Mother Nature as an Oil Polluter". *Bull. Amer. Assoc. Petrol. Geol.,* **57** (4), 637–641, (1973).
13. R. Lebkuchner *et al.,* "Asphaltic Substances in Southeast Turkey". *Bull. Amer. Petrol. Geol.,* **56**, 1939–1964 (1972).
14. W. K. Link, "Significance of oil and gas seeps in world oil exploration". *Bull. Amer. Assoc. Petrol. Geol.,* **36** (8), 1505–1542 (1952).
15. S. T. Ovnatanov and G. P. Tamrazyan, "Thermal studies in subsurface structural investigations, Apsheron Peninsula, USSR". *Bull. Amer. Assoc. Petrol. Geol.,* **54** (9), 1677–1685 (1970).
16. P. H. Phizackerly and L. O. Scott, "Major Tar Sand deposits of the world". *7th World Petrol. Congr., Mexico, PD* 13(1), (1967).
17. M. F. Ridd, "Mud Volcanoes in New Zealand". *Bull. Amer. Assoc. Petrol. Geol.,* **54** (4), 601–616 (1970).
18. H. R. Ritzma, "Oil impregnated sandstones". *Bull. Amer. Assoc. Petrol. Geol.,* **57** (5), 1961 (1973).
19. H. H. Suter, "The General and Economic Geology of Trinidad", 2nd Ed., HMSO London, (1960).
20. E. N. Tiratsoo, "Natural Gas" 2nd Ed. 1972, Scientific Press, Beaconsfield.
21. US Bureau of Mines, "Surface and Shallow Oil-impregnated Rocks and Shallow Oilfields in the USA". *Monograph* 12 (1965).
22. J. S. Wells and K. H. Anderson, "Heavy oils in Western Missouri". *Bull. Amer. Assoc. Petrol. Geol.,* **52** (9), 1720–1731 (1968).

GEOLOGICAL SURVEYING, AIR PHOTOGRAPHY & REMOTE SENSING

23. Amer. Soc. Photogrammetry, "Manual of Photogrammetry". (3rd ed.) 1966.
24. D. Bannert, "Afar tectonics analyzed from space photographs". *Bull. Amer. Assoc. Petrol. Geol.,* **56** (5), 903–915, (1972).
25. A. Barringer, "Remote-sensing techniques for mineral discovery". Paper 20, *9th Commonwealth Min. Met. Congr.,* London, (1969).
26. L. F. Dellwig, H. C. MacDonald and J. N. Kirk, "The potential of radar in geological exploration", *Proc. 5th Symposium Remote Sensing of Environment,* Univ. of Michigan, Ann Arbor, 747–763, (1968).
27. W. Domzalski, "Airborne Techniques in Petroleum Exploration". *J. Inst. Petrol.,* **48**, 459, 55–67, (1962).
28. S. C. Freden, *et al.,* Symposium of Significant Results Obtained from ERTS–1, Abstracts, NASA/Goddard Space Flight Center, Greenbelt, Md., March 509, 1973.
29. B. J. Gorham, "The Laser in Surveying". *Hydrographic Journal,* 1, 3, 43–47, (1974).
30. J. G. W. Greenwood, "Role of satellite photographs in geology". *Nature,* **224**, 506–508, (1969).
31. International Assoc. of Geodesy, "Electromagnetic Distance Measurement". Hilger and Watts Ltd., London (1965).
31a. F. Lahee, "Field geology", Chapters 16, 21, 5th Ed., McGraw Hill, 1952.
32. J. Lintz, "Remote sensing for petroleum", *Bull. Amer. Assoc. Petrol. Geol.,* **56** (3), 542–553, (1972).
33. P. A. Merrifield, "Space photography in geologic exploration". *Bull. Amer. Assoc. Petrol. Geol.,* **56** (5), 916–924, (1972).
34. R. G. Ray, "Aerial photographs in geologic interpretation and mapping". *Prof. Paper US Geol. Surv.,* 373, (1960).
35. J. K. St. Joseph (Ed.), "The Uses of Air Photography". (Cambridge Committee for Aerial Photography), John Baker, London (1966).
36. B. A. Tator *et al.,* "Photo-interpretation in Geology", pp 169–342 in "Manual of Photographic Interpretation", Amer. Soc. of Photogrammetry, Falls Church Va., USA, (1960).
37. E. N. Tiratsoo, "Petroleum Geology", Chapter 12, Methuens, London, 1951.
38. R. S. Wing and H. C. MacDonald, "Radar geology—Petroleum exploration technique, Eastern Panama and Northwestern Colombia". *Bull. Amer. Assoc. Petrol. Geol.,* **57** (5), 825–840 (1973).

CHAPTER 6

Subsurface Exploration for Petroleum

GEOPHYSICAL SURVEYS

PETROLEUM, even when in an accumulation, forms only a small proportion of the total fluids present in a rock section, and none of its properties differ sufficiently from those of the salt water continuum in which it is normally found as to make it susceptible of discovery at a distance. However, since rocks can vary considerably in their physical natures—notably their densities, magnetic properties, electrical conductivities, and the velocities with which they transmit elastic shock waves—it has proved possible to use these variations in rock properties to assist in the location of subsurface structures which are favourable for the accumulation of petroleum. The only geophysical techniques which are believed to be directly related to the properties of petroleum itself are the geochemical and radioactivity surveys, the validities of which are still the subjects of considerable controversy. All the other techniques concentrate on the discovery of "anomalies" in the rocks which overlie or surround possible petroleum accumulations.

Geophysics was first applied to petroleum exploration in the United States in the early 1920's, by which time it had become evident that surface surveys alone could no longer be relied upon to discover and define petroleum accumulations which might lie at considerable depths or be obscured by unconformable shallower structures. Undoubtedly, the dominating position of the United States as an oil and gas producer has been largely due to the extent and thoroughness of the geophysical exploration operations that have been carried out over the past half-century, most notably in the Gulf Coast area.

During this period, geophysics has played an increasingly important role in oil exploration in nearly every country in the world where the search for oil and gas has taken place; and nowadays geophysical surveys of one type or another are generally considered to be standard prerequisites before an exploration drilling programme is embarked on. Thus, initial aeromagnetic surveys may be flown to determine the thickness of the local sediments and the depth and configuration of the crystalline "basement"; gravity surveys

150

may then be conducted to determine the position of local structural "highs", and these will generally be investigated by seismic reflection surveys before a decision is made as to the location of the first exploration well. Further detailed seismic surveys will probably then be made to delineate the most promising structures, while geochemical investigations may help to differentiate between those which are hydrocarbon-bearing and those which are barren.

While most geophysical surveys are carried out by specialized contractors who use their own field parties to collect the data, which is subsequently processed and interpreted in appropriately equipped centres, the petroleum geologist should understand the scope and limitations of the techniques used, and must be able to relate the resultant sections and maps to the surface geological evidence and the subsurface data furnished by well samples and cores. Properly applied, geophysics is an invaluable exploration tool, but one whose balance, application and limitations must be properly understood if it is to be successfully utilized.

The descriptions of the various geophysical techniques which follow are designed to provide only a conspectus of what has now become a highly specialised subject. Details are given on p. 175 of various standard textbooks and case histories, to which reference should be made for further information [8, 11, 16, 23, 26].

GRAVITATIONAL ("GRAVITY") SURVEYS

Gravitational methods have been in use in petroleum exploration since 1922, although the basic principles involved were utilized in geodetic work almost two centuries earlier. Gravitational surveys were particularly successful in discovering the salt domes often associated with oil accumulations, notably in the Gulf Coast area of the USA. In general, however, this type of survey is employed as a preliminary to the seismic survey, enabling areas of maximum interest to be delineated, rather than as an independent exploration technique, since it has the drawback of lack of certainty in interpretation.

Three different varieties of instrument have been used for field gravity determinations—the torsion balance, the pendulum, and the gravity meter. Of these, the gravity meter in one form or another has almost completely replaced the others, which are therefore mentioned here only for their historical interest.

The Torsion Balance

The Eötvos Torsion Balance was invented in 1890 but not used for petroleum exploration before 1922, when it achieved some notable successes, particularly in the Gulf Coast area.

It measures the gravity gradient, i.e. the *rate* at which the force of gravity changes horizontally and also the direction in which this change is a maximum. It does not measure the vertical component of gravity, as does the gravity meter.

The instrument is basically composed of two small masses supported at different levels at either end of a light aluminium beam which is suspended by a torsion wire. The angular deflection of this torsion wire can be measured by the deflections on a scale of a beam of light reflected from a mirror attached to the wire, and this can be recorded photographically.

The principle upon which the instrument operates is that if the gravitational field is not uniform there will be a very slight difference in the direction of the gravitational force on the two weights, which will tend to rotate the system about the torsion wire. The amount of rotation is measured by the optical system and is a measure of the local distortion of the earth's gravitational field caused by geological features, after due corrections have been applied.

The instrument records its results automatically on a photographic plate and requires three to four hours for a complete set of readings at each station.

The Pendulum Survey

Relative gravity measurements have also been effected in geophysical surveys by the use of a compound pendulum. These depend upon the fact that if a pendulum has a period T_1 at a base station where the gravity is g_1, the gravity g_2 at any other station can be determined from the period at that station T_2, since $g_1/g_2 = (T_2/T_1)^2$.

This method has fallen into disuse, since it is both lengthy and subject to many corrections. Even the best purchasable pendulums have a probable error of $\frac{1}{2}$ milligal (defined below), and since the gravitational anomalies characteristic of oil structures suitable for oil accumulations may only be of the order of $1\frac{1}{2}$ milligals, this type of instrument is clearly insufficiently sensitive.

The Gravity Meter

The gravity meter is designed to measure the difference of gravity between points on the earth's surface, and accomplishes this by weighing the same object with great precision at different stations. Then, since the weight recorded is a product of the mass (which is constant) and the acceleration due to gravity, any variations detected in the weight of the object will, other factors being equal, reflect variations in the gravitational acceleration.

The variations noted in g in any particular area on the earth's surface may be due to the local uplift of more or less dense underlying rocks; hence, gravity variations may be used to make approximate deductions about subsurface structures (Fig. 46).

The gravity variations in question are very small, of the order of one part in ten million of the total value of gravity, and the design and construction of gravity meters therefore requires great skill and precision. The smallness of the anomalies to be expected at the surface from subsurface structures is shown by the calculation that a sphere of 1,000ft diameter with its centre 1,000ft deep will produce a gravity anomaly at the surface of only 0·001 gal if it differs in density by 1 gm/ml from the surrounding rock. (The gal is the unit used in geophysics to denote an acceleration of 1 cm per second². The average value of the earth's gravity is therefore about 980 gals; due to its oblate spheroidal shape this varies with latitude, from 978 gals at the equator to 983 gals at the poles.)

Gravity meters used for prospecting may be either *static* or *astatic* instruments.

The static type contains a weight attached to an elastic spring, the elongation of which varies according to the local gravitational pull. This variation is magnified by an optical system and can be measured from station to station in the field. An alternative arrangement is to measure the auxiliary force necessary to counterbalance the elastic restoring force of the elongating spring.

Astatic gravity meters achieve high sensitivity by a system of balancing controls near a position of unstable equilibrium.

The instruments must compensate in some way for variations in temperature and imperfect elasticity of the spring material, and also allow for the damping-out of external vibrations.

For field use, the gravity meter may be mounted for mobility on any suitable vehicle. Stations are selected and marked out, usually about one mile apart in lines traversing the area to be examined. For detailed work, closer stations will be chosen. It is usual for the lines to be run in "loops" of 20 to 30 miles, so that on closing a check can be made and surveying or instrumental errors detected. The meter truck is driven over each station in turn and the gravity meter lowered through the floor on a tripod to the ground. After levelling, the gravity value is read off a calibrated scale in terms of the increase or decrease in milligals relative to the gravity value at a selected datum point. Each reading may take only a few minutes.

The measured difference of gravity cannot, however, be attributed entirely to geological variations in the subsurface. Corrections must be applied for latitude, topography and level differences. The latitude correction due to the oblate spheroidal shape of the earth can be applied from a general formula, but the topography correction is more difficult to assess, since it is a summation dominated by the effects of the nearby surface features. However, this factor is usually small compared with the other corrections, since the gravity stations are chosen to be as far as possible from marked topographical irregularities.

The elevation correction takes into account the variation in gravity values resulting from the variation in elevation between individual stations. It can be calculated as the arithmetical sum of the "free air" and the "Bouguer" corrections, which are opposite in effect. The former factor is to compensate for the variation in gravity attraction due to varying distance from the centre of the earth. Over the range of levels of the earth's surface this is equivalent to a decrease in gravity of 0·1 milligal per foot, and assumes that the layer of rock between the field station and the datum station has no gravitational effect, i.e. is filled with free air. In the Bouguer correction, this layer is taken into consideration and assumed to have a gravitational effect.

When the observed values of gravity at each station have been duly measured and corrected, it is then possible to construct a gravity profile or a contour map showing isogravity lines.

Marine Gravity Surveys

Marine gravity surveying is a difficult operation, due to the wave effects. In shallow waters, the gravity meter can be mounted in a pressure-tight container and lowered to the sea floor, readings being taken photographically.

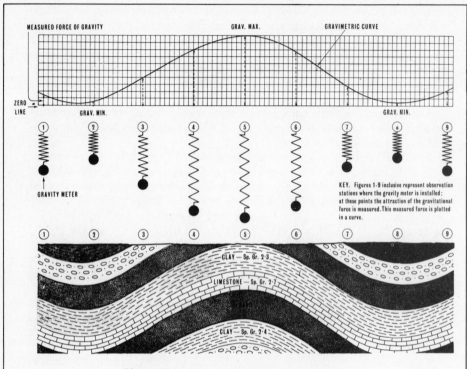

Fig. 46. Principles of gravitational surveying.
(Shell International Petroleum Co.)

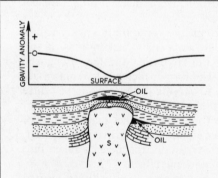

Fig. 47. Negative gravity anomaly over a subsurface salt dome, with associated oil accumulations. C=cap rock.

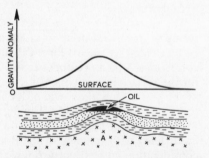

Fig. 48. Positive gravity anomaly over an igneous uplift or "buried hill", with an oil accumulation in the overlying sediments.

However, this is a slow and expensive process. A stabilized platform has been developed which is said to enable gravity data to be obtained at sea at a survey speed of about 8 knots; if successful, this may be expected to be widely used.

In deep-sea operations, submarines have been successfully used as vehicles for gravity meters, but are obviously impractical for normal exploration work.

The Askania-Graf shipborne sea gravity meters have been used with some success for reconnaissance in Continental Shelf marine exploration. Due attention must be paid to the corrections necessary for both instrumental and environmental factors—the latter being caused by the effects of navigation over the surface of the rotating earth and errors due to the accelerations of the vessel (pitch, yaw and roll).

The accuracy of shipborne gravity surveys is still limited by the effects of swell and the navigation accuracy possible to 2–3 milligals, but this will improve with the increasing availability of sophisticated electronic position-finding equipment. However, this order of accuracy is still sufficient to outline major geological trends[5].*

Interpretation of Gravity Results

The geological interpretation of results obtained from gravity surveys is a complex matter. It is true that once the appropriate corrections have been made, the isogravity lines obtained represent points of the same gravity value, and it is convenient to assume that they represent gravity contours on some subsurface bed and in this way reflect its structure. Unfortunately, there is generally no evidence to show that the same bed is always concerned. Well marked gravity "highs" can be explained as being due to dense rocks having been raised nearer to the surface, while conversely, gravity "lows" may be due to light rocks replacing dense rocks. But even so, the shape and depth of the subsurface structure involved cannot be estimated from gravity results alone, because the data are not susceptible of a unique interpretation. By laborious calculations in which the gravity values which might be expected over various theoretical structures are calculated and compared with the actual results obtained, it may nevertheless be possible to indicate likely structures.

In general, although the quantitative or even qualitative interpretation of gravity surveys is difficult, they can still give valuable results. This has been shown particularly in the exploration for salt domes. Where the salt is relatively deep, a marked gravity "low" is generally obtained (Fig. 47), owing to the density of the salt being low compared with the surrounding sediments. However, where the salt is shallow, the effect of the dense cap rock (if present) may outweigh that of the low-density salt, with the result that a gravity "high" results (e.g. the *Spindletop* and *Nash* domes). Gravity "highs" are usually taken to be indications of buried anticlines, although there is always doubt as to whether the cause of the anomaly is indeed the arching of the underlying sediments or is due instead to an upthrust of deeper "basement" rocks (Fig. 48).

A decrease in density, besides being perhaps due to salt, may also clearly be due to the presence of other types of rocks, e.g. a thick bed of light dia-tomaceous shale at depth[24].

Using multi-sensor systems, it is now practicable to conduct underwater gravity, seismic and magnetic surveys simultaneously from a single vessel.

Gravity surveys are subject to so many complicating factors that no quantitative assessment of subsurface structure should be based on them without corroborative geological or seismic evidence. For successful gravity assessment, the structure examined must be one of marked relief, since the limiting factor in the use of the gravity meter is its sensitivity.

An important development in this respect is the borehole gravity meter, said to be capable of very high precision (± 0.015 to ± 0.005 milligal) and rapid operation in wells, thus providing a means by which the gravitational effects of the different beds in a rock column can be compared and the results of surface gravity meter surveys checked.

MAGNETIC SURVEYS

The first application of magnetic surveying for mineral exploration was the use in Sweden as long ago as 1640 of variations in the direction of a compass needle for the location of iron ores.

In petroleum exploration, advantage is taken of the fact that whereas the earth's normal magnetic field is uniform over areas of homogeneous magnetic material, it is locally distorted where concentrations of rock materials of high magnetic susceptibility are present, e.g. iron minerals such as ilmenite and magnetite. Since crystalline rocks contain much higher proportions of these minerals than do normal sediments, the magnetic anomalies which result from the presence of intrusions of "basement" material may show the existence of arching and structural traps in the overlying sediments, and by indicating the depth of the "basement" may also serve to measure the thickness of the sedimentary section present.

The local magnetic distortions of the Earth's field can be measured either absolutely, or relatively. The total magnetic intensity at any point can be resolved into horizontal and vertical components; the angle between the direction of the total magnetic intensity and the horizontal is termed the magnetic dip.

Many types of magnetometer have been developed for measuring these factors—earth inductors, dip circles, the sine galvanometer, the compass variometer, the magnetic gradiometer, and various iron induction instruments. For measuring local as opposed to absolute anomalies, dip needles, deflection magnetometers, magnetic torsion balances and magnetic field balances have been used. Mention must also be made of the sensitive micro-magnetometer and magnetron instruments.

The vertical component of the Earth's magnetic field varies from a maximum of 0·6 gauss at the earth's magnetic poles to zero at the magnetic equator, while the horizontal component varies similarly from zero to 0·3 gauss. For geophysical prospecting, the gauss (by definition equivalent to a force of 1 dyne on unit pole) is subdivided into 100,000 *gammas*. The accuracy achieved in most surface magnetic exploration methods is of the order of ± 1 gamma, i.e. about 1/50,000 of the earth's field.

As with the gravitational survey, certain basic corrections must be applied to magnetometer field results before they can be used as indicators of sub-surface conditions. These corrections are for latitude (i.e. for the variation of the normal component of the earth's field with distance from the magnetic poles); for diurnal variations in the strength of the Earth's field; and for occasional electromagnetic storms.

In certain parts of the world, as for example the Gulf Coast of the United States, normal magnetic surveying methods have had little success. The crystalline basement may be too deep and magnetic anomalies associated with salt domes are usually not sufficiently definite to make such surveys serious competitors with other geophysical techniques. On the other hand, notable success has been achieved in some parts of the United States, particularly in Kansas and Oklahoma, where subsurface structures have been produced by the arching of sediments over buried granite ridges. Magnetic surveys have also been particularly valuable in southern Oklahoma, where the Wichita and Arbuckle mountains and the intervening basin are related to "basement" upwarpings with resultant magnetic anomalies, and in Texas, where serpentine plugs give strong magnetic anomalies and are often associated with fault-trap oil accumulations. Magnetic surveys have also had considerable success in the USSR.

However, as an independent method of exploration, the ground magnetic survey cannot usually detect subsurface structures with acceptable reliability, since the results obtained are nearly always imprecise in depth. Furthermore, there is no constant quantitative relationship between a particular type of structure and its surface magnetic expression—e.g. a buried anticline may under different local conditions register either a magnetic' 'high' or a "low" at the surface.

Aeromagnetic Surveys

Wartime research into the problem of detecting submerged submarines led to the development of methods of magnetic prospecting from aircraft which have since had considerable success in petroleum and mineral exploration. (Aeromagnetic surveys, together with the airborne radioactivity surveys mentioned below, can be classed as "remote sensing" techniques).

The airborne instruments initially used included single or coupled magnets, stationary or rotating coils, magnetrons and saturated core reactors, all of which have had some success in the detection of subsurface magnetic anomalies. More recently, however, a new generation of much more sensitive instruments has been developed which can be carried on aircraft wing-tips, in a special non-magnetic section of the aircraft's tail, or separately in a streamlined "bird" towed on a cable behind the aircraft. These include the flux-gate, proton-precession, metastable-helium, and alkali-vapour instruments[4].

The flux-gate instrument has a sensing head in which three mutually perpendicular flux-gates are mounted, two of which maintain the third unit in the direction of the total magnetic field. The voltage required to compensate for changes in the ambient magnetic field is recorded continuously on a chart.

The proton-precession magnetometer uses a sensing element filled with water around which an induction coil is energised by a direct current. The hydrogen protons in the water align themselves with the induced polarising field, and when this is switched off, precess with a frequency proportional to the total magnetic field at the point of observation.

The metastable-helium and alkali vapour instruments operate on the principle of atomic spin resonance. Initially, such instruments could measure the total magnetic field with an accuracy of only ± 1 gamma, but the development of "optically-pumped" rubidium and caesium vapour instruments with a "noise" level of only 0·01 gamma has led to an effective precision of $\pm 0·1$ gamma in surveys. Greater sensitivity is probably not possible, or justified, because of the practical problems of navigation and altitude keeping.

In aeromagnetic surveying, as with air photography, the operating aircraft must be accurately flown at a constant height (usually 1,500–5,000ft) and along pre-set flight lines—usually 1–2 km apart, although for reconnaissance, lines of 10–20 km apart may be used. Various semi-automatic navigational systems have been employed which use transmissions from ground stations to fix the aircraft's position. The Doppler Navigator device, on the other hand, operates on radar principles but without ground stations (p. 145).

The advantages of the use of the airborne magnetometer over ground instruments lie in the speed of survey (an average of 8–10,000 miles of traverse per month is feasible), and the possibility of reaching otherwise inaccessible areas. Furthermore, local influences which would affect the accuracy of the ground instrument are avoided. On the other hand, the costs of aeromagnetic surveys, as of all forms of air survey, are high, since they inevitably involve many hours of expensive aircraft flying time, and such surveys are therefore uneconomic for the investigation of small areas. However, the aeromagnetic survey now provides a rapid and effective method of estimating the depth and shape of the crystalline "basement" in unprospected areas, and hence the approximate thickness of the overlying sedimentary material—essential knowledge in assessing potentialities for petroleum generation and accumulation.

Exploration using the "gradiometer", i.e. two magnetometers suspended at different levels, while the survey aircraft is flown at only 500ft above the ground, has also achieved considerable success in detecting "micro-anomalies" caused by shallow subsurface structures. The sensitivity of measurement achieved by this technique enables the geologist to determine the distribution and shape of igneous intrusions and the characteristics of deep-seated basement structures. The dual sensor configuration of the gradiometer also counteracts diurnal changes in the earth's magnetic field.

The general end-result of the aeromagnetic survey may be either a magnetic profile, or more usefully a contour map of the total magnetic field, preferably keyed to air photographs taken during the same survey flights. However, the quantitative interpretation of such a map is difficult unless surface data are also available, since a number of unknown factors will otherwise be involved. Close co-operation is always necessary between the petroleum geologist and the geophysicist in studying aeromagnetic survey results.

In recent years, digital processing techniques have been increasingly used in aeromagnetic mapping. Digital recording in flight enables contoured maps to be rapidly produced, and two-dimensional digital filtering can be used to accentuate or suppress near-surface effects as may be required.

SEISMIC METHODS

The seismic methods of geophysical exploration fundamentally involve the measurement of the travel times of shock waves, artificially created at the Earth's surface, passing through subsurface strata. Such shock waves can be initiated by various chemical or mechanically explosive means, and have periods of between 0·01 and 0·1 sec (in contrast to normal earthquake waves which may have periods as long as 60 sec).

The degree and direction of reflection and refraction of the waves in the subsurface depend upon the nature, depth and dip of the various beds penetrated. Seismic prospecting for petroleum is essentially a method of locating subsurface geological structures by measuring and defining reversals of dip in subsurface beds; it does not in general differentiate between those structures which contain oil and those which do not.*

When an explosive shock is set off on the surface, three types of wave radiate outwards from the point of explosion. These comprise a longitudinal wave ("P" wave), a transverse wave ("S" wave) and a surface wave ("Rayleigh wave"). Each of these travels with its own special velocity, which varies with different rocks.

In seismic prospecting, the "P" waves are used, which are the fastest. When a "P" wave strikes a discontinuity between two subsurface beds of differing elastic properties, four new waves will be set up (Fig. 49)—a reflected longitudinal wave (PP_1), a reflected transverse wave (PS_1), a refracted longitudinal wave (PP_2), and a refracted transverse wave (PS_2). If the angle of incidence is critical, the refracted longitudinal wave (PP_2) will travel along the interface between the two beds.

Measurements are then made of the time interval that elapses between the original explosion and the receipt at detectors on the surface of the refracted or reflected longitudinal waves. In refraction techniques, the waves which travel by minimum time paths are refracted back to the surface and recorded at points some distance from the explosion; in reflection techniques, the waves which are reflected more or less vertically back to the surface arrive considerably later than the direct wave at detectors near the explosion point.

Fig. 50 illustrates how for a simple subsurface section of two horizontal layers, three sets of waves are received at the detection point D at the surface —the direct wave along OD, the reflected P wave along OBD, and the refracted P wave along the path OACD.

The equipment used in seismic prospecting comprises essentially:—

(a)   a means of generating the shock wave;
(b)   detectors for picking up the resultant ground motion;
(c)   an amplifier-filter-recorder system which records a selected portion of the detector responses and filters out extraneous "noise";
(d)   an accurate timing system sensitive to intervals of the order of ·001 sec for measuring the time that elapses between the explosion and the receipt by the detector of the desired waves.

---

*Amplitude anomalies produced by the large reflection coefficients of certain low-velocity reservoir rocks (particularly in the US Gulf Coast area) have been used in the recently-developed "bright spot" or "hot spot" techniques to define the presence of hydrocarbon-containing rock sections under favourable circumstances[20, 21].

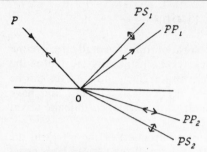

Fig. 49. Reflection and refraction paths
of seismic waves.

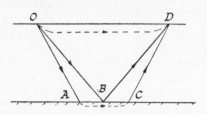

Fig. 50. Diagram illustrating paths of
surface, reflected and refracted waves.

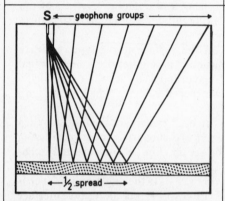

Fig. 51. A single-ended spread, with geo-
phone groups extending in one direction
from S, the shot-point.

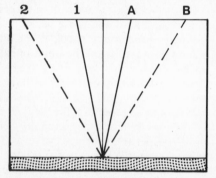

Fig. 52. Diagram illustrating principle of
200% common depth point seismic reflection shooting. 1, 2 are shot-points; A, B are
geophones.

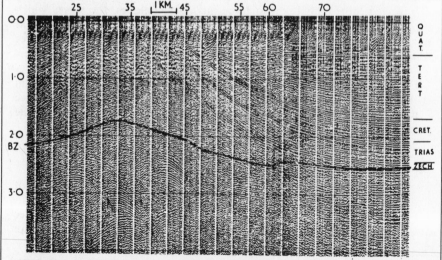

Fig. 53. A seismic section in the North Sea (BZ=Base Zechstein, vertical scale in
seconds). (after J. T. Hornabrook "Seismic Interpretation problems in the North Sea",
"Offshore Europe", p. 125, 1968).

TABLE 33

"P" WAVE SPEEDS IN DIFFERENT MEDIA

	ft/sec
Alluvium	1,000- 2,000
Sand	2,000- 6,000
Clay	3,000- 9,000
Sandy Clay	6,000- 9,000
Shale	6,000-13,000
Sandstone	8,000-13,000
Limestone	8,000-20,000
Rock Salt	14,000-17,000
Granite	15,000-19,000
Metamorphic Rocks	10,000-23,000

The chemical explosive used is generally nitroglycerine, suitably packaged and electrically detonated when required, after being tamped into shotholes previously drilled on a prearranged pattern by a portable rig. The depths of the holes may be 50–200ft, according to the thickness of the weathered layer that must be avoided.

A successful flexible energy source system* uses a long, thin strand of plastic-coated detonating fuse which is ploughed into a shallow linear furrow only a few inches deep. It is claimed that the efficiency of the explosion produced in terms of the elastic waves generated is enhanced in this way, and that by using suitably arranged explosive arrays, the amount, quality and direction of the energy released can be controlled as required. A variation of this technique is to combine the detonating fuse with flat packs of explosive laid on top of it on the surface of the ground at intervals of about 20ft.[22]

A variety of mechanical "surface sources" of shock waves has also been developed to replace the expensive and potentially dangerous chemical explosives, which must be handled and stored with special care. These include the dropping of 3-ton weights ("thumper" survey) or the use of the "Vibroseis"† technique, wherein multiple vibrators operate in synchronism to generate a controlled energy input.

The detectors used ("geophones") comprise an inertia element suspended or supported in such a way as to allow movement between it and its case which is placed on the ground. The relative movement which occurs when a disturbance reaches the detector is then suitably amplified.

The small electrical impulses from each detector are passed by cable to an amplifier-filter which is designed to cut out as far as possible unwanted external influences, and to magnify and transmit the desirable frequencies— usually in the 25 to 60 cycle/sec range. The output of the amplifier is connected to an oscillograph-camera system, which passes sensitized paper on a rotating drum under the beams of light reflected from galvanometer elements, and thus makes a record of the variations in the impulses received. Usually 6 to 12 oscillograph elements are used, and the sensitized paper is

*The I.C.I. "Geoseis" system.
†"Vibroseis" is a registered trademark of Continental Oil Co. This process has been used very successfully in sand-dune areas. It uses a controlled range in the vibration train, in addition to the output from a series of vibrators and sweeps.

rotated on the drum by an electric motor. (Magnetic tape recording is nowadays increasingly replacing the older photographic types of record, allowing deeper reflecting horizons to be identified.)

The timing mechanism employed is based on a carefully calibrated, electrically-actuated tuning fork, which is caused to make time lines 0·01 sec apart on the photographic record. This permits intervals of 0·001 sec to be interpolated.

Seismic refraction techniques were developed during the 1914–18 war for ranging enemy artillery. They were first applied in petroleum exploration operations in the United States in 1921, and during the following decade were successful in locating a number of salt domes with which hydrocarbon accumulations were associated in the Gulf Coast area, using several different techniques. Although also employed with great success in Iran for detecting anticlines, the refraction process was thereafter increasingly replaced by reflection methods of survey, which gave greater penetration in depth and were also capable of producing more detailed information from considerably smaller energy sources.

Seismic reflection techniques are now used considerably more extensively than all other geophysical methods put together. They differ in practice from the refraction techniques, which are nowadays only used for reconnaissance purposes or in special circumstances, in the disposition of the detectors relative to the explosion point. The refraction methods utilize detectors at distances from the explosion point that are several times greater than the estimated depths of the beds to be examined, whereas with reflection methods the furthest detector points are generally less than 1,000ft away.

The geophones (usually nowadays in 24 or 48 groups) are set out in a straight line through the shot-point, which may be either in the centre or at one end of the "spread". Either the "dip method" or the "correlation method" of operating can be used. The "dip method" uses the different arrival times of a particular wave front to describe the differences of depth of points along a reflecting surface, and hence to measure its inclination relative to the horizontal. (A variation of this technique—Fig. 51—is "continuous profiling" and this is widely used in exploration operations. In it, the shot-point is moved one-half the length of the "spread" before each explosion is set off, so that the reflecting horizon is fully covered). The "correlation method" of reflection shooting, on the other hand, relates the reflections coming from the same reflector bed so as to depict the varying depths of this surface below a datum plane.

The reflection survey is particularly useful for depth estimations whenever there is a continuous surface available to act as a good reflector. Suitable limestone beds, for example, have been most successfully mapped in this way over considerable areas. Where, as in the Gulf Coast, there are few continuous reflecting horizons, the reflection survey cannot delineate a continuous bed, but, by recording general dips, even though these dips may be in a variety of reflecting strata, a "phantom" picture of the structure can often be obtained. Faults can also be mapped with accuracy, using the displacement of the pattern of reflections on the two sides of the fault. Excellent results have been obtained by the reflection seismic survey to depths as great as 15,000ft or more. However, in common with all geophysical investigations it has certain limitations, and these are pronounced in areas which are much faulted, or where there are very steep dips or much blanketing alluvium.

The *common depth point* (otherwise called *common reflection point* or *horizontal stacking*) technique, which has been developed since 1959 as a means of eliminating multiple reflections, enhancing weak signals and attenuating background noise, provides a means whereby seismic surveys can be carried out with success in areas which previously had proved too difficult, either by reason of the depth or the nature of the strata[27a].

In this method, which is designed to emphasise the reflected energy while filtering out unwanted "noise", the shot-point is moved linearly while the positions of the geophones remain constant (Fig. 52). Shooting is carried out 6, 12, or 24 times, and the resultant consumption of explosives is obviously high. However, so successful has this technique been proved, that nowadays almost all seismic reflection surveys use "stacking" methods, and many areas which had previously been surveyed by older techniques have since been resurveyed, with strikingly improved results.

[In the processing centre, a number of treatments are usually applied to the recorded seismic data. By moving the traces up and down, so-called "static" corrections may be applied for the effects of ground weathering and variations in local topography. The "normal move-out" or "offset" is removed by a "dynamic" process in which the longer traces are shortened to the length of the trace nearest to the shot-point. Band-pass filtering removes unwanted background "noise" from the records, and the distortion of the transmitted seismic energy by earth characteristics (multiple reflections and "ringing") is removed by a "deconvolution" operation. "Dip migration" is applied by manual or computer methods to allow for the markedly non-vertical reflection paths from beds which dip at angles of more than about 10°.].

After the completion of processing, a series of record sections is obtained, on which dominant peaks and troughs on the traces are aligned and identified as coming from common reflecting surfaces. Such sections can reveal important structural information (Fig. 53). A contour map may eventually be produced which seeks to show the structural attitude of one or more of the reflecting surfaces at depth. The precise identification of such surfaces must generally await drilling, but the lack of this information need not affect the validity of the structural picture obtained. It is usual to contour such maps in terms of time—i.e. the intervals between the release of the shock-wave and its return to the surface, but even an approximate knowledge of the likely velocities of transmission in the underlying beds will enable a rough estimate of the depth in feet of the reflecting surface to be made. In this way, the form of the beds which lie fairly close to the sought-for petroleum reservoir can be shown, and an approximate idea obtained of the amount of vertical closure and the position of the structurally highest point of any potential structural trap which is present. Such information, coupled with the general geological picture, may enable a location to be made for the first exploratory well, and as this progresses, formation velocity surveys will be made. The samples and cores obtained, together with perhaps a further more detailed seismic survey along selected lines of section, will enable a more accurate structure contour map of the subsurface structure to be compiled.

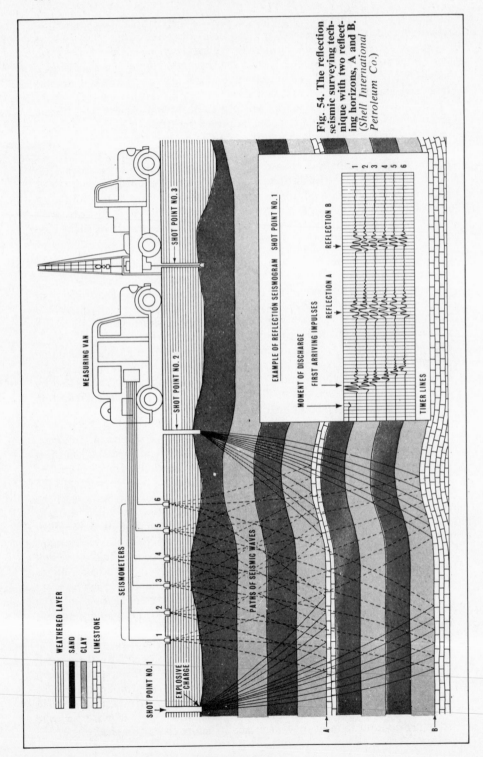

Fig. 54. The reflection seismic surveying technique with two reflecting horizons, A and B. (*Shell International Petroleum Co.*)

The use of the digital computer, utilizing coded numerical representations of the seismic data, permits subsequent processing to any degree of precision. This provides the most effective way of recovering and utilizing very low-energy signals and also of processing great volumes of seismic data. As yet, much seismic information is still being processed by digitizing analogue-type signals, but the trend is towards the "total digital system" wherein analogue signals from the seismometer are passed through a converter and directly recorded in digital form on magnetic tape for subsequent processing by a digital computer.

Lasers are being used to scan seismic profiles to permit the attenuation of random noise over a wide range of apparent velocities, and since variations of frequency are highly critical in the search for reefs by seismic exploration, notably in areas such as Alberta, "laser scanning" has become an accepted technique in these operations.

Seismic surveys can provide important information about the lithology of subsurface strata in favourable circumstances. Thus, since rocks absorb seismic energy to varying degrees, depending on their nature, measurements of the attenuation (reduction of amplitude) of the seismic signals propagated through layers of rock, coupled with a knowledge of the respective velocities of transmission, can together provide a comparative index of the lithology. In general, carbonates are characterized by a higher velocity of signal propagation than sandstones, which in turn show higher velocities than shales; on the other hand, sandstones show greater degrees of attenuation of seismic signals than both carbonates and shales[27].

Signal velocity analysis can also be used to predict the existence of abnormally high-pressured zones in subsurface strata before drilling commences. Since wave velocity increases with bulk density, rocks containing high-pressure fluids are likely to be less compact than normal strata, and may thus be detected by the relatively low seismic velocity that results.

Seismic reflection techniques have also had limited but encouraging successes in recent years in locating some specialized types of "stratigraphic" trap. Thus, indirect evidence of the presence of a bioherm may sometimes be obtained by amplitude ratio analysis of the reflected waves, using a calibrated gain-control unit. Examples are known where the disappearance of a high-amplitude reflection coincides with the presence of a reef mass, which exhibits unusual absorption characteristics.

The structural "draping" of sediments over a reef body, due to differential compaction effects may, of course, also be shown by seismic reflections if sufficiently pronounced: such structures may be distinguished from normal folds by the tendency towards thinning of the strata over the "high".

Some success has also been obtained, particularly in Canada, in the delineation of other types of stratigraphic trap by seismic methods. For example, it is now possible to construct a synthetic seismogram using the continuous velocity log obtained from a well, which in suitable circumstances can be used to help identify phase changes arising from variations in lithology and thus to recognize sand truncation points and "wedge-outs".

It is likely that with the development of improved instrumentation and techniques, the resultant high-quality field data obtained will permit greater refinement of analysis of geophysical data, and hence greater success in the search for the stratigraphic traps in which a large proportion of future oilfields will undoubtedly be found.

Offshore Seismic Surveys

Seismic surveying at sea is considerably more rapid and therefore proportionately cheaper than on land, since there are no boreholes to drill or networks of geophones to lay out by hand—they are simply trailed behind the shooting vessel or a companion ship.* It is thus possible to fire perhaps 1000 seismic shots a day at sea, compared with a score or less on land. However, it has been found that shallow-water dynamite explosions have a low efficiency factor for the production of seismic waves, while at depths of 40ft or more where the efficiency is higher, the formation of bubbles produces unacceptable amounts of "noise" for the recorders. As a result, seismic operations conducted offshore usually employ sources of low to medium energy shock-waves such as "sparker" transducers (electric spark discharges), magneto-strictive devices, or the explosion of combustible gases (acetylene or propane and oxygen) just below the water surface, so that no oscillating gas bubbles are produced to complicate the records.

The "shallow water airgun" system is used in marine areas too shallow for conventional techniques, and is designed to provide energy in the portion of the frequency spectrum that is most effective for deep penetration. The "guns" are trailed from a boom over the stern of the shooting barge on which is a suitable air compressor, while the recording barge follows, with its hydrophones laid on the sea floor.

For marine refraction surveys, it is possible to place the receiving units on the floor of the sea, rather than at its surface. Hollow steel spheres equipped with hydrophones, tape recorders and high-precision clocks are lowered to the sea bed and there record the times of arrival of seismic waves resulting from explosions at the surface up to 50 miles away. Later, when the survey is completed, the sea-bed spheres are made to release their ballast and float to the surface by the transmission to them of a coded signal.

Marine geophysical surveying depends for success on the availability of accurate position-fixing systems, several of which can give a continuous and precise record of the position of the survey vessel relative to marker buoys or to shore stations. Examples are the Decca and the Loran systems, which are essentially methods of hyperbolic radio position-fixing. The Decca system, for example, consists basically of a chain of four ground-based transmitters, a central "master" station and three "slaves" sited in a star formation. These radiate low-frequency radio signals which produce a hyperbolic pattern of reference lines: each such line represents a path along which the phase difference between the "master" and one of the "slave" stations is constant.

A receiver on board a moving vessel (or aircraft) can then utilize a phase comparison technique to establish two (out of three) intersecting position-lines which, when plotted on a chart overprinted with the radial pattern, determine the moving unit's position within an accuracy of a few metres.

---

*Geophones used in marine operations are called "hydrophones".

A light-weight, mobile position-fixing system ("Hi-Fix") has also been developed which can give position references accurate to within 1 metre on the "master"/"slave" base-line at ranges of up to 150 miles from the transmitters in daylight. "Hi-Fix" operates on the 2 Mc band and the transmitter chain consists of a "master" and two "slaves" sited in such a way as to give the best possible coverage of the area under survey. The survey vessel and the "slaves" transmit a radial position-line pattern which requires no special charts as radials can be easily drawn on ordinary charts from known wavelengths.

Satellite position-fixing systems have now become practical techniques, particularly for offshore work. The *Transit* system depends on the fact that the path of a satellite can be defined by the analysis of the changing Doppler shift of a steady system broadcast by the satellite. Three transmissions are needed, together with the orbital data, and when this information is fed into a suitably programmed computer a print-out of latitude and longitude of the ground receiver can be made every $\frac{1}{2}$–$1\frac{1}{2}$ hours, depending upon latitude, with an optimum accuracy of 0·1 mile.

Another technique (*"Navstar"*) uses two or more satellites in geostationary orbits with an attainable accuracy which is said to be as close as 60ft.[14]

When satellites are only infrequently available, an electromagnetic logging system can be used which measures both the fore-and-aft and sideways speeds of a vessel with an accuracy of 0·1 knot, thus permitting more precise dead-reckoning calculations between "fixes" than have previously been possible.

The fact that water has excellent properties for the transmission of low-frequency sound waves has been utilized in subsurface sonic surveying techniques which depict a profile (or "sonogram") of the irregularities of the sea floor, and hence may provide useful geological data. Pulses of sound of frequencies of the order of $1\frac{1}{2}$–6 kHz may be generated in the same way as the energy required for seismic surveys, i.e. by spark discharge or the explosion of combustible gases, and submarine sound detectors have been developed which are capable of acoustic directional sensitivity.

The *transit sonar* or *side-scan sonar* technique uses an acoustic transmitting and receiving system which is sensitive to signals arriving along a two-dimensional vertical plane, enabling a plan to be produced in analogue form which shows reflections received from irregularities on the sea floor caused by rock outcrops. A further development utilizes a system directionally sensitive to signals from narrow sources, which enables pictures to be presented on a cathode-ray tube screen[7].

The sonic *spectral analysis* technique uses both the incident and reflected acoustic signals of a low-frequency sub-bottom profiling system to identify and correlate types of sediments to maximum depths of some 200ft.

ELECTRICAL SURVEYS

Electrical methods of geophysical surveying have had considerable success in the location of metallic lodes and ore bodies. They suffer from the drawback of lack of penetration in depth when used for petroleum exploration, and are now little used commercially.

Most electrical techniques seek to measure the effects of artificially genera-ted electric currents applied under controlled conditions to the subsurface via surface electrodes, and attempt to detect variations that may be due to structure, lithology or the presence of hydrocarbons[17, 18].

Conductive Methods

In these, an electric current, either direct or low-frequency alternating, is applied to the surface of the ground by two electrodes, and the potential difference resulting is measured between two other electrodes. The figure obtained is a measure of the resistivity of a volume of the ground and is dependent on the spatial arrangement of the electrode system. A common configuration of electrodes used in practice is the "Wenner" arrangement, in which the four electrodes are equally spaced in a straight line, the input electrodes being on the outside. The further the electrodes are apart, the thicker is the zone vertically in which the bulk of the current flows, and the resistivity recorded becomes a weighted average of the resistivities of all the beds down to the deepest bed which the current penetrates. If a succession of resistivities are measured corresponding to increasingly wide spreads of the electrode system, it is possible to gain information about the depths of beds of varying resistivity.

In practice, the separation of the input electrodes is usually best fixed at between three and six times the depth desired to be reached. With direct current sources, non-polarizing electrodes must be used, or the Gish-Rooney system, which reverses the current flow about thirty times a second, can be utilized to eliminate the polarization effect. In reconnaissance work, station spacing is usually at 1-mile intervals.

Another method of resistivity exploration known as the "Elflex" system uses low-frequency commutated currents, by which measurements of the average resistivity of a section 800 to 1,200ft thick, centred at desired depths of 3,000 to 9,000ft, have been obtained. It is claimed that portions of the sections measured which contain oil may be distinguished from barren portions by their appreciably higher resistivities.

Induction Methods

A number of electrical surveying methods have been designed on the principle of inducing a current in the ground by a primary alternating current, measuring the resulting electromagnetic effect at the surface, and comparing the strength and shape of the field recorded with the field that could theoretically be expected if the subsurface were uniform and homogeneous. On this basis, it may be possible to make deductions concerning the nature and conductivity of the subsurface strata, and these techniques have been of special use in detecting faults.

The application of induction methods to prospecting below a depth of about 200ft is limited by the fact that for this purpose very low-frequency alternating currents are necessary whose generation and measurement are difficult.

The Electric Transient ("Eltran") Method

This prospecting technique is based on the fact that if a sharp current pulse is introduced into the ground, the spread of the current through a bed of high resistivity will be much faster than through a bed of low resistivity, i.e. the current transient in the former will be steeper than in the latter. The relative resistivities of the strata in the path of the current can therefore be determined from observations of the time constants and wave forms resulting.

The current pulse is applied by closing or opening an electric circuit in contact with the ground through buried electrodes. Measurements of the rate of variation of the resultant potential are made through two electrodes placed outside the first electrodes, connected to a cathode-ray oscillograph and high-speed recording camera of the revolving-drum type.

There has been controversy regarding the relative merits of such direct resistivity surveys. It seems that the direct-current, steady-state resistivity measurements give more constant results owing to the absence of inductive and current redistribution effects, but that the Eltran technique may have the advantage of greater depth range.

Radio-Wave Surveys

Methods of using radio broadcast energy as a power source for subsurface surveys have been developed, for which some success has been claimed in oil exploration. The transmitters used typically operate on a frequency range of 1600–1700 Kc; higher frequencies lessen depth penetration.

The receiving set-up usually consists of an amplifier, a manual phase-shifter, a second amplifier and receiver and an output-meter. The techniques used may be summarized as follows: (1) variations in the strength of surface propagated waves are plotted; the results, it is claimed, can reveal mineralization "haloes" caused by seepages; (2) the electrical conductivity of the earth transmitter and receiver is determined; this can also indicate the presence of "haloes"; (3) the time displacement of waves received from the ground compared with waves directly from the transmitter is measured, indicating possible changes in structural elevation or faulting effects.

Telluric Surveys

It has been known for a considerable time that natural currents flow through the earth both locally and over wide areas. Local natural currents may result from electrochemical, meteorological or man-made factors, whereas the so-called telluric currents cover the whole globe and are thought to be due to astronomical causes. The direction and intensity of the telluric currents at a given spot vary with time, but at a given moment the currents have approximately the same direction over a wide area[6, 12].

An exploration technique has been established on this basis, since many geological anomalies appear to cause a local variation in the telluric currents. Salt domes, folds, buried ridges, etc. have been investigated by this means, using non-polarizable electrodes and a photographic recorder to measure the telluric current variations.

The telluric methods have many of the same advantages and disadvantages as the gravity and magnetic methods of surveying, but can sometimes provide an additional type of information by making use of a range of frequencies.

The major problems lie in the methods of analyzing and interpreting the measurements. Digital computers are currently being used for analysis, and under the most favourable conditions, semi-quantitative depth estimates with ±10% error can be made, which can possibly be further improved upon by more refined techniques.

GEOTHERMAL SURVEYS

Surface rock temperature measurements have been used with occasional success to locate and identify underlying structures and faults. A sensitive electric temperature detecting device, i.e. a thermocouple capable of detecting temperature changes of the order of 0·01°C, must be used. The thermocouple is constructed so that the hot junction can be lowered into a series of shallow holes, while the cold junction is kept at constant temperature in an insulated flask full of melting ice at the surface. In this way, accurate readings of local earth temperatures can be made.

The depth of hole used must be sufficient to eliminate diurnal variations in temperature; usually holes of about 3 to 4ft in depth are sufficient for this purpose. Stations are chosen from 100 to 500ft apart, depending on local conditions and the structure to be examined, and measurements are usually taken the day after a particular hole has been made, to eliminate the local heating effect of drilling.

In general, it has been found that areas overlying buried structures tend to be warmer than those over undisturbed areas, if the measurements are corrected in each case to allow for the variation of temperature with seasonal changes. A positive temperature anomaly curve may therefore be expected when an anticline is traversed.

The geothermal survey has also been used to detect salt domes, since the ground temperature above such domes is often slightly higher than normal, due to the fact that the thermal conductivity of salt is higher than that of the surrounding sediments. This type of survey may also be used in some cases to delineate a granitic contact.

GEOCHEMICAL SURVEYS

Geochemical methods of exploration attempt to link the presence of very small quantities of hydrocarbons in surface soils—only detectable by analytical techniques—with oil and gas accumulations in the subsurface[19].

They depend on the theory that hydrocarbons from a buried accumulation will tend to percolate upwards in the course of geological time, and may therefore be detected in surface soils by a sufficiently sensitive technique. The mechanism of this percolation is unknown: it has been suggested that the hydrocarbons may be carried upwards in the form of minute gas bubbles, or that the movement may be due to somewhat obscure processes of capillary transport. However, the coefficient of diffusion of methane gas through water-saturated clay is so low that to penetrate some thousands of feet of compact sediments would take a period of several million years. Since the rate of diffusion of liquid oil components would be several thousand times less than that of gases, it is clear that to envisage the diffusion of crude oil on this basis would require such enormous periods of time that it can be considered as virtually impossible where an unfractured reservoir is concerned.

We have seen that the escape of mobile petroleum constituents towards the surface along faults and fissures to form seepages and surface "shows" has probably been an intermittent process, periodically ceasing and restarting over lengthy periods of time. It has been suggested that instead of—or in addition to—this process of *macro-fissuring* and hydrocarbon escape, there may be present a less apparent system of *micro-fissuring,* which permits the escape of gas and oil in molecular quantities only. Given sufficient time, detectable hydrocarbon concentrations could then reach the surface and may be detectable there if sufficiently sensitive analytical techniques are employed.

It certainly seems possible for a sealed subsurface hydrocarbon reservoir to be subjected to periodic earth movements which result in the opening of temporary channels to the surface. These may be subsequently re-sealed after the escape of some oil and gas—possibly as a result of the clogging of capillary passages by the heavier hydrocarbon constituents, or by the circulation of meteoric waters resulting in base-exchange effects in the clays. The whole process could presumably be repeated many times during the geological history of the accumulation.

The curious type of surface "show" called "paraffin dirt" was mentioned earlier. It is widely dispersed rather than only in the immediate neighbourhood of seepages, which seems to indicate that there is a widespread percolation of gaseous hydrocarbons in these areas; it is reasonable to assume that some of the gases will be retained in the surface soils and perhaps subjected there to biochemical processes. Various types of surface survey have been developed on this supposition.

In the *"bitumen" survey* (mainly practised in the USSR), samples are taken of the soil in closely spaced grids across the area to be examined, and from depths of from $\frac{1}{2}$in to 10ft, depending on local circumstances; these samples are then subjected to laboratory analysis to determine the content of "bitumen".* The total "bitumen" content of the soil is not a reliable index, since in whatever way it is measured it will necessarily include soil bitumens having no relation to petroleum, so that it has been found preferable to use instead the "bitumen" proportion of the organic constituents of the soil.

In view of the often confusing nature of the results obtained, it is clear that the reliability of this technique is obscured by extraneous factors, such as the organic composition of the surface soils[12a].

A variation of the surface bitumen survey is the *bacterial survey*. This depends on the use of bacterial cultures for determining the presence of diffusing gases. The capacity of various types of bacteria to oxidize methane and higher hydrocarbons has been used in a technique to detect the presence of these gases in shallow surface boreholes. It has been suggested that when methane-oxidizing and propane-oxidizing bacterial anomalies coincide in the subsoil, this is an indication of subsurface oil or gas concentrations. Methane-oxidizing bacteria are said to predominate over gasfields and propane-oxidizing bacteria over shallow oilfields. The technique is time-consuming, but is in standard use in the USSR.

---

*The term "bitumen" has unfortunately been used in a number of different ways, but signifies here a natural organic substance soluble in neutral organic liquids. Plant and soil bitumens are related to the surrounding rocks and are chemically different from the oil bitumens which, in the form of asphalt, ozokerite, etc., are secondary deposits.

In the *soil gas analysis survey,* holes of 10–30ft in depth are made in a close grid pattern across the area to be surveyed, constructed in such a way that a bell-shaped sample container can be introduced, into which the soil gases can be drawn by means of a vacuum pump. A number of variations in the techniques employed in the USSR include: the "delayed gas survey", in which samples are not taken directly after the drilling of the test holes, but only after a period allowing equilibrium to be established, i.e. 12–24 hours; the "deep gas survey" in which several samples at different depths are taken to avoid contamination by surface soil gases; and the "water gas survey", in which ground waters are examined for dissolved hydrocarbon gases[15].

The soil gas obtained contains a number of combustible components, including methane, carbon monoxide and various volatile organic substances, as well as a proportion of the atmospheric gases. There is considerable fluctuation in the content of the various gases entering the survey holes, depending on the season, and even the time of day. It is therefore generally desirable to take samples from as deep as possible.

Errors of analysis can obviously vitiate the whole process, but assuming that a sound technique is employed, it is claimed that valuable gas anomaly maps can be compiled in this way. Maps based on the detection and measurement of the heavier hydrocarbon gases are, of course, more likely to be reliable than those based on methane only.

By measuring variations in the fluorescence spectra of samples taken by augering several feet down into the soil over the Nottinghamshire oilfield areas in Britain, it was shown that "isofluor" lines could be reliably obtained which were frequently parallel to the structural axes in the underlying coal measures; while over the oil-bearing structures these "isofluors" formed haloes along the flanks of the anticlines[9].

In the United States, it has been found in some cases that if the higher hydrocarbon gas content is plotted across the section tested, a typical "halo" pattern is obtained surrounding a "bald spot" vertically over an oil area. It has been suggested that this result is to be expected as a result of uplift of structure and resulting compaction, which might cause the upper part of a dome to be less permeable than the flanks; and also that hydrocarbons are likely to escape from the oil-water boundary zone. On the other hand, it has been reported from the USSR that the methane content of soil gas samples is likely to be at a peak over the crest of a structure, coinciding with a heavy hydrocarbon gas "low". The significance of these results is not clear.

The French "sniffer" technique is claimed to have had success in defining which structures contain hydrocarbons and which are "dry". The collection of samples is done systematically on a grid pattern from a constant depth of 10ft, and the samples are sealed in polythene bags for transport to the laboratory, where they are subjected to acid digestion under vacuum at 80°C. The gases liberated are then analysed on a gas chromatograph and saturated hydrocarbons are distinguished from unsaturated hydrocarbons. The interpretation of results depends on noting anomalies in the proportions of various groups of saturated hydrocarbons in relation to the lithology of the sediment samples.

A variation of this technique is a recently developed underwater "marine sniffer" or "seep detector" which is claimed to be capable of analysing sea-water for saturated and unsaturated hydrocarbons containing up to four carbon atoms, and thus detecting prospective areas for subsurface petroleum accumulations. However, some observations by a chromatographic technique which can reproducibly measure less than one part per million of individual components in the vapour obtained by heating soil samples, indicated that the paraffinic and olefinic hydrocarbons detected were largely derived from plant root remains. Thus, components derived from a deeper source may be overshadowed by hydrocarbons from small amounts of vegetation distributed in the soil. It is probably therefore incorrect to ascribe the presence of saturated hydrocarbons in the soil only to a petroleum source, and very careful background work is necessary to distinguish between vegetable and petroleum contributions[13, 25].

Even where petroleum hydrocarbons are detectable at the surface, it would often be wrong to assume that their presence indicates the existence of a vertically-underlying subsurface accumulation. Most strata are more permeable along their bedding planes than across them, so that gas is more likely to diffuse to the surface along the bedding planes than vertically, with the result that any anomaly detected may be at a considerable distance from the vertical projection of the hydrocarbon concentration from which it is presumed to have been derived.

Another essentially geochemical technique which has been developed principally in the USSR is the *oxidation-reduction ("redox") potential survey*. This depends on the measurement by a suitable instrument of the electrical potential developed between rock or soil samples and a standard hydrogen electrode. Since hydrocarbons are relatively reduced substances compared with most sediments, their presence may be expected to lower the local redox potential of the environment. Certain complicating factors have been noted, however; thus hydrogen sulphide gives a strong reducing effect, whereas heavy oils have little effect.

The soil redox potential survey has been used to detect subsurface hydrocarbons, either by measuring the electrical properties of soil samples taken over a pre-arranged grid or by direct measurements in a series of shallow pits. If hydrocarbon gases have escaped to the surface from an underlying reservoir, it may be expected that a pattern of negative redox potential anomalies is likely to occur in the surface soils overlying the reservoir. However, the reverse case has also been reported—abnormally high redox potentials over oil pools. This has been explained by the presence of micro-organisms which utilize the hydrocarbons in their own life-processes. The method is therefore still very restricted in its application as an exploration tool.

Soil-chloride surveys have been used from time to time as geochemical indicators of the brines associated with underlying hydrocarbon accumulations, based on the supposition that hydraulic currents move upwards from the bottom of sedimentary basins.

Soil-iodine anomalies have also sometimes been proposed as evidence of underlying oil and gas; it is claimed that a concentration of as much as 100 times the normal iodine content may sometimes be detected in soils over a hydrocarbon accumulation, and this has been explained by the preferential concentration of iodine salts in oilfield waters.

An increase in the *calcium sulphate* content of soils has been taken as a sup-

plementary surface indication, based on the supposition that calcium chloride
derived from oilfield brines will react with sodium and magnesium sulphates
in the soil, producing gypsum. However, rising hydrocarbon gases would also
presumably reduce such sulphates to sulphides.

The analysis of the elements present in local waters which may be of deep-
seated origin is another technique which can be considered as a method of
geochemical exploration. *Bromine* and *boron* are elements whose concentration has been related to association with subsurface petroleum. A high *benzene*
content is also said to be indicative of an oil accumulation, since benzene is a
hydrocarbon relatively highly soluble in water (p. 20).

RADIOACTIVITY SURVEYS

The only other method of surface exploration which endeavours to make
direct use of a physical property distinguishing petroleum from the surrounding rocks is the radioactivity survey. It is based on the hypothesis that most
crude oils contain radioactive material, some of which—notably dissolved
radium salts or radon gas—may be carried to the surface by percolation and
thus reveal the areas under which oil may lie.

Tests for radioactivity may be made on gas samples drawn from shallow
surface holes or on soil samples. Both types of sample are recovered along
closely-spaced profiles or grids covering the area under test, and the relative
radioactivity of each sample is then measured and plotted against its map
position. At one time it was thought that by such measurements radioactivity
"haloes" could be defined over oilfields which were similar to those resulting
from geochemical surveys. Unfortunately, the results obtained have generally
been complicated by extraneous factors, and relationships have not been
found which have undisputedly led to the discovery of an oil or gas accumulation not previously known to exist[1, 3].

However, radioactivity surveys of this type have been reported to have been
of use when conducted prior to drilling extension wells in "shoestring" sand
fields. By taking closely-spaced profiles on the surface across such sands it has
been claimed that their limits could be defined with exactness.

Radioactivity measurements have also been used with some success to find
the trend of a fault after it has been cut by a well, or the strike of a fault buried
beneath surface alluvium, on the basis that radon-bearing gases are likely to
migrate more rapidly along fault lines.

By using a portable instrument (usually a scintillation counter), sensitive
to the beta- and gamma-rays emitted from the radioactive elements in mineral
particles contained in surface soils, it is often also possible to distinguish the
boundaries between different sedimentary formations, e.g. high-radioactivity
shales and low-radioactivity limestones, even though these are covered by
several feet of soil and vegetation cover, and in this way to provide data
capable of structural interpretation. It is important, however, that the soil
should be autochthonous, since otherwise the instrument readings will not
reflect the radioactivity of the underlying rocks[28].

Airborne radioactivity (or "aeroradiometric") surveys, using sensitive
scintillation counters and a flight altitude of a few hundred feet, have been

successful in locating areas of relatively high radioactivity and hence concentrations of uranium minerals. When flown over known oilfields, such surveys have shown that distinct gamma-ray patterns can be found, but predictably these are generally related to surface geological, pedological and hydrological factors, rather than to the presence of oil at depth.

REFERENCES

1. J. S. Adams and P. Gasparini, "Gamma-Ray Spectrometry of Rocks". Elsevier Publishing Co. (1970).
2. A. M. Alexeyev et al., "Application of electromagnetic methods in oil and gas exploration in the USSR". Proc. 7th World Petrol. Congr., Mexico, v.2, 697–708, (1967).
3. F. E. Armstrong and R. J. Heemstra, "Radiation halos and hydrocarbon reservoirs: a review". Bur. Mines Info. Circular 8579, Washington, (1973).
4. A. Barringer, "Remote-sensing techniques for mineral discovery". Paper 20, 9th Commonwealth Min. Met. Congr., London (1969).
5. R. C. Browne, "Gravity surveying at sea". Journ. Inst. Petrol., 55, 541, 22–26 (1969).
6. L. Cagniard, "Basic theory of the magneto-telluric method of geophysical prospecting". Geophysics, 18, 3, 605–635 (1953).
7. H. Charnock, "Ocean fine structure". Nature, 239, 36, Sept 1, 1972.
8. M. B. Dobrin "Introduction to Geophysical Prospecting", McGraw-Hill, New York (1952).
9. W. D. Evans, B. S. Cooper, D. W. Corbett and K. Gough, "A geochemical survey of the Nottinghamshire oilfields and related sediments". Q. Journ. Geol. Soc., 118, 23–38 (1962).
10. A. F. Gregory, "Analysis of radiometric sources in aeroradiometric surveys over oilfields". Bull. Amer. Assoc. Petrol. Geol., 40 (10), 2457–2474, (1956).
11. C. A. Heiland, "Geophysical Exploration", Prentice-Hall, New York (1940).
12. W. B. Heroy (Ed.) "Unconventional Methods in Exploration for Petroleum and Natural Gas". 87–104, SMU Symposium, Dallas (1969).
12a. G. D. Hobson. "The status of geochemical prospecting". Petroleum, 24, 255–258 (1961).
13. L. Horvitz, "Vegetation and geochemical prospecting for petroleum". Bull. Amer. Assoc. Petrol. Geol., 56 (5), 925–940 (1972).
14. International Symposium on Marine Navigation, Sandefjord, Norway, Sept. 1969.
15. A. Kartsev et al., "Geochemical Methods of Prospecting and Exploration for Petroleum and Natural Gas" (translated by P. Witherspoon and W. Romey), Univ. California Press (1959).
16. J. J. Jakosky, "Exploration Geophysics, 2nd Ed., Los Angeles (1950).
17. G. V. Keller, "Electrical Prospecting for Oil". Quarterly of Colorado School of Mines (1968).
18. G. V. Keller and F. C. Frischknecht, "Electrical Methods in Geophysical Prospecting". Pergamon Press, London (1966).
19. H. Kroepelin, "Geochemical Prospecting". RP3 7th World Petrol. Congr., Mexico (1967).
20. J. P. Lindsey and C. I. Craft, "How hydrocarbon reserves are estimated from seismic data". World Oil, 23–25, Aug. 1973.
21. J. P. Lindsey, "Bright spot: a progress report and look ahead". World Oil, 81–83, April 1974.
22. T. A. Magub, "The Geoseis system". APEA J. 12 (1), 28–35 (1972).
23. L. L. Nettleton, "Geophysical Prospecting for Oil", McGraw-Hill, New York (1940).
24. L. L. Nettleton, "Gravity and magnetics for geologists and seismologists". Bull. Amer. Assoc. Petrol. Geol., 46 (10), 1815–1840 (1962).
25. G. H. Smith and M. M. Ellis, "Chromatographic analysis of gases from soils and vegetation". Bull. Amer. Assoc. Petrol. Geol., 47 (10), 1897–1903, (1963).
26. Soc. Explor. Geophys., Tulsa, Oklahoma, "Geophysical Case Histories", Vol. I (Ed. L. L. Nettleton) (1948); Vol. II (Ed. P. L. Lyons) (1956).
27. M. T. Taner and N. A. Anstey, "The problems of corrections and display", Geophys. Soc. Houston Sympos. "Lithology and direct detection of hydrocarbons", Oct. 1973.
27a. L. H. Tarrant, "Geophysical methods used in prospecting for oil". Pp. 67–107, "Modern Petroleum Technology", Applied Science Pub. Ltd. (1973).
28. E. N. Tiratsoo, "Radioactivity measurements as an aid to geological mapping". Proc. 18th Internat. Geol. Congr., 5, London (1950).

CHAPTER 7

Formation Evaluation

WELL SAMPLES

FORMATION Evaluation may be defined as the procedure of determining the composition and physical properties of formations drilled and the nature and amount of the fluids they contain[17, 27, 30]. The operations involved vary depending on whether drilling is being undertaken for exploration, appraisal or development purposes. In the first case, the principal objective is the location of hydrocarbon-bearing beds; in the second case, it is the establishment of parameters for log analysis; and in the third case, the aim is the determination of the type and volume of the hydrocarbons present.

It is clearly desirable that every well drilled should furnish the maximum of subsurface information, irrespective of its success or otherwise as an oil producer. Hence, a close geological check is nowadays kept on all drilling wells—particularly those whose purpose is exploratory.

The "resident" or "well-site" geologist has the responsibility for the collection of all possible evidence concerning the rocks penetrated, and the subsequent compilation of an accurate *geological well log* (sometimes called a "strip log"). During the drilling operations he will also be expected to advise the drilling staff of any expected oil-, gas- or water-bearing beds likely to be penetrated, and of the nature of the formations to be expected, with particular emphasis on any unusual strata, such as heaving shales, caving formations, or extra-hard rocks. He will also advise on the selection of suitable formations as casing seats, which are chosen in dense, unfissured formations to give a firm and impermeable support for the casing. (Compact limestones, anhydrite beds, dense shales, hard sandstones, and salt beds are suitable formations for casing seats, while fissured or water-bearing strata should be avoided.)

To facilitate the collection of geological evidence such as formation boundaries, the well-site geologist is usually empowered to ask for cores to be taken where necessary; and he will be responsible for obtaining the maximum information from each core, and preserving both cores and well cuttings in such a way as to permit subsequent worth-while re-examination.

The functions of the resident geologist can be summarized by saying that his primary objective is to piece together from all available sources of information the sedimentary and structural history of the area, by using all the evidence that can be obtained by drilling, together with whatever surface information is available. A further objective is to assist the drilling operations by providing correlations between wells and prognostications ("prognoses") of the depth, nature, and thickness of important formations.

The Drilling Fluid

Throughout the period of drilling the well, the well-site geologist is responsible for the examination of material brought up by the drilling fluid or "mud". This mud is itself nowadays a subject so complex as to fall outside the scope of this work; but it is necessary, however, briefly to describe its constitution and functions in order to understand its relevance to subsurface operations.

In the early rotary wells, it was the practice to use water to lubricate the drilling bit and to carry away the cuttings. The water was pumped down the drill-pipe, emerging at pressure through an opening in the bit, and carrying the débris from the bottom of the hole up through the borehole annulus round the drill pipe to surface. In soft formations, mud was formed in the course of circulation by the action of the water on the cuttings, and it was found that it had a beneficial effect on the drilling process by plastering the walls of the hole and preventing crumbling and caving. From this beginning, the practice of circulating mud instead of plain water developed, and proved so advantageous that it soon became standard. Conveniently situated native clays were originally the source of the solid matter in the drilling mud, but nowadays artificial fluids are usually employed, made by mixing special clays with various chemical additives. These generally have properties which are far superior to the older "natural" muds.

There have also been developments in the substitution of other fluids for fresh water as the liquid medium. Thus, oil-base and salt-water muds are quite often used, while dry air and air-mist have been the subject of some interesting applications.

The responsibility for the quality and performance of the drilling fluid was originally that of the driller. Subsequently, the responsibility often fell to the resident geologist, but nowadays the subject has become so complex that specialist mud engineers are generally needed[25].

Good mud is of great importance for the successful drilling of a well; in fact, experience in oilfields all over the world has shown that the mud used is probably the most important single factor governing the success of rotary drilling operations. The rationalization of mud techniques enables greater depths to be reached than were previously possible, with consequent increased drilling hazards; yet the careful mud conditioning and control of modern practice enables a high proportion of even the deepest holes to be completed safely, and eliminates many stoppages due to sticking tools, twist-offs and "fishing" jobs. Even the highest pressure wells may be brought-in under complete control if adequate supplies of the correct heavy mud and suitably powerful mud-pumps are available.

The principal functions of the drilling fluid may be summarized as follows:

(a) During drilling, the walls of the hole have a tendency to crumble and cave, particularly in soft or fractured strata. The mud column tends to prevent this caving by exerting hydrostatic pressure, while the coating of mud on the walls of the hole tends to prevent loss of fluid and helps to consolidate loose strata.

(b) A back pressure is exerted on any permeable formation penetrated which contains fluids,thus preventing "blow-outs" and allowing desirable productive formations to be brought in smoothly.

(c) The revolving bit and tools are continually cleaned and cooled by the flow of liquid through the hole, while the drilling tools are lubricated.

(d) The presence of the mud column reduces the weight of the drill-pipe and casing which the hoist has to lift.

(e) The jet of high-pressure mud emerging from the bit has a considerable erosive effect on soft strata, and the very rapid footage made when drilling in Tertiary rocks in such areas as California is often due to this factor.

(f) Bit cuttings are removed and deposited at the surface in the ditch, and when circulation is temporarily stopped, these solids are held in suspension in the mud by the formation of a gel, thus obviating "freezing" of the tools. Upon circulation being restored, the gel turns rapidly back into a sol and the mud becomes liquid again.

It might be thought from the above summary that a viscous fluid of high specific gravity would be an ideal drilling mud. But such a fluid would have the disadvantages of requiring excessive power for pump operation, and also of taking too long to deposit its detritus in the surface troughs, even with the aid of screens.

On the other hand, a low-viscosity fluid would permit the cuttings to settle in the hole when circulation was temporarily suspended, and the drilling tools would tend to stick as a result. The ideal mud must therefore be able to increase its viscosity soon after being brought to rest, although it possesses a comparatively low viscosity while in motion. This is the characteristic of a *thixotropic* substance, and the ideal drilling fluid must therefore be a reversible hydrosol or thixotrope. That is to say, while in motion it behaves as a liquid, and is a sol; and when at rest it becomes a semi-solid mass or gel, the two states being unrestrictedly interchangeable.

The optimum drilling fluid has been defined as one that is a "low-cost, low-viscosity, high-density, stable fluid, of which no portion will pass through a filter, independent of the flocculating action of salt-water or the influence of temperature variations, and having the unusual characteristic of setting as a gel when allowed to remain quiescent, but becoming extremely fluid when agitated".

Most muds used nowadays consist essentially of a mechanical mixture of several substances, chiefly clays and colloid-rich clays (bentonites), with water to which various chemicals and other materials are added for special purposes. These muds are thus essentially single-phase fluids with solid additives. (The more sophisticated two-phase, or emulsion, muds should also be mentioned: it has been found that by emulsifying up to 20% of oil with a water-base mud the resultant fluid then has improved lubricating properties.)

The chief mud additives used comprise:–

(a) Weighting material, such as barytes, haematite, or galena, added to increase the specific gravity of the mud.

(b) Organic or inorganic chemicals, used for improving some special property of the fluid, usually its plastering power, or for decreasing its viscosity. Thus, alginates reduce viscosity, phosphates and tannates are deflocculating agents and soluble silicates reduce the gel strength. Various inhibitors and preservatives are also used.

(c) "Plugging agents", usually fibrous materials of various kinds, which are used when circulation of the mud is lost in fissured strata.

In making-up the mud in the conditioning pit it is reduced to as homogeneous a condition as possible by repeated pumping through jets, the conditioning agents being added in a powdered state. The common practice is to have separate storage, conditioning and suction pits, which can be interconnected as required, and from the last of which the fluid, when it is of suitable consistency, is drawn by the mud-pumps and circulated down the well. Upon emerging, laden with cuttings, the mud is passed over one or more vibrating screens and then along a system of long troughs with slats and baffle boards, during the passage of which much of the detritus is dropped. Care is taken not to agitate the mud too much in transit, as the consequent absorption of atmospheric oxygen tends to cause corrosion in the drill-pipe, besides adversely affecting the quality of the mud. Special equipment may be installed to eliminate gas brought up by the mud, and when deep wells are being drilled coolers are often necessary.

As drilling progresses, increasing volumes of drilling fluid will be required to fill the borehole, and in addition a small loss of drilling fluid by evaporation and absorption, as well as loss on the derrick floor, is to be expected; the circulating mud column must therefore be periodically augmented from the storage pit. Where the drilling fluid loss is greater than normal, but circulation remains partial, the leakage may be stopped by chemical means, interacting reagents being added to precipitate an impervious "filter cake" on the walls of the hole. Typical combinations of such chemicals are sodium silicate with calcium chloride or aluminium sulphate. Larger fluid losses are generally counteracted by increasing the colloid content of the mud, and hence its plastering properties, by the addition of some form of bentonite.

However, circulation may be completely lost when the bit passes through fractured, fissured, or cavernous formations, through gravel, or when very permeable strata are penetrated. On such occasions, fibrous material must be added to the mud, to plug the fissures through which the fluid is being lost. Many substances have been used for this purpose: straw, chopped rope, feathers, cottonseed hulls, cane stalks, beet pulp, sawdust, oyster shells, hay and peat moss, and various flaky organic materials.

Mud weights are generally expressed in terms of lb/cu ft. Optimum densities vary from field to field and also depend on the depth of the well being drilled, but the practice is to keep mud weights as low as possible consistent with safety and drilling speed. If the fluid does not have to carry much inert material, a light mud of about 70lb/cu ft, with a viscosity of 36 sec API is usually adequate. Much heavier muds than this are used in special circumstances, as, for example, in Trinidad where 110lb muds are common, and 120lb mud is used in heaving formations.

During drilling operations, a close watch is always kept on the specific gravity of the drilling fluid, often with the use of continuous mud weight recording instruments; a sudden drop may be due to the entry of gas and a "gas-cut" mud must be treated at once by the addition of barytes or the admixture of fresh mud.

Any traces of oil or gas in the mud must be noted. Crude oil will generally show as a fine dark film best seen by reflected light, but even very small traces of oil may be detected by their characteristic fluorescence under ultra-violet light of wavelengths down to 2,000 ÅU, while the so-called "near ultra-violet" light of about 3,650 ÅU causes an even brighter fluorescence. The quartz mercury arc lamp releases much of its light at 2,536 ÅU and is convenient to use. A glass filter is employed to cut out visible light rays, and samples can be examined at the wellhead in a hooded box. This is best done after allowing the mud samples to stand for some time in wide-topped dishes so that any droplets of oil they contain can rise to the surface. By comparison with known concentrations of oil, the test can be reasonably quantitative. However, contaminating oil from bearings or pipe couplings may sometimes give a similar fluorescence. In the USSR, the reported fluorescence of colonies of hydrocarbon-oxidizing bacteria is also used as a method of detecting the presence of traces of oil and gas in cores or drill cuttings.

When a formation containing much heavy oil or bitumen is being drilled through, characteristic black spots may appear in the mud, while a reservoir rock when entered may produce a notable admixture of liquid oil, so that the drilling fluid actually becomes dark brown or black in colour. It is important to note that the mud weight and the consequent pressure exerted on the subsurface formations will in fact be the critical factor controlling the appearance of oil "shows" in the mud when oil-bearing formations are penetrated. The pressure exerted on the mud at its point of exit from the bit may be equivalent to double the hydrostatic head of the fluid column; and in a deep well with a heavy mud, this is likely to prevent efflux of fluid from the formation. In fact, the mud filtrate may actually enter the strata, and drive back any contained fluids from the walls of the hole. (If this occurs to an extensive degree irreparable damage may be caused to the reservoir horizon.) Various additive chemicals are employed to minimize the caking tendency of the mud on the surface of the hole, but continuous control of the weight of the drilling fluid is always essential.

Gas bubbles seen on the surface of the emergent mud are important evidence, and samples should be taken, wherever possible, for laboratory analysis. It is sometimes possible to confuse air trapped in the drill pipe, which is released when this has been run into the hole, with subsurface gas. Such spurious gas "shows" will, however, be purely temporary phenomena.

The appearance of any authentic gas or oil show in the mud must be the signal for cautious progress. Samples of the bit cuttings should then be taken with increased frequency, and if the mechanical condition of the hole allows, coring should be substituted for drilling. The object must always be not to pass through the petroliferous zone without adequately identifying it, and if necessary casing it off from possible interference by adjacent waters. The examination of many old boreholes has shown that potential producing formations have quite often been passed through unbeknown, or have even been "mudded-off" and lost, owing to too heavy a mud having been used during drilling or insufficient attention having been paid to small traces of oil and gas in the mud returns.

The presence of natural gas in the drilling mud (or in the bit cuttings) was customarily detected by a *hot wire analyser*, which can provide a semi-continuous log of the mud and is sensitive to about 200 p.p.m. of combustible gas; but this instrument is now being replaced by more sensitive techniques.

Partition gas chromatography is being increasingly used in analysing oil-well gases, particularly to indicate the approach of possible oil horizons. If mud samples are examined as drilling progresses, it has been noted that in some areas there is a pattern of "first" detection of hydrocarbons in sequence of their molecular weight, commencing several hundred feet above the reservoir. Methane can commonly be detected throughout the section, with no traces of ethane; but CO_2 is invariably associated with methane, and there is a possible "first" appearance of CO_2 before methane, if ethane appears later in the sequence.

The *infra-red analyser* can also analyse almost continuously, but for only one hydrocarbon component at a time, generally methane. It has been found that the best general technique is to monitor the mud flow regularly for the presence of methane by this device, and when a positive "show" is recorded to confirm and extend the analysis by using the chromatograph.

To improve the efficiency of gas sampling, a reflux unit has been devised in which mud samples can be treated with steam so as to release almost 100% of their content of C_1–C_5 hydrocarbons for analysis.

Bit Cuttings

The mud which returns to the surface up the annular space between the drill-pipe and the casing will bring with it in suspension the small fragments of rock which the rotating bit has cut from successive subsurface formations, and will deposit these "cuttings" in the settling trough at the well-head before recirculation into the hole. The outflowing mud is generally passed through a mechanical shaker before it enters the settling ditch, and consequently deposits on the meshes of the shaker a large proportion of the suspended solids.

At regular intervals (usually after drilling every 5ft or so, but sometimes more frequently when detailed information is being sought) a sample of the cuttings is collected from the shaker, washed, packaged and labelled with the depth of the well at the time of sampling. (A mechanical sample catcher has also been developed which permits the semi-automatic collection of selected portions of the cuttings as they fall off the shaker screen. They are directed over the top plate of the device, which consists essentially of two cylinders, the outer one serving as a reservoir into which a continuous water supply is fed, the inner one serving to catch the separated samples.)

In practice, it is usual to take each sample as representative of the formation at the depth (measured by the drill-pipe) reached when it is collected, although some time will have elapsed since it was actually cut, and during this period the bit will have progressed further. For deeper holes, however, the time-lag that occurs in practice for the cuttings to reach the surface can be estimated by adding a small quantity of marking material (e.g. red mica) in the mudstream and noting when it reappears in the ditch. The time taken for a given sample of drilling fluid to pass from the top to the bottom of the hole can be calculated and subtracted from the "round trip" marker travel time, to give the approximate upward travel time of marker and cuttings. There is, of course, always a certain amount of mixing of cuttings in the upward mud flow and a differential speed of movement depending on their size, so that a lithological log based only on cuttings has obvious limitations in its accuracy. A rule-of-thumb for estimating the time for cuttings to reach the surface is about 1 minute per 90ft—i.e. with a hole 9,000ft deep the time-lag would be about 100 minutes.

Samples of all the formations through which the bit passes are obtained in this way; cuttings are collected even when coring is in progress, since complete core recoveries are seldom obtained. In the laboratory, the bit cuttings are then examined after further washing and sieving. Portions of each sample may be extracted with solvents (e.g. chloroform) to detect any trace of oil or bitumen. A binocular microscope with a magnification of about $\times 5$ and a wide field of view is generally used for visual examinations. In an oilfield where drilling is in progress, the resident geologist will have to examine a large number of samples daily, so that a technique that entails minimum fatigue is required, and the binocular microscope is well suited for this purpose.

The following minimum information about each bit cuttings sample is systematically recorded:—

(1) Type of rock—limestone, sandstone, shale, etc.
(2) Physical description—colour, texture, grain size, hardness, etc.
(3) Presence of minerals, organic remains or other special features.

The physical description of a sample of bit cuttings will inevitably vary subjectively according to different observers. It is therefore advisable that a terminology which is as nearly standard as is possible should be established for each area.

The presence of more than one type of rock in specimens of bit cuttings may be evidence that a lithological and perhaps a stratigraphic boundary has been crossed. However, in some formations rapid variations in lithology occur, while there is always a tendency for bit samples to be contaminated by fragments caving from higher up the uncased portion of a borehole.

Bit cuttings may sometimes be used for approximate porosity determinations where suitable cores are not available. For this purpose, a weighed quantity of dried, washed cuttings from which any bituminous matter has been extracted is saturated with water, and the bulk volume of the cuttings is then obtained by removing excess water by suction on a porous plate, after which the volume they displace is measured. The largest cuttings available should be picked for this determination, as far as possible representative of the formation to be studied.

The presence of gas in bit cuttings may be confirmed by subjecting them to pulverizing action, either in a vacuum or under water, with or without heating, analysing any gas evolved for hydrocarbon constituents[19]. (This has been called *gas-in-cuttings* or *microgas* analysis.)

The *geochemical log* has been developed as a technique for detecting the presence of potential source rocks. Measurements are made on cuttings obtained from selected horizons of the total organic carbon content, and this is compared with average values (1·14% for shales, 0·24% for carbonates and 0·05% for sandstones). In addition, measurements are made of the absorbed light hydrocarbon gases present in the cuttings, and the ratios of methane to ethane, propane, etc. derived. Based upon this information, it is believed that both the geochemical evaluation of a prospective area and geochemical correlations between neighbouring wells can be carried out.

Another variety of geochemical log depends upon the use of an automatic instrument for measuring and recording the increasing pressure and hence the rate of CO_2 formation resulting from acid attack on a carbonate cutting sample in a constant volume cell. A rapid differentiation can be made in this way between limestone and dolomite samples.

Drilling-time Record

From the driller's log of each well it is possible to compile a time-footage chart showing the rate of drilling during the period the bit is on the bottom of the hole. Due allowance has to be made for bit wear, shut-downs, "fishing" and other delays, and if this is done a rate of drilling index results which is directly related to the hardness of the formation being drilled.

The drilling-time record can be of great assistance to the resident geologist in compiling a well log. By comparing the relative hardness of the formations penetrated (as shown by the time required to drill each successive foot) with the composition of the samples obtained over the same range, a reasonably close estimate of lithological boundaries may often be obtained.

A difficulty traditionally encountered in the past in estimating drilling times was the driller's tendency to underestimate in his log the actual depth reached. The development of automatic recording devices (geolographs) has removed this source of error.

The recording instrument of the geolograph installation may be placed at any convenient place at the well-head, or even in the resident geologist's laboratory. It is connected by means of a flexible cable to the swivel or travelling block in the derrick. Recording is automatic and continuous on a strip chart marked to indicate 5-min intervals, which rotates on a drum operated by a clock mechanism. Two pens keep a parallel record on the chart, one making a mark each time a foot has been drilled, and the other making a mark whenever the bit is raised off the bottom of the hole. A continuous and accurate check of the drilling time required for each foot of the well's progress can thus be obtained.

The driller's own log can also be utilized to assist in the location of formation boundaries. Experienced drillers can often tell by the "feel" of the drilling string that they are in a particular type of formation, and are aware when the bit passes into a bed of a different lithological type.

Core Analysis[14]

In place of the ordinary drilling bit, a core barrel and annular bit can be used to obtain subsurface evidence. The earliest types of core barrel consisted of a hollow tube with a sharpened edge which was driven into the formation under pressure of the string of tools, and carved out by rotation a solid cylinder of rock which was held by springs until the barrel had been withdrawn to the derrick floor. The modern rotary type core barrel has some additional refinements. For example, there is usually an outer barrel for carrying the tool joint, and an inner barrel for holding the core, with an arrangement of valves which allows the mud to circulate freely while coring is in progress. The cutting edge is arranged in steps, which gives a better recovery. Sometimes two types of cutting head are combined—one with cone teeth for hard rocks, and a second with scraping blades for soft formations. When it is desired to withdraw the core barrel, increased weight is applied and the resulting torque breaks off the core.

Coring is a slower and consequently more expensive process than normal drilling, but it provides much more comprehensive evidence than cuttings can give. The well-site geologist must establish the optimum relationship between footages cored and drilled in consultation with the drilling staff. Cores are taken for the following purposes:

(a) To furnish information about subsurface conditions and in particular to measure dips, provide evidence of faulting, and generally to elucidate the local structure.

(b) To determine important formation contacts.

(c) To establish horizons suitable for casing seats prior to cementation.

(d) To examine and to determine the properties of all reservoir rocks, in particular their porosity and permeability.

(e) To provide rock specimens for special study, e.g. mineral and micro-fossil examination.

In exploratory wells, 50-ft long cores are taken relatively frequently, since the maximum information is required about the formations being penetrated. Occasionally, periods of continuous coring may be resorted to at depths which have been forecast as being near to an important horizon. However, because of the financial considerations involved, it is evident that coring policy must be dependent not only upon geological considerations, but also upon relevant economic and mechanical factors.

Devices have been developed whereby small-diameter cores can be taken in rotary wells while the bit continues to drill without interruption. In principle, a small core barrel is lowered on a wire line down the drill stem, and by a latching arrangement rotates with the bit, cutting a pencil of core (about 1½in in diameter) through a special aperture. An overshot can later be run down the stem to recover the barrel. This procedure, is of course, far quicker and more economical than full-size coring, but has the disadvantage that the cores obtained are correspondingly small.

Another increasingly used device is the "wall corer" which enables a triangular strip of rock material to be taken from the wall of the hole when drilling is suspended.

It is also possible to take samples from the uncased sections of a hole after the well has been completed. This technique must be used when, for example, it is desired to re-examine an old well whose drilling records have been lost, or when it is suspected that an important horizon has been passed. Using the "gun" type sample-taker, for this purpose, it is lowered into the well to the required depth, and a hollow bullet is then fired into the walls of the hole. During propulsion, the bullet bottom cap is forced off so that the mud cake on the wall of the hole, which is first collected, passes through, and a true sample of the formation is trapped in the bullet. The bullets are mounted in batteries in a "gun", and when this is withdrawn from the well the bullets can be recovered with their contents, since each is attached to the "gun" by wire springs.

When a core is recovered for examination at the surface it undergoes a rapid pressure and temperature decline. Any gas present therefore expands and tends to escape, and also some of the oil. The interstitial water may also be contaminated by the entry of drilling fluids, unless the precaution has been taken of using an oil-base rather than a water-base drilling mud when drilling through the potential reservoir formation. Techniques are available for minimizing the loss of a core's fluids by freezing it quickly at the surface, or by sealing it in the subsurface in a special container under approximate reservoir conditions.

Any gas tending to be evolved when fresh cores are broken at the surface may be retained for analysis by using special pressure storage techniques for transporting the cores to the laboratory. Here, the cores may be degasified completely by steam heating or solution in acids. With rocks of low permeability, there may be no apparent evidence of oil when cores are first examined, but drops of oil will appear subsequently on fresh surfaces as the oil is forced out under surface conditions. These are called "weeping" or "bleeding" cores.

For the detection of traces of oil in cores the *ultra-violet lamp* is a very useful field instrument whose application was described on p. 180. A few parts in a million of oil can be easily detected by this means, and ultra-violet light is also useful for revealing planes of cleavage and dip in apparently compact cores.

The use of core samples in place of or in addition to bit cuttings does not entirely remove the risk of contamination, since drilling mud is liable to permeate the surface of a porous core. It is for this reason that the outer surface of the core sample must be scraped clean of mud, and it is always advisable for the core to be split open so that internal portions can be selected for detailed examination.

The unused parts of the washed core should be stored in suitable metal or wooden troughs, both ends being carefully labelled with their respective depths to obviate later confusion. It is usually advisable to keep these trays on suitable shelves in a separate core house. The length of core obtained is measured in each case so that the recovery can be calculated, due allowance being made for any portions removed. This is, of course, the ratio of the core recovered to the length of the core bored, expressed as a percentage. In firm formations and under favourable conditions recovery may be nearly 100% but it is usually of the order of 50 to 60%, and in shattered zones, poorly-cemented rocks or under difficult drilling conditions, it may approach zero. Hence, the value of the bit cuttings taken while coring is in process.

The lithology of the core specimens is recorded along the appropriate section of the geological well log in a similar way to that employed for bit cuttings; that is to say, a description is given of colour, texture, hardness, composition, and any special features observable, such as bedding, lamination, dip or cleavage. The latter features are of special importance, as they afford evidence of the conditions of sedimentation. Photographs of the core under both white and ultra-violet lights are often made for record purposes.

Clay-bearing cores may be handled under controlled-humidity conditions to obviate excessive drying of the clays; the clay content may be determined by centrifuging a sample-water slurry over a non-miscible liquid. The clay fraction is recovered in the water and heavy minerals in the denser liquid. The presence of swelling clay layers is important in evaluating fresh-water sensitivity.

Core samples may be subjected to a variety of laboratory tests. Thus, portions are extracted with carbon tetrachloride or similar solvents to detect traces of oil or bitumen, while other portions are examined for microfossils and heavy mineral residues (Chapter 8.)Any particularly porous sections are noted, and porosity and permeability measurements are made on samples which appear to indicate promising reservoir rocks.

The effective porosity of a solid core sample can be measured by determining the volume of a suitable fluid held in its pores under known conditions. One method of doing this is by *liquid absorption,* a technique in which a convenient unit of rock is caused to absorb the maximum possible volume of a non-reactive liquid, such as benzene. This volume of liquid is then equal to the volume of the connected pores. If the rock sample is of regular dimensions, its overall volume can then be determined by mensuration; if not, by liquid displacement after coating the surface. Another effective technique of measuring the porosity of a consolidated rock is by *gas expansion,* while a gas compression method is also sometimes used.

In order to measure the permeability of solid specimens of compacted rock, suitably homogeneous samples are chosen from the cores, and "plugs" of these are cut with a diamond drill to the desired cross-section, and then cut into pieces about 1in long. The prepared samples are then extracted with carbon tetrachloride, dried at 110°C to evaporate any moisture, cooled, and their dimensions exactly measured. If water is to be used as the test fluid, the cores are first soaked in distilled water for several hours before mounting, to ensure complete saturation.

There are several varieties of Permeameter. In a typical instrument, the rock is held in a rubber stopper, wedged tightly into the core holder by means of thumb-screws attached to the brass riser connected to a flow line from the fluid reservoir. Connections lead to a manometer which records the differences of pressure between the two ends of the core; the exit tube leads to a container in which liquid test fluids can be weighed.

Each sample should be tested with both air and water, over a series of pressure heads, the volume of fluid flowing through the sample in a given time at constant head being measured in each case. The pressures used should be the lowest range possible to give reasonable flow rates under non-turbulent conditions.

Where clay is present in a sandstone, the permeability of the core to water will decrease with time, due to the swelling of the clay which gradually obstructs the pores. But the permeability will remain constant if the sample has been well saturated for a considerable period before test; the permeability remains constant when dry air is used. These measurements are of course only measurements of the permeability of samples of the reservoir material relative to air or water, rather than to the actual reservoir fluids. The data for air, however, do provide an index which is comparable as between different specimens, while the permeability to water is of great importance in any consideration of secondary recovery by water-flooding processes.

In general, the rock permeability for a gas is higher than for a liquid, due to "slippage" of the gas along the rock pore walls. The measurements may be related by extrapolating the curve obtained by plotting air permeability values against the reciprocal of the mean pressures to a theoretical infinite pressure point at which a gas would behave as a liquid.

Permeability estimates are needed before any forecast can be made of the productive capacity of a reservoir. As many direct measurements as possible of both vertical and horizontal permeabilities are therefore made on suitably-spaced core samples, and as much supporting evidence as possible is also obtained from instrumental well logs (see below).

(However, it must be remembered that examination of such core samples can be of only limited value in estimating the overall permeability of a reservoir, because of the relatively small proportion of the reservoir bed that can be examined in this way. Once oil production has started, the estimated permeability of the producing formation can be checked from the output of a well under a known pressure draw-down or from the interpretation of pressure build-up data.)

Interstitial water saturation can be measured at the surface by heating solid core samples and distilling off their water contents. After condensation and weighing, a percentage saturation can be deduced, related to the effective porosity of the samples. Such computations can only be approximate, since it is nearly inevitable that the core will have been contaminated to some degree by the invasion of water from the drilling fluid and there may also have been the release of some constitutional water. More accurate calculations of the *in situ* water saturation (as well as the saturations of oil and gas, if present) can be obtained from the analysis of the appropriate electric log (see below).

Where a core has been taken of a reservoir bed, it is possible to run an oil recovery test on a suitable portion of it, in order to make some estimate of the potential recoverable oil in the formation. To do this, the prepared sample is mounted in a pressure vessel and water allowed to enter, driving the oil out into a weighed receptacle. After complete flooding, the water-saturated core is removed and broken up, and the residual oil removed with solvents. In this way, an approximate estimate can be made of the proportion of recoverable oil in the reservoir. Obviously, a number of cores, suitably spaced out, would have to be examined.

A technique of impregnating carbonate rock specimens with plastic or resin, which is forced under pressure into the void spaces in the rock and hardens in position before the carbonate matrix material is dissolved away by acid treatment, has been used with some success to study the porosities and pore forms of important reservoir rocks.

Other physical properties of core samples which are also sometimes measured include their electrical resistivity, radioactivity and direction of magnetization.

The *bacterial log* (or "bio-log") has been used in the USSR as a standard well-logging technique, using core samples from known depths. It depends on the same principles as the surface bacterial survey, but there must remain a doubt as to the presence of indubitably indigenous microflora at depth.

The *petrographic* ("Thin Section") examination of compact core samples is of great value in studying the fabric of a rock and elucidating its history and relationships. Thin sections, when viewed under the petrological microscope by transmitted normal and polarized light, can give useful information about the grain sizes of the constituent minerals and the type and degree of cementation present.

A small piece, a few millimetres thick, of the core specimen to be examined is sawn off and its flattest side polished on a succession of progressively finer lap wheels before being mounted, flat side downwards, on a glass slide. The mounting medium used is usually Canada balsam, which has to be "cooked" to a suitably tacky consistency. After cooling, the reverse side of the rock specimen is ground in several stages, until ideally a uniform thickness of about 0·03mm is attained. This can be checked by optical methods as the grinding operation proceeds. The thin section is then covered by a glass cover slip, cemented in place again by Canada balsam, and the slide is then ready for optical examination by standard petrographic techniques.

Several methods have been devised for classifying sedimentary rocks—particularly sandstones and carbonates—on the basis of their microscopic appearance, or on their content of granular and cementing materials. For carbonate rocks, in particular, the most effective classifications make use of the traces of the original depositional features which can usually be seen in thin sections, in spite of later diagenesis (reference to such classifications was made in Chapter 3).

In addition to optical examination, the technique of *stain analysis* of thin sections (in this case using uncovered specimens) has proved of value, using a variety of staining agents, which permit the rapid distinction between such apparently similar materials as calcite and aragonite, calcite and dolomite, and the three groups of clay minerals—kaolinite, illite and montmorillonite.

Fluid Samples

The fluids recovered from the well itself provide an important source of subsurface information. Correlations can often be based on the fluid contents of different wells—bituminous zones, petroliferous sands, waters of similar salinity and chemical contents, high-pressure gas horizons—all can furnish significant "marker" levels.

Before development drilling commences, the levels at which fluids are likely to be met are calculated, and the casing programme provisionally arranged on this basis. Subsequently, as drilling progresses, these estimates must continually be revised so that high-pressure zones are not encountered unawares. Discretion must of course be used in attempting to correlate by fluid content, and the possible effects of faults and of local fluid pockets must be borne in mind.

Fluid samples are collected from the subsurface by a "formation" or "drill-stem" tester—a device which consists in principle of a tube which can be set against the formation it is desired to test, and from which the drilling fluid can be excluded by a system of valves. The general practice is to obtain fluid samples from any formation entered by the bit which appears to be "making" liquid into the hole.

Chemical analysis may often be helpful in arriving at a decision, particularly in relation to oilfield waters, whose nature and geological significance[12] were referred to in Chapter 1.

Most oilfields possess a central chemical laboratory, to which the subsurface geologist can refer water samples for analysis. Contact between such waters and oil-bearing formations has always to be avoided, since the entry of water into the pores of a reservoir rock may obstruct the subsequent flow of oil, form troublesome emulsions, and, if the water contains dissolved sulphates, chemically damage the crude.

The simplest way to compare the salt contents of different brines is to express them in terms of milligrams/litre or parts of solute per million of solution (ppm). The relationship is $\text{ppm} = \dfrac{\text{mg/l}}{\text{specific gravity of solution}}$ and hence for low concentrations the terms are virtually interchangeable. Table 34 gives the analyses of a few typical oilfield waters and shows the widely varying amounts and proportions of their constituent solids. The total salt content is usually more than 6% by weight and may be as much as 20%.

The use of water analyses and the related drawing of "iso-concentration lines" can provide some interesting structural information. For example, there is a normal increase of salt concentration with depth, possibly due to gravitational concentration, but more probably related to the effects of longer

TABLE 34

COMPOSITION OF SOME OILFIELD WATERS

(Quantities expressed as parts per million)

Field and source of water	$Na^+ + K^+$	Ca^{++}	Mg^{++}	SO_4^{--}	Cl^-	HCO_3^-	Total Solids
Rangely, Colorado, Weber	37,725	3,509	568	973	65,000	565	108,053
Lander, Wyoming, Tensleep	34	41	17	13	31	230	249
Kawkawlin, Mich., Dundee	66,280	25,740	4,670	155	161,200	60	258,105
Bay City, Mich., Salina	21,383	206,300	7,300	0	403,207	1,208	642,798
East Texas, Texas, Woodbine	24,540	1,388	282	278	40,958	569	67,649
For comparison: Average Sea Water (approximate)	11,000	420	4,300	2,690	19,350	140	35,000

Source: G. D. Hobson "Some Fundamentals of Petroleum Geology", 1954, p.19.

burial and thus longer exposure to mechanisms of concentration. In some cases, however, where more dilute waters are found at greater depths than more concentrated waters, the explanation is the existence of an unconformity which marks an erosion surface permitting dilution of the older waters before reburial.

Water analyses may also be used for local correlation purposes between neighbouring wells and for the identification of particular reservoirs when several exist. (The significance of water and other fluid pressure measurements is discussed in Chapter 8.)

INSTRUMENTAL ("WIRE-LINE") WELL LOGS[18, 21, 29]

A number of specialized devices have been developed for measuring and recording important properties of the rocks through which the drill has penetrated. Some of these methods of "well logging" have proved so successful that they are nowadays used on every exploration and development well as a part of normal drilling practice, and new techniques are being continually developed to provide more information and greater accuracy. In principle, instrumental well logging methods have the same objective as the geological log, i.e. the eventual production of an accurate subsurface map on the basis of correlations, together with the provision of information about the porosities, permeabilities and fluid contents of the strata penetrated. Since these logs are produced by quasi-direct physical contact with the layers of rock adjacent to the walls of the well, they may be expected in general to be more precise than logs which depend only on surface information. They all involve lowering a special instrument on a calibrated wire-line down the hole, for the most part after drilling operations have been at least temporarily suspended and the drilling string of tools removed, but while the hole is still filled with fluid. Some instruments can be used even after the hole has been cased; none, however, can detect the presence of oil or gas directly but only by inference. They may nevertheless be used to evaluate reservoir rocks[3, 7].

Electrical Logs[32]

The method originally devised by C. and M. Schlumberger for electrically surveying a borehole was used first in the *Pechelbronn* field in 1929,[24] and since—with many modifications and improvements—has become a part of standard drilling practice. The well logs obtained by the electrical survey are of great assistance to the well-site geologist, since they frequently enable him to locate lithological boundaries and to determine correlations between wells. Furthermore, oil and gas zones which might otherwise have been overlooked may often be detected by this technique.

The two parameters which are determined can be considered for practical purposes to be (i) the apparent resistivity of the formations close to the hole, and (ii) the spontaneous potentials (S.P.) set up by the physical interaction of the drilling mud and strata of the subsurface. These potentials are related to the permeability of the rocks, and therefore this second parameter is to a certain extent an indication of permeability—although to speak of it as a permeability or porosity survey is incorrect.

The Resistivity Log

The *resistivity* of a rock is defined as the electrical resistance of a cylinder of rock of unit length and unit cross-sectional area, expressed in ohms–m^2/m (ohm-metres). Unless the rock contains a metallic ore, its conductivity is almost entirely electrolytic, i.e. dependent on the presence of fluid in the pore space. Any measured resistivity will therefore vary with the porosity of the rock and the degree of saturation and nature of the salt content of the contained fluid. In other words, the specific resistivity of a rock is inversely proportional to the amount of fluid contained in each cubic metre of the rock and to the conductivity, i.e. the salt content of the contained fluid. Thus, a bed saturated with salt water has a low resistance, but if filled with fresh water its resistance is relatively high, and if filled with oil (a poor conductor) its resistance is very high indeed.

The "formation resistivity factor" (F) is defined as the electrical resistivity of a rock saturated with a conducting electrolyte (R_t), divided by the resistivity of the electrolyte (R_w). As might be expected, the formation resistivity factors of most reservoir rocks increase as the porosity (and hence the degree of cementation) decreases, and *vice versa*. Thus, $R_t = FR_w$, and also $F = P^{-m}$ where P is the rock porosity and *m* its cementation factor (lying between 1 for non-cemented, and 3 for highly-cemented rocks).

In order to measure this important formational resistivity, an instrument assembly containing several electrodes is lowered down the hole on a special cable. (The three-electrode system is most commonly used, but systems of one, two or four electrodes have also been employed.) The holes must be filled with water-base mud, to ensure good electrical contact between the electrodes and the formations making up the walls of the hole. The cable used includes conductors, each connected to an electrode, passing at the top of the hole over a measuring pulley on to a large drum, whose rotation can be conveniently controlled by a special gearing in the motor truck on which the apparatus is carried. It is generally operated by specialized logging contractors.

In Fig. 55, the electrode A of a typical three-electrode assembly is shown connected by cable to one pole of a source of direct or alternating current of voltage E. The other pole of the current source is connected to earth near the mouth of the hole—usually the top string of casing. The current coming from A may be considered as spreading spherically outwards, giving rise to a series of equipotential surfaces centred on A. The other two electrodes, M and N, are suspended close to A as shown. The distances AM(r) and AN(r^1) are constant and are relatively large compared with the hole diameter. The potential difference measured between M and N is equivalent to the potential difference measured between any other two points on the same equi-potential surfaces, i.e. it corresponds to the potential difference inside the formations between points distant r and r^1 from A.

The potential difference between M and N may be measured by a potentiometer at the surface. Suppose it is $\triangle v$. If the current flowing from A is I, and the resistance between the two shells of rock radii r and r^1 is R, then $R = \triangle v/I$.

Assuming the rock has a specific resistivity ρ, the resistance across a thin shell of thickness $\triangle r$ will be $\rho \triangle r/4\pi r^2$.

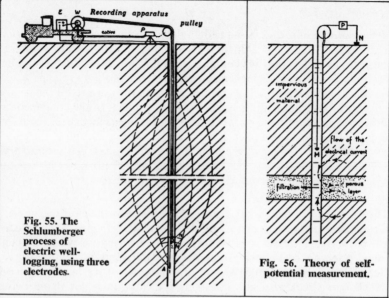

Fig. 55. The Schlumberger process of electric well-logging, using three electrodes.

Fig. 56. Theory of self-potential measurement.

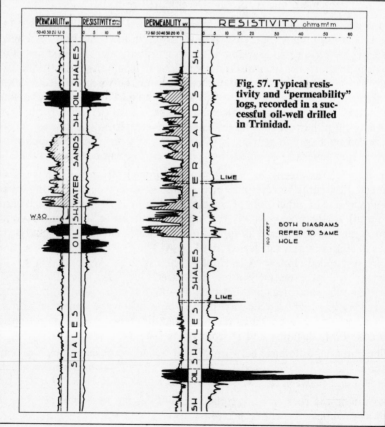

Fig. 57. Typical resistivity and "permeability" logs, recorded in a successful oil-well drilled in Trinidad.

BOTH DIAGRAMS REFER TO SAME HOLE

The resistance between shells of radii r and r^1 will therefore be:-

$$R = \frac{\rho}{4\pi} \int_{r}^{r^1} \frac{\triangle r}{r^2} = \frac{\rho}{4\pi} \cdot \frac{r^1 - r}{r \, r^1}$$

Hence $\rho = K. \triangle v/I$ where K is a constant.

Since r and r^1 are known from the construction of the instrument, K is known. The values I and $\triangle v$ may be measured at the surface and a continuous oscillograph trace of the resistivity ρ of the formations can therefore be made as the electrodes move from the bottom of the well upwards at the rate of several thousand feet an hour. (Readings are taken as the electrode moves upwards to allow for the elastic elongation of the wire.)

The resistivities recorded in this way are not, as we have seen, the true resistivities of the strata but only their *apparent resistivities*. Various factors such as the diameter of the hole, the nature of the mud, and the thickness and homogeneity of the beds would have to be taken into account before true resistivities could be obtained. However, the apparent resistivity can be used as a comparative index of the resistivity of different types of rock, and hence as a comparative measure of the saturation and salinity of the fluids they contain.

The Spontaneous Potential Log

The potentials spontaneously generated at the surface of strata traversed by a mud-filled drill hole are believed to be due to electro-filtration and electro-osmosis phenomena.

It was discovered as long ago as 1859 that electric potentials could be generated by the passage of water through a porous layer. This electromotive force is proportional to both the amount of the liquid filtering through and to its resistivity. A current flows in the same direction as the fluid, and its magnitude is evidently related to the permeability of the rock.

This effect is particularly important where the mud used is made relatively heavy to avoid caving and blow-outs, and thus filtrate from it tends to penetrate the walls of the hole. In the opposite case, when the fluid flows from the walls of the hole into the well, the electro-filtration current will be reversed in direction.

Quantitatively, the difference in potential can be expressed as:-
$$E = K\rho \, (p^1 - p)$$
where p^1 and p are the pressures of drilling fluid and formation fluid, ρ is the resistivity of the percolating liquid, and K is a constant.

Electrochemical potentials are generated at the plane of contact between solutions of dissolved salts, and hence at the wall surfaces of the hole where drilling and formation fluids of different salinities come into contact. The electromotive force resulting is proportional to the logarithm of the ratio of the resistivities of the two electrolytes:- $E^1 = K^1 \log \rho_1/\rho$.

When the rock fluid is more saline and consequently has a lower resistivity than the drilling fluid, this potential will be negative.

In the Schlumberger system of measurement, an electrode made from a material having a low electrochemical contact potential, such as lead, is used, in the form of an electrode enclosed in a porous container filled with lead acetate or lead nitrate solution. This is lowered into the hole on an insulated cable, the other end of which is connected to one side of a recording potentiometer at the surface. The other side of the potentiometer is connected to a second electrode grounded at the surface where the potential is assumed to be zero. The readings of the potentiometer as the electrode is moved up the hole are therefore a measure of the relative potential existing at the surface of the formations traversed (Fig. 56).

The potentials recorded are the algebraic combination of the electrofiltration and electrochemical effects and can be used as a qualitative indication of the permeability of the strata contiguous to the walls of the hole; but they are not a quantitative measure of permeability. The greatest value of the S.P. log is for comparison, differentiating in particular impermeable clays from permeable sands.

Other Electric Logs

Many variations of these basic electric logs are now available. The *conventional electric log* generally used combines an S.P. log with three standard resistivity curves—"short-spacing" (two-electrode) to define bed boundaries; "long-spacing" (two-electrode) to determine resistivity; and a "long-lateral" (three-electrode) curve. It is most effective with fresh mud and fairly thick formations.

The "ultra-long spaced" electric log (ULSEL) is obtained by the use of two potential-measuring electrodes mounted on a 5,000ft bridle lowered down the well and held at various distances above the current electrode. It is claimed that the long spacings pick up anomalies in resistivity differences, which when examined by computerised techniques provide a means of looking beyond the well-bore to a distance of as much as 1,800ft, and thus locating salt domes, isolated oil sands or faults.

The *induction log* energizes the formation by creating an alternating magnetic field between the instrument and the rock. It has a number of advantages, notably its ability to record resistivity curves in gas-filled holes, or where a non-conductive oil-base mud has been used. The combination called the *induction-electric* log, i.e. the induction log combined with the normal "short" and S.P. curves, is nowadays the most widely used well-surveying combination where normal fresh-water drilling fluids are employed.

With salt-water fluids, which are more conductive than most rock formations, the *laterolog* is of value, the current applied being sharply focused to eliminate the effect of the mud column. The *microlog* and the focused *microlaterolog* measure the resistivities of small sections of rock around the borehole. They can give a broad indication of the permeability by measuring the resistivity of the rock segment that is penetrated by the mud.

Interpretation of Electric Logs

In practice, both resistivity and S.P. logs are taken simultaneously and automatically recorded in the surface apparatus as continuous graphs with a common depth ordinate. Apparent resistivity (in ohms–metre^2/metre) and spontaneous potentials (in millivolts) are plotted as abscissae on the two sides of the depth ordinate. The pattern produced is as shown in Fig. 57.

TABLE 35

GENERALIZED RELATIONSHIP BETWEEN ROCK LITHOLOGY
AND ELECTRIC LOGS

Lithology	Resistivity	S.P.
Clay, shale	Low	Low
Sand, salt water	Low	Very high
Sand, fresh water	High	Medium
Sand, oil or gas	Very high	Very high
Limestone, compact	High	Low
Limestone, porous	Low	Very high
Limestone with oil	Very high	Very high

Exact marking and calibration of the cable used are necessary, since the depths must be accurately and continuously known. The scales used are usually 1/1,000 for general surveys, rising to as much as 1/150 for specially detailed work.

Electric logging runs are usually made several times in the course of drilling, in uncased sections of the hole, with the drilling tools withdrawn. A variation of the usual technique has also been developed which enables electric logs to be taken while drilling is actually in progress. This method utilizes the bit as an electrode, insulating it from the rest of the drill column and connecting it electrically to a potentiometer at the surface.

The interpretations of the electric logs produced are in general simple and positive. Correlations can be made between neighbouring wells using peaks on the curves (Figs. 59, 60) and in some cases these correlations can be extended over substantial distances, and are of great assistance in structural evaluations. Faults are also often revealed by the logs. In regions of rapidly changing lithology and lensing formations correlation is often difficult, but "electrical marker beds" may frequently be noted whose characteristics persist over a sufficiently wide area to be of value for correlation purposes.

The interpretation of the individual log is always done in conjunction with the other evidence available—i.e. that of cores, bit cuttings, etc. It can generally be assumed that a high resistivity indicates either a non-porous formation or a porous formation containing a non-conducting fluid (i.e. oil and gas, or to a lesser degree fresh water). A low resistivity shows a porous formation containing a conductive fluid (i.e. salt water). Similarly, a high S.P. value shows a permeable formation, and vice versa (Table 35).

The precise lithological significance of the log can be determined only in relation to experience and local geological evidence. In particular, the properties of the drilling fluid being used must be borne in mind. The effects of mud penetration upon the S.P. curve are more important than the resistivity effects, since the latter may depend on a larger volume of rock going some distance back into the formation, whereas the former is measured at the walls of the hole. When water from the mud filters into the formation and pushes back the formation fluids, the magnitude of the current generated will depend not only upon the permeability of the formation, but also on the time of contact and the salinity of the mud. A poor mud will lose its water rapidly to the formation because it does not form a good "filter cake" and hence the E.M.F. recorded will therefore be high. The use of a good mud would result in a lower porosity being recorded for the same rock. The specific gravity and

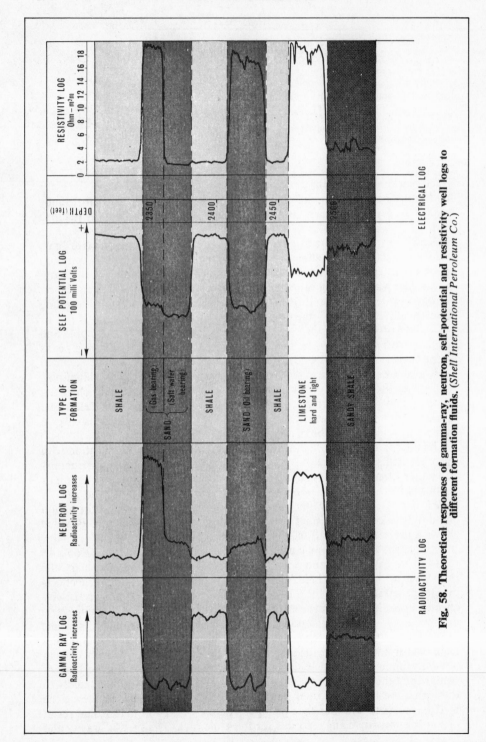

Fig. 58. Theoretical responses of gamma-ray, neutron, self-potential and resistivity well logs to different formation fluids. (*Shell International Petroleum Co.*)

hence the pressure of the mud column will also affect the rate at which fluid filters into the formation. Long contact of mud with the formation tends to lower the recorded "porosity", since the capillary spaces will become gradually choked and the rate of fluid flow consequently diminished.

An important application of electric logs is their semi-quantitative use for the calculation of the fluid contents of a rock section.

The relationship involved for the salt water content of a sandstone reservoir (S) may be written as follows:

$$R_t = \frac{R_w}{P^m S^n} \text{ or } S = (FR_w/R_t)^{\frac{1}{n}}$$

where R_t is the true resistivity, derived from the log, expressed in ohm-metres;
  P   is the porosity obtained from core analysis, the S.P. curve or other logs;
  R_w is the resistivity of the formation water, obtained from the S.P. log or by analysis; this is also a useful index for comparing different waters;
  m   is the cementation factor for the particular reservoir rock; and
  n   is the saturation factor (usually between 1·9 and 2·0).

Where S is less than 40% of the total pore space, it can be assumed that the remainder of the pore space is filled with hydrocarbons, and hence that the reservoir will probably be an economic oil or gas producer.

Specific permeability can also be found directly from electric log data by the use of various empirical formulae. The following equation has been claimed to give results for water-wet, shale-free sands which are close to those obtained by core analysis[16]:-

$$K = 3·3 \times 10^3 (R_w^2 R_t)/R_o^3$$

where K = specific permeability, md
  R_w = resistivity of formation water, ohm-metres
  R_o = resistivity of formation saturated with formation water
  R_t = true resistivity of non-invaded formation containing irreducible water saturation, ohm-metres

Nuclear Logs[4, 13, 28]

Since sedimentary rocks contain varying amounts of radioactive elements derived from the crystalline rocks from which they were formed, or from their organic matter, the *relative radioactivity* of the sediments encountered in a well can be used as a comparative index of rock properties.

The *gamma-ray log* measures a physical parameter of the sediments concerned, and not, as is the case with the electric log, a product of the interaction between the sediments and the drilling fluid. The apparatus used consists essentially of a sensitive gamma-ray radiation detector with an amplification system, which when passed up the hole produces a record at the surface of the gamma-radiation outputs of the strata.

Gamma-ray logs can be made through casing and are little affected by the fluid in the hole. Correlation between electric logs run in open holes, and gamma-ray logs run through casing may be difficult; peaks can sometimes be correlated (Fig.58), but they do not always coincide in depth. Thin shale

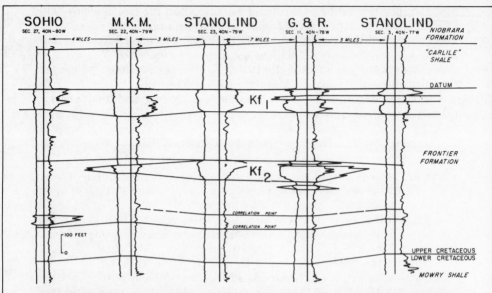

Fig. 59. Example of correlation by electric logs: west-east cross-section of Salt Creek field, showing correlation of First and Second Frontier sandstones (Kf¹ and Kf²).

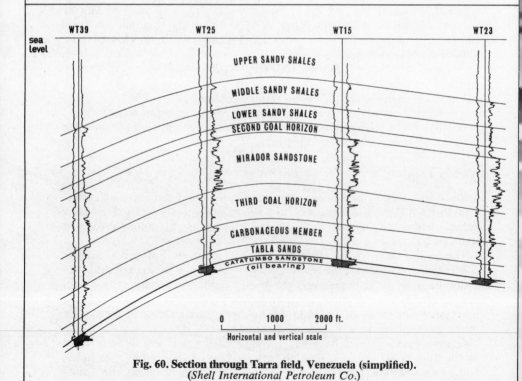

Fig. 60. Section through Tarra field, Venezuela (simplified).
(*Shell International Petroleum Co.*)

breaks in a thick sandstone are reliable gamma-ray markers, and shales in general are more reactive than sands, and sands than limestones.

The *gamma-density log* uses collimated sources of radiation such as cobalt-60 and caesium-137, with a scintillation counter to receive the radiation rebounding from the first few inches in depth of the formations surrounding the hole. If calibrated in the well against a bed of known density, this log can give a valuable indication of porosity, and has been used in combination with the electric induction survey for locating gas-saturated sands in a gas-filled hole.

In the *neutron log* technique, the subsurface formations are themselves exposed to a source of primary (neutron) radiation contained in the instrument, which can also detect the secondary radiations generated by the interaction between the neutrons and the atomic nuclei of some elements present in the formation fluids.

The instrument used contains a strong neutron source,* suitably shielded from an ionization chamber and amplifier. The fast neutrons strike the hydrogen atoms of the formation fluids or undergo inelastic collision with the constituent rock atoms, giving rise to secondary, or slow neutrons. These slow neutrons are captured by the atomic nuclei of some of the elements present in the rocks, and gamma-rays are produced which can be detected as in the gamma-ray technique. The neutron log in practice chiefly measures the amount of combined hydrogen present in the formations and is of value for its capacity to distinguish limestones from sandstones and shales from other materials mixed with shales[22].

Considerable success has been achieved by using the gamma-ray and neutron curves (Fig. 58) in conjunction for defining porous zones containing fluids. Dense formations such as anhydrites or compact limestones are logged as high-intensity peaks on the neutron curve, while porous formations show low readings. The gamma-ray log can distinguish between porous formations on the neutron log and shales, which also log low on the latter.

Since the fluid content of the formations will be either salt water or hydrocarbons, the neutron log is essentially a porosity log and does not differentiate between the different fluids. However, this type of log is influenced by the presence of certain elements with high neutron capture cross-sections in the fluid or surrounding rocks, and also responds to the presence of chemically bound water (e.g. in gypsum) or water adsorbed in clays or shales.

The *neutron-epithermal neutron log* has certain advantages, since it uses a lithium iodide scintillation counter to measure neutrons just below the thermal level, where the probability of a capture reaction is small, and it is therefore insensitive to variations in the chemical constitution of the rock and contained fluids.

The *neutron-gamma log* is particularly influenced by the capture cross-section of chlorine and can produce a chlorine (i.e. salt water) saturation curve, which by comparison with the neutron-neutron hydrogen (i.e. hydrocarbon) curve can be used to separate oil and water zones. "Dry" gas formations will give high values on both curves, due to the low hydrogen content, but the chlorine curve will be the more influenced of the two.

---

*Usually a radium-beryllium mixture. Californium-252 has also been used experimentally; it has 300 times the emission of conventional sources of neutrons.

The *neutron lifetime log* uses a pulsed high-energy neutron source and measures the rate of decay of the neutron flux during the interval between pulses, in this way largely eliminating borehole effects. The recorded "capture cross-section" of the elements present relates chiefly to the chlorine atoms in the formation water; information as to the presence of hydrocarbons is given by the reducton of apparent water salinity.

Other logs of this type used in modern practice include the *oxygen* and *silicon activation logs,* in which a very high-energy neutron source is employed to activate silicon and oxygen atoms in the strata penetrated, the effect being measured by the gamma-rays produced. Lithological variations or hydrocarbon saturation may be detected in certain circumstances by this technique.

The *nuclear magnetism log* uses the magneto-mechanical properties of the hydrogen nucleus to measure the quantity and quality of the fluids in a formation. It is claimed that measurements are obtained representing the pore space which is filled with fluid that is free to move. A subsidiary test of the thermal relaxation time of intervals of special interest may then distinguish zones filled with oil and gas from zones filled with water. The instrument contains a coil system which is lowered into the well on a wire line and applies a magnetic field to the formation; on collapsing the field, precession of the protons in free fluids generates small signals which are detected, amplified, and recorded.

The system is, of course, only responsive to the element hydrogen, since the other elements met with in subsurface logging have weak or nil magnetic spins. The material containing the hydrogen must be in the fluid state, so that materials that are bound to the surface of the reservoir mineral particles do not give a measurable signal. However, no distinction is made between hydrocarbon gas and fluid oil or water.

Acoustic Velocity ("Sonic") Logs[15]

The velocity of sound through a subsurface formation is a function of the elastic properties of the rock and also of its porosity and fluid content. The modern compensated type of acoustic logging device, drawn up the hole on a calibrated wire line, emits a pulse of sound energy every 1/10th sec from two transmitters set above and below the receivers. A surface computer averages the transit times for the receiver pairs and divides by the span to give the transit time, and the transit time is also integrated to give total travel time. The resultant acoustic velocity log affords an excellent way of estimating porosity within the ranges of 5–30%, and in combination with the electric resistivity log, of measuring oil or gas saturation in "clean" formations.

The transit times through oil- and gas-saturated sands are different, so that the sonic log has been used with some success for locating gas-oil interfaces and differentiating between oil and gas reservoirs.

The *acoustic borehole televiewer* may be used to scan the walls of a borehole with bursts of sonar energy, analysing the reflected "picture" by a ceramic acoustic lens and transmitting it to the surface by continuous telemetry. The log is obtained even in the presence of drilling fluid or crude oil and is reported to indicate fractures, vuggy porosity and bedding planes in uncased holes.

A slim borehole televiewer has been developed which is only 1·75in o.d., compared with the usual 3⅜in o.d., and can thus be lowered even through 2in tubing installed in a production well, to observe fractures, perforations, or the condition of the hole below the casing.

Thermal Logs

Sensitive recording thermometers lowered into the borehole permit continuous records of temperatures to be made. These instruments depend on the variation of the resistance of an electric conductor with temperature, and record these variations at the surface in the form of potential differences, with an accuracy of the order of 0·1 °C, and a logging speed of up to 5,000ft/hr.

The instruments actually record the temperature of the drilling fluid column at a given point, rather than the formation temperature. Two sets of circumstances are possible, dependent upon whether the drilling fluid in the well is in thermal equilibrium with the formation or not.

Conditions of complete thermal equilibrium are rarely met with in practice, since not only must the column of fluid in the well stand undisturbed for a considerable time to take up the formation temperatures, but the formation itself in the neighbourhood of the well will need time to recover its thermal equilibrium after the disturbing effects of drilling. If such an equilibrium were attained, the temperature gradient in the subsurface would be found to depend on the thermal conductivity of the rocks and the local tectonics, and its value would vary between different formations, each gradient change corresponding to the boundary between major lithological and thermal zones[11, 23].

The state of affairs much more commonly encountered, however, is that existing when wells are in various degrees of thermal evolution. During drilling, the temperature of the drilling fluid differs at most points from that of the formations penetrated. When circulation is stopped there is an exchange of heat between the formation and the mud. This exchange will vary with the nature of the formations, and also with the depth at which it is measured. Supposing that there is a steadily rising formational temperature from the surface to the bottom of the hole, it is likely that the mud will be losing heat to the formations in the top part of the hole and absorbing heat from the lower beds. The rate at which heat is gained or lost depends upon the local lithology, and the convection currents which can flow in permeable beds tend to keep such strata cooler than a relatively warm contiguous mud and vice versa. A valuable means thus becomes available for detecting *water sands* on the geothermal log, and where a stratum contains oil and gas which escape even in small quantities into the well, another thermal anomaly results. The expansion of the escaping gas causes a local lowering of temperature in the mud column, and in limestone fields where production is related to reservoir fissuring, such thermal evidence is particularly valuable. The temperature drops involved are very small—of the order of ½°C—but the instruments used are capable of detecting them.

Temperature measurements can also be made in cased holes, although the casing will exert a damping effect. The level of the top of the cement collar set in the annular space between the casing and the walls of the hole can also be found by the evolution of heat during the setting process.

Other Well Logging Methods

A number of other instruments for examining boreholes are available, none of which are as widely used as the devices already described. Various types of *cameras,* including televiewer devices, have been devised for visually scanning the walls of a borehole; their application is obviously limited in opaque fluids[33]. A *photoelectric instrument* can locate water flowing into a well by the increased translucence of the drilling fluid near the point of entry. The points at which fluids (including gases) are entering or leaving a well can also be detected by the use of a propeller-like *mechanical spinner,* which can be lowered down the hole on a cable and transmits the rate of revolution of the spinner blades to a recording device at the surface.

The diameter of a drilled well will vary locally (quite apart from the question of decreasing bit diameter) due to differences in the lithology of the beds traversed. The diameter of the hole can therefore be used to afford some information about the subsurface strata, and to measure this parameter, various forms of *automatic calipering devices* have been devised, which when passed down the hole record at the surface the variations in diameter. The logs obtained are of value for geological correlations as well as for cementing processes[26].

Modern instruments can simultaneously measure as many as 11 separate parameters and often use four measuring arms in planes 90° apart acting in opposing pairs, a caliper curve being recorded by each pair. Data are presented digitally and are then suitable for computer evaluation, the geometry of the borehole being reconstructed, together with high-resolution data on beds which may be only a fraction of an inch thick.

Borehole gravity meters (p. 156) have been developed primarily to assist in the interpretation of surface gravity data and may also prove useful for measuring the coarse porosity of subsurface formations and perhaps distinguishing gas from oil.

To measure the direction and amount of dip of the strata penetrated, various types of continuous logging *inclinometers* and *dipmeters* are available which record several micrologs by electrodes pressing directly on the walls of the hole, and also indicate the direction and amount of the deviation of the well from the vertical.

The results of such dip measurements may be of special value in determining the orientation of local sand bodies, barrier bars and reefs, the presence of cross-bedding and current bedding[6], as well as serving to confirm the general structural picture obtained by previous ground or seismic surveys.

Conclusion

It will be noted that none of the instrumental logging methods described in this section can *on their own* give a clear-cut differentiation between oil and hydrocarbon gases, and this is not after all surprising, since these two fluids are so closely interrelated in the typical petroleum reservoir. While techniques for analysing gases from the drilling mud are available, it is often important to be able to distinguish "dry" gas zones from oil zones after drilling has ceased. In a series of sands having similar porosity, it is sometimes possible to detect gas-filled beds by running two neutron curves with different spacings between sources and detectors. Considerable success has also been obtained in distinguishing gas from oil zones by using neutron and formation density logs, which in the correct combination can give a quantitative evaluation of the average gas saturation in the sector being investigated.

In general, however, the presence of gas as opposed to oil is best detected by a normal neutron log run in conjunction with a different type of log, such as the sonic log or electric microlog.

Each well-logging technique has its advantages in different circumstances, and the resident geologist and petroleum engineer must select the most appropriate methods to be used. In any case, the great volume of information that is provided through the use of modern techniques and instruments is nowadays necessitating new measures being taken to record, process and retrieve the data obtained[5, 10].

In recent years, there has been a marked increase in the use of computers for well data storage and retrieval operations, and their use has made possible the handling of the large amounts of information recently made available about earlier drilling operations, particularly in the USA. An example[8] is the Permian Basin Well Data System comprising 135,000 wells and their logs. Such data normally includes details of ownership, location, drilling and completion activities, formation boundaries, core descriptions, etc., and it is relevant to note that the study of the Muddy Sandstone in the Powder River basin area by such well-data computer techniques has revealed previously untested productive trends on which subsequent drilling has discovered at least 250 million brl of oil reserves[9].

A bottleneck in the application of computer programmes for applications such as these is the preparation of the data, and this is being overcome by the increasing use of fully automatic machines which can optically scan a log print as frequently as ten times per foot of depth. However, since readings are obtained at fixed depth increments, it is possible that errors may occur due to the fact that many readings will be located on the slopes of peaks on the logs. For comparing logs for correlation purposes this is unimportant, but in calculating data such as oil content the discrepancy might be serious. Various techniques are being developed for overcoming this difficulty, and in any event a number of semi-automatic digitizing procedures are available, mostly of the "pencil-follower" type, with output on punched cards, paper tape or magnetic tape[1, 2]. Information retrieval systems are also increasingly necessary, and here again the translation of the data into digital computer language is an essential prerequisite[20, 31].

The radio transmission of well logs for quick assessment at a shore base has been common practice for some years in US Gulf Coast offshore operations. Tests have also been made on the possibility of transmitting digitized well data between remote drilling locations and computing centres by communications satellites, and the feasibility of such an operation has been proved. The availability of additional satellites is necessary, however, before such transmissions could become routine.

REFERENCES

1. E. A. Breitenbach, "Computer evaluation of logs". *Journ. Petrol. Technol.*, **18** (4). 493–501 (1966).
2. E. A. Breitenbach and D. W. Peterson, "Procedures for the Use of Digital Well-Log Data in Exploration", Canadian Well Logging Society, 4th Formation Evaluation Symposium (May 9–10, 1972).
3. A. A. Brown, "New Methods of characterizing reservoir rocks by well-logging", PD 3 (2), *7th World Petrol. Congr., Mexico* (1967).
4. R. J. S. Brown and B. W. Gamson, "Nuclear magnetism logging". *Journ. Petrol. Technol.*, **219**, 201–209 (1960).
5. C. F. Burk Jr., "Computer-based geological data systems: an emerging basis for international communication". R.P. 5, *8th World Petrol. Congr.*, Moscow (1971).
6. R. Campbell, "Dip measurements in sandstones". *Bull. Amer. Assoc. Petrol. Geol.*, **52** (9), 1700–1719 (1966).
7. E. T. Connolly, "Production logging, a resumé and current status of the use of logs in production". SPWLA Symposium, Dallas (1965).
8. C. C. Cooper, "Experience with the Permian Basin well-data system". *World Oil*, 82–86, April 1967.
9. J. M. Forgotson and P. H. Stark, "Well-data files and the computer; a case history from northern Rocky Mountains". *Bull. Amer. Assoc. Petrol. Geol.*, **56** (6), 1114–1127 (1972).
10. J. W. Harbaugh and D. F. Merriam, "Computer applications in stratigraphic analysis". Wiley and Co., London (1968).
11. H. A. Hedemann, "Geologische Auswertung von Temperaturdaten aus Tiefbohrungen". *Erdol. u. Kohle,* **20**, 5, 337–371 (1967).
12. B. Hitchon and M. K. Horn, "Petroleum indicators in formation waters from Alberta, Canada". *Bull. Amer. Assoc. Petrol. Geol.* **58** (3), 464–473, (1974).
13. W. G. Hoyer and R. C. Rumble, "Field experience in measuring oil content, lithology and porosity with a high-energy neutron-induced spectral logging system". *Journ. Petrol. Tech.* **17** (7) 801–807 (1965).
14. D. K. Keelan, "A critical review of core analysis techniques". *J. Can. Petrol. Technol.*, **11** (2), 42–55 (1972).
15. F. P. Kokesh *et al.*, "New approach to sonic logging and other acoustic measurements". *Journ. Petrol. Tech.*, **17** (3), 283–286 (1965).
16. J. Kumar, "New chart offers fast permeability estimate". *World Oil*, 38–39, Feb. 1971.
17. E. J. Lynch, "Formation Evaluation". Harper and Row, New York (1962).
18. S. J. Pirson, "Handbook of well-log analysis". Prentice-Hall Inc., Englewood Cliffs, USA, (1963).
19. B. Pixler, "Formation evaluation by analysis of hydrocarbon ratios". *Journ. Petrol. Technol.*, **21** (6), 665–670, June, 1969.
20. P. Podmaroff, "The application of computer techniques to exploration for oil and gas". *Journ. Inst. Petrol.*, **55**, 541, 13–21 (1969).
21. J. Riboud and N. A. Schuster, "Well-logging techniques". RP 7, *8th World Petrol. Cong., Moscow* (1971).
22. W. L. Russell, "Interpretation of neutron well logs," *Bull. Amer. Assoc. Petrol. Geol.*, **36** (2), 213–341 (1952).
23. R. J. Scheppel and S. Gilaranz, "Use of well-log temperatures to evaluate regional geothermal gradients". *Journ. Petrol. Tech.*, **18**, 667 (1966).
24. C. and M. Schlumberger, *Trans. AIMME(Geophysical Prospecting)*, **110**, 237–289 (1934).
25. J. P. Simpson, "What's new in mud engineering". *World Oil*, 135–139 (April) and 118–22 (May) (1967).
26. W. Tapper, "Caliper and temperature logging"; In "Subsurface Geologic Methods", 2nd ed., Colo. Sch. Mines, 439–449, (1951).
27. D. J. Timko, "Recent trends in formation evaluation". *World Oil*, 97–106 (June 1968).
28. C. W. Tittle and L. S. Allen, "Theory of neutron logging", *Geophysics*, **31**, 214–224, (1966).
29. M. P. Tixier, "Modern Log Analysis". *Journ. Petrol. Technol.*, **14**, 1327–1336 (1962).
30. R. M. van der Graaf, "Recent advances in formation evaluation", RP 5, *7th World Petrol. Congr., Mexico* (1967).
31. R. M. van der Graaf, "Application of computers in formation evaluation". *Journ. Inst. Petrol.*, **54**, 540, 380–384 (1968).
32. M. R. J. Wyllie, "The Fundamentals of Electric Log Interpretation", 3rd Edn. (1963).
33. J. Zemanek, E. E. Glenn, L. J. Norton and R. L. Caldwell, "Formation evaluation by inspection with the borehole televiewer". *Geophysics*, **35** (2), 254–269 (1970).

CHAPTER 8

Delineation of the Reservoir

PRINCIPLES OF CORRELATION

WHEN a deposit of a solid mineral is extracted by mining, the position and trend of the ore body is continuously exposed as the operations develop, so that by the normal methods of surveying, maps of each level can be made, which can eventually be consolidated into a three-dimensional model of the subsurface. Furthermore, unless faulting causes the ore body to be temporarily lost, close physical contact is always maintained with it.

With a fluid mineral such as petroleum, it is obviously impossible to keep contact by mining methods except in very exceptional cases. The drill is the only connecting link between the mineral and the surface, and the samples obtained by drilling and the information from the borehole itself are the means by which the geography, geometry and lithology of the oil reservoir bed can be established.

However, an accurate map of the subsurface is required just as much in oil operations as when mining solid minerals, since only this will provide the information needed for the optimum location of future wells and the efficient and economic development of the oil resources of the area. Furthermore, since the initial "discovery well" will necessarily have been positioned on the basis of limited surface geological and geophysical data, it is only from the subsequent results obtained by drilling a series of "outstep" and "appraisal" wells that the soundness of the preliminary assumptions can be established.

Obviously, if the first wells are successful, subsequent wells will be drilled with the specific purpose of finding the same productive reservoir bed, and the maximum subsurface information must therefore be collected from the early wells to ensure that later wells are properly sited. Even a well which is "dry" and produces no hydrocarbons may prove to have been well worth drilling if the geological evidence that it produces is collected and properly assessed. In fact, the true economic value of an early well may often lie much less in the oil that it discovers than in the geological data that it provides. Similarly, a "wildcat" may produce oil in plenty, but unless the geological evidence that it can make available has been collected throughout the course of its progress and properly interpreted, the full potentialities of the area may never be properly tested. In any case, when a hydrocarbon-bearing reservoir bed has been located, the maximum information is needed about it so that estimates can be made of its dimensions, nature and the volume and producibility of the fluids it contains.

Thus, it is always necessary to accumulate subsurface data and to compare and correlate the information made available from successive wells. As has been discussed in Chapter 7, all possible information should be recorded about the strata penetrated during drilling operations, and drilling samples, cores and logs must be carefully labelled and stored in such a way that they can be retrieved easily for subsequent examination. The fact that this was not done during the early years of the industry had the consequence that much valuable information was irretrievably lost and many needless holes drilled.

The collection of drilling data and the making of correlations between wells serve the following purposes[23]:

(a) by the identification of "marker" beds and the measurement of their depth in different wells, taken in relation to a common datum plane, the subsurface structure can be defined. It may be notably different from the structure at the surface; in particular, repetition or omission of marker beds may be the first evidence of the intersection of faults.

(b) a check is obtained on drilling progress in future wells, especially the depth at which the oil reservoir bed may be expected in each case.

(c) the lithology of the reservoir and its horizontal and vertical extent can be determined, and estimates made of recoverable hydrocarbon reserves.

(d) the most economic spacing of future wells can be planned.

All methods by which correlations can be made depend upon the identification of special features in the subsurface beds pierced by one well and the subsequent recognition of the same features in following wells. Thus, the examination of bit cuttings and cores will identify lithological "marker" beds, the drilling-time log will show hard and soft layers of rock, the electric survey will mark zones of greater or less resistivity or self-potential, and so on. Subsequently, when sections in two wells are found to have similar lithological parameters, a tentative correlation between the equivalent beds can be established, due allowance being made for differences in the elevation of the derrick floors above sea-level. The more distinct and specific a correlation, the more useful will it be. The ideal, of course, is to use several different methods of correlation which can substantiate and support one another. Subsurface geologists therefore take into account for correlation purposes the continuous geological log derived from cuttings and cores, coupled with the drilling-time record from the geograph; and the various instrumental logs which will be taken at suitable intervals, depending upon local circumstances and the availability of equipment. Various items of additional information, such as the occurrences of permeable sections, faults, saline waters, and even gas or oil "shows" may be used as correlation evidence.

The importance of sound correlations cannot be overestimated in oilfield development work. They enable drilling and casing programmes to be properly planned, and possible high-pressure zones to be prepared for in advance, so that gas-cutting and "blow-outs" can be avoided by having suitable muds available; while from the structural point of view, accurate correlations are essential to delineate the structure and in particular to discover possible faulting, with omission or repetition of beds.

All correlations when first made are tentative, and the resultant forecasts of the depths at which particular beds are likely to be encountered are approximations which must continually be revised in the light of new evidence. The drill is the final arbiter of the subsurface, and it is therefore often desirable to drill test holes with portable rigs purely for geological information. Such rigs can reach depths of 8,000—10,000ft, using small-diameter drill-pipe, and test holes can be drilled wherever subsurface geological information is desired, particularly to assist in the delineation of structure by the close correlation of samples.

All methods of correlation assume *continuity of deposition* of the strata. It is, in fact, generally true that any formation that is continuously traceable over a limited area is *synchronous* throughout that area. However, some formations may persist laterally for miles, while others may "finger out" within a relatively short distance. Even within a synchronous horizon, variations of lithology will occur according to the distance from the shore line of deposition. Hence, it follows that beds that might make good markers are not always sufficiently widely deposited or lithologically constant to be suitable for this purpose.

Correlations may be local or regional, depending on the extent and uniformity of the marker horizons involved. Complications arise when there is a facies change in sediments of the same age, or put another way, if dissimilar facies are time-equivalent. The boundaries of what appear to be the same rock units do not necessarily coincide with the time-units established on the basis of palaeontology—which represent all the sediments deposited in a given time-interval. In other words, boundaries based on rock-property correlations may not always coincide with time-rock or time-stratigraphical boundaries, which involve the correlation of different rock-types accumulated during the same time-interval.

In general, the lithogenic units of the field geologist are limited to relatively small-scale correlations, whereas the time-rock units of the palaeontologist are useful for regional correlations. Sometimes the relationship between the two boundaries is impossible to evaluate.

Correlations between neighbouring wells can be made by using similar physico-chemical characteristics of the rock units, such as peaks on the electric and nuclear logs—which sometimes are recognisable over considerable distances, depending on the existence of electrical "marker" beds. This is nowadays a standardized technique, which is very widely employed. Other characteristics of the rocks penetrated can also sometimes be used for correlation, including the presence of trace elements—non-detrital elements may have been incorporated into sediments in constant proportions over a basin of sedimentation.

The fluid contents of porous rocks can also be used for correlation purposes. In the case of brines, the spectrographic analysis of the salts obtained by evaporating samples of oilfield waters can sometimes be used to correlate the waters with regard to their source. Any sharp changes of salinity can be indicative of the existence of an unconformity or the sudden development of a restriction to free marine circulation.

Evidence of polarity reversals (p. 247) may sometimes be used for defining stratigraphical boundaries in certain circumstances—e.g. the Palaeozoic-Mesozoic boundary—or for making regional correlations.

Methods of correlation which have now become so detailed and widely used that they amount to specialised scientific studies in themselves are the uses of palaeontology, in particular micro-palaeontology and palynology, and heavy mineral analysis. A brief discussion of the principles involved in these techniques as they affect the petroleum geologist follows, but the reader is referred to the specialised text-books for a fuller exposition of these subjects.

PALAEONTOLOGY AND MICROPALAEONTOLOGY[7, 12, 14, 21]

A study of the fossil content of an area is always essential in the investigation of its stratigraphical history. Fossils can be classified into genera, species and time groups, and can be used as a relative index of the age of rocks in which they are found. There are anomalies and gaps in the story, but generally speaking, it is found that there has been a gradual increase in the complexity of any particular genus with time. The gradual processes of evolution can be traced by the changes that have occurred in fossil types and the relative position in the time-scale of a particular specimen can often be defined with considerable precision. All fossils do not, however, lend themselves to stratigraphical classification. Some species remained almost unchanged over very long periods of time, while others were only local in their distribution. Varying environmental conditions affected the development of the living creature: so that while in one part of the world a certain stage of evolution was relatively rapidly attained, elsewhere, the same stage was only reached much later. Changes of environment may have driven one species away from a locality and brought in another, thus giving a false appearance of evolutionary connection.

The ideal "zone-fossil" is one that has the maximum distribution in space combined with the minimum distribution in time. The oilfield stratigrapher seeks "zone-fossils" in the surface rocks and in the cores and samples available from drilling wells, which can serve as markers for correlation purposes. The frequency of occurrence of different species is noted, and this will generally reveal a number of species which persist through most of the succession, as well as species which are so rare as to be useless, since the time consumed in finding them and the likelihood of missing them in a sample are prohibitively high. Between these extremes, there will probably be found species of relatively short vertical (stratigraphical) range and wide lateral (geographical) distribution, which can be used as key specimens to which similar samples found in other wells can be related. Abrupt variations in faunal frequency can also be utilized as marker levels. The reliability of the correlations made will clearly increase with the narrowness of the range of each individual correlation level, and the greater the number of corresponding faunal changes that can be found. Similar sequences of faunal changes are not likely to have occurred independently in different places, and therefore when these sequences can be detected in two sets of well samples, the correlations obtained will have the optimum degree of reliability. Regional correlation is sometimes possible over large areas, provided that conditions of life were reasonably uniform.

In a developing oilfield, when the "zone-fossils" or "index-fossils" have been ascertained, samples are photographed and distributed to the laboratories controlling the various drilling wells. In this way, the local resident geologists can compare fossil samples obtained from each new well with key specimens, and thus achieve rapid and accurate correlations.

Macrofossils, mostly lamellibranchs, are sometimes found in cores, and can be removed by a chisel or dental drill. Corals, brachiopods and gastropods are also relatively common oil-well fossils. Most work, however, is done on the very small fossils—*microfossils*—which are recoverable not only from cores, but also by careful treatment of bit cuttings.

Micropalaeontology has nowadays become a science of its own, and the micropalaeontologist occupies a valued place in oilfield laboratories. The study of microfossils does not differ in principle from that of macrofossils; but their minute size calls for special methods of collection and their abundance necessitates statistical analysis.

Micropalaeontology was first employed for correlation purposes in the 1920's in Texas and California, and today there is scarcely an oilfield in the world, particularly where Tertiary strata are dominant, where the systematic examination of microfossils is not standard practice.

The foraminifera are easily the most important group of microfossils for the petroleum geologist. Second to them in stratigraphical value are the ostracods, but these are more difficult to classify and identify, and since the families are generally of long time-range they cannot generally be used as index fossils, except locally[1]. Other microfossils also sometimes used include the radiolaria, conodonts, otoliths, calcareous algae (nannoplankton)[19], diatoms, etc.

Although micropalaeontology provides an excellent and often unique means of correlation, it should not be forgotten that it is also of great value in providing evidence of the environmental conditions that prevailed at different times during the course of local sedimentation.

Variations in the environment of sedimentation produce changes in the ecological and depositional pattern of the resulting sediments and their fossil contents, which can broadly be summarised as variations in "facies", and each type of facies will be broadly associated with a different group of animals and plants. Conditions on the sea floor, the rate at which solid matter accumulates from the inflowing rivers, and the local wave and current action are all factors which influence both the depositional facies and the content and nature of the life which exists there.

Thus, it is possible by a study of the nature of the microfossils and their occurrences elsewhere to deduce the water conditions that prevailed in the course of their lifetimes—i.e. whether the environment was deltaic, neritic, (i.e. extending from the low tide line to a depth of about 600ft), bathyal (600–6,000ft depths of water) or abyssal (below 6,000ft); and whether they flourished in warm or cold waters.

Since the conditions for oil generation and primary migration were probably shallow-water marine, some valuable indications of whether favourable conditions prevailed during the formation of a particular bed can be derived in this way, and source rocks perhaps identified. Methods of rock classification have been derived based on this principle (e.g. the Dunham environmental index).

Collection and Preparation of Material

The field geologist will only rarely be able to observe the occurrence of microfossils in hand-specimens, but must collect rock samples in orderly sequence across the section he is examining, and will rely on subsequent laboratory examination to disclose the microfaunal content. (About 1 lb of rock is sufficient for each sample, and the exact location of derivation must be marked on the map of the area, using a letter or number code which is repeated on the sample bag label.)

Microfossils are most easily obtained from soft, friable rocks; bit cuttings, properly washed and sieved, may provide suitable samples, but care must be taken to eliminate extraneous microfossils which may have been introduced from the clays and shales used in the mud fluid.

Cores are sampled after washing by breaking off transverse slices from portions which appear promising, or by chipping off pieces over a length of several feet—thus affording an average sample of the microfauna present.

Care must be exercised in the labelling and storage of samples, since large numbers can accumulate during the course of drilling a few wells, and the transposition of samples or depths may lead to serious errors in correlation and subsequent structural interpretation.

To release the contained microfauna from rock specimens obtained in the field or from cores, a hammer and chisel can initially be used, but to reduce samples below a cubic inch in size, continuous pressure must be applied by some variety of vice, rather than by percussion.

Most clays and marls can be broken down by boiling, preferably in water to which has been added a little washing soda. For harder rocks, such as shales and calcareous marls, more concentrated alkaline solutions may be necessary to achieve disintegration. Rapid freezing and thawing, the use of an autoclave or oxy-acetylene torch, and the crystallization of sodium sulphate solution impregnating the rock are other processes which have been employed for breaking down rock samples and releasing their microfossil contents.

When the specimens have been broken down by one or other of these methods, the microfossils can be separated out by washing the material with a jet of water through a set of sieves of increasing fineness, usually with 40, 80 and 200 meshes to the inch. If plenty of water is used (say 2–3 gallons per sample) the microfossils will be virtually held in suspension during passage through the sieves and will consequently not be damaged. If foraminifera with air-filled chambers float off on the surface of the water used for washing, they must be recovered by filtration.

Microfossils can be separated from the dried residues of the sieving and washing processes by flotation in a small dish filled with bromoform or carbon tetrachloride. The heavy mineral residue sinks, while most of the foraminifera float and can be decanted off and dried. Another method of separation, which will separate foraminifera from light minerals as well as from heavy grains, is to add a soap solution and then blow a stream of air through the liquid. The resulting foam lifts the foraminifera to the top, while the quartz, etc. remain at the bottom of the vessel.

A photoelectric process has also been developed for separating dark, phosphatic microfossils from light-coloured material[15].

The microfauna obtained are examined initially against a dark background, using a binocular microscope. Specially prepared trays are often used, embodying an enamel base divided into small squares by white lines. Such trays enable systematic counts to be made of the number of specimens obtained from any given rock sample, a process greatly aided by the use of a divided grid in the eyepiece of the microscope. The best specimens can be picked out by the use of a moistened brush of fine red sable and mounted on slides for preservation and future detailed examination. Slides can be made either of complete assemblages or of individual species.

Where a rock cannot be conveniently disintegrated, it will be necessary to make a thin section of it for microscopic examination. This can be done by grinding one side of a small chip with carborundum or emery powder of increasing fineness, and mounting this face in properly "cooked" Canada balsam on a glass slide. The free surface is ground down until the specimen is thin enough to be translucent; it is then cleaned and covered with a plastic glass slip on which a little balsam has been spread.

The identification of microfossils is a process essentially based on comparison, specimens being first compared with each other and then divided into groups of the same species; these species groups are then compared with previously defined types. When identified, a record must be kept, cross-referenced to the slide number, of the genus and species of foraminifera, its relative abundance, lithological and stratigraphical context and probable age. On the basis of such data, a comparative record of the microfauna of a given area can be systematically built up, which can be used as a source of correlation material.

Under the stress of intensive work in many laboratories, it is sometimes necessary to identify microfossils simply by temporary code numbers for purposes of comparative identification. Computer techniques are being developed to facilitate the storage and retrieval of the large volumes of data that become available during the exploration and development of an oilfield and to enable statistical analyses to be carried out[6, 20].

Semi-automatic diffraction pattern analysis has been used successfully to differentiate and classify microfauna. Thus, it has been found that the diffraction patterns of foraminifera exhibit a typical star-shaped structure consisting of broken rings. Foraminifera with structured rather than smooth surfaces, produce speckled diffraction patterns that are basically ringed star shapes. The characteristic structure of the diffraction pattern for a specific shape is not dependent on the size of the object or its location.

A photo-detector can pick up the diffraction patterns, which are electronically evaluated, using a series of revolving slits to select characteristics of the pattern. When incorporated in an automatic evaluation system, individual objects may be separated and passed on for size analysis; sorting into rod-shapes, spheres, helices and others may be performed.

PALYNOLOGY[2, 4, 11]

The science of palynology—the examination and comparison of pollens and spores—although initiated as long ago as 1916 and employed extensively thereafter in the coal industry in the examination of Late Palaeozoic strata, was not used in petroleum geology until about 1936. Since the 1939–45 war, it has been increasingly employed as a method of providing correlations, particularly in the otherwise difficult to differentiate Tertiary strata of tropical areas such as Venezuela, Borneo and Nigeria. (Relatively few data are available as yet on Cretaceous palynology). During this period, it became generally realised that fossil pollens and spores ("palynomorphs") when examined under very high magnification (usually × 1000) were eminently suitable for stratigraphic differentiation and environmental studies, and could supplement and complement the evidence provided by micropalaeontology.

The first system of comparison used was based on the setting up of a number of "morphological species", related to the pattern of furrows and pores on individual pollen grains. These groups did not bear a simple relationship to the natural species observable in living plants, since some plants may produce more than one morphological variety of pollen, while on the other hand the same "morphological species" may be produced by different living plants of varying species, genera or families. Subsequently, a comprehensive decimal code system was devised, whereby groups of digits could be employed to describe the varieties of shape, aperture, sculpture and structure of pollen grains, and this has since been extended to include spores, dinoflagellates and other organisms. The parameters used are classified under the headings of dissociation, symmetry, vertical and horizontal differentiations, elementary bodies and detailed features.

This system has of course proved extremely suitable for computer data processing techniques, which are now extensively employed in palynostratigraphy. In particular, statistical analysis can be relatively rapidly carried out by this means to test such factors as the percentage variations of common species (lowest in marine environments where the pollen deposited has been well mixed and highest in a terrestrial environment); statistically significant associations of species with a common source area (enabling different topographical source environments to be differentiated); and variations in the occurrence of rarer pollen types.

In general, adequate palynomorph samples can be obtained from well cores, side-wall samples or bit cuttings, and analysed within a few hours of receipt. The chemical and physical methods used for extraction vary in different areas, but in general depend on the acid resistance. Neogene material is treated with hydrofluoric acid followed by bromoform separation, while Palaeogene samples may need to be treated with an oxidizing solution. Clays or shales, with silty or carbonaceous material, are selected for sampling, limestones, conglomerates or sandy sediments being avoided as far as possible. If the sediments have been subjected to oxidation soon after deposition or as a result of subsequent exposure and weathering, or if they have been subjected to heating as a result of deep burial or igneous or tectonic activity, determinable pollen grains may be scarce or absent. It has been found that a count of the order of 100 marker species with a sampling distance of 70–100ft is adequate in practice for correlation purposes. Since Tertiary pollen floras are very rich and prolific, it is not generally difficult to select 100–200 for stratigraphical marker purposes out of the perhaps 1000 or so which are commonly available in a particular area[5].

In setting up time-stratigraphical zones from pollen grain counts it is necessary to discriminate between stratigraphically reliable and unreliable pollen types and to make allowances for the primary factors of evolutionary change, migration and climatic variation which have controlled the presence of the plants from which the pollen was derived; and also the secondary factors which have affected the dispersal of the grains before fossilization. Fossil pollen species usually appear suddenly in the stratigraphic column, presumably as a result of a climatological changes of environment, but delay may occur in their lateral dispersal. Secondary factors also influence the dispersal and mixing of the grains: the duration of the phase of air transport and the environment of their deposition in water—rivers or the sea—and the degree and effectiveness of waterborne transportation before settling must also be taken into account. There are obvious complications in the possible recycling and reworking of pollen grains as a result of the erosion of the sediments in which they were originally deposited. By suitable statistical analysis, and due consideration of all the factors involved, it has been possible to classify marker pollen species into three groups—a few "pantropical" species, some markers occurring in both South American and West African areas, and the largest number which are significant only within a single botanical province. This classification has, of course, been assisted by botanical consideration of related Recent plants and the additional evidence of palaeontology. Environments of deposition may be broadly classified as upper coastal plain (fresh water), lower coastal plain (brackish water) and marine. In individual oilfield areas, subdividing the local sedimentary column enables correlations to be made between different wells which can supplement the correlations derived by other means and also provide valuable environmental evidence of the conditions of sedimentation.

Palynology can also provide specific information about temperature conditions in the sediments which are reflected in the palynomorph colour or translucency, and this is an important contribution to petroleum geology, inasmuch it can indicate the degree of local eometamorphism.

HEAVY MINERAL ANALYSIS[14, 15, 22]

Heavy mineral analysis provides a valuable method of correlation in certain areas. It was first used in petroleum geology in Trinidad in 1916, when notable success was achieved by the application of this technique to the correlation of Tertiary sediments.

Since clastic rocks have been basically formed by the weathering and breakdown of pre-existing crystalline materials, they may be expected to retain some of their original mineral contents. The degree to which this occurs in fact depends on the nature of the individual minerals; the lighter, softer constituents are usually completely altered by weathering, aqueous transport and chemical solution, but some types of heavier mineral are often preserved and may be diagnostic of the source-type of the igneous material, and hence are of correlative value.

Heavy minerals can be used for correlations between localities if they have a common source or "provenance", i.e. igneous or metamorphic rocks, or older sediments which have been reworked and redeposited. If the parent rock was homogeneous, the sediments derived from it will also be homogeneous, but if it varied in its constitution, i.e. if for example, different sections of rock were exposed by progressive erosion—then this variation will be reflected in the sediments derived from it and their constituent minerals.

An assemblage or "suite" of minerals can be used as a means of correlation only if it is distinctive and if there is a relative abundance of the constituent minerals, since in general only the sand-grade portion of a sediment can provide such assemblages, and the heavy minerals may form less than 1 % of this. Clay and silt facies do not as a rule provide useful heavy mineral suites. However, a similar technique of insoluble residue analysis has been applied with success to some carbonates, where the material remaining after digestion in hydrochloric acid is composed of sand grains, secondary minerals and chert particles.

The principle upon which correlation between wells is based is the identification of relatively thin sections which represent near-simultaneous deposition from a common source. However, it should be remembered that detrital minerals may reappear in successive zones, unlike fossils which become extinct, and comparison between correlations achieved by micropalaeontology and heavy mineral analysis can therefore only be made with caution. When attempts are made to correlate by heavy mineral analysis over relatively long distances, the difficulties increase, since the variations in local conditions of weathering, transport and deposition become more significant.

The standard laboratory technique for the separation of heavy minerals comprises the floating off of unwanted lighter constituents in a liquid of suitable specific gravity. The two liquids most widely used for this purpose are acetylene tetrabromide ($C_2H_2Br_4$) and bromoform ($CHBr_3$). These have specific gravities when pure of 2·96 and 2·89, respectively, and are colourless and chemically inert.

A variety of other liquids have been used to isolate particular heavy minerals after the lighter minerals have been separated. Among these are methylene iodide, thallous formate and thallous formate-malonate $CH_2(TlCO_2)_2$. Concentrated aqueous solutions of the latter salt (Clerici's solution) have a specific gravity as high as 4·4.

For normal separation, the following procedure is usually adopted: a quantity of the disintegrated rock sample is sieved through $\frac{1}{2}$mm mesh, and a weighed quantity transferred to a filter funnel full of bromoform or acetylene tetrabromide and fitted with a pinch-clip over the exit. The liquid is stirred and then allowed to stand, and when no more settling takes place the pinch-clip is opened and the heavy mineral residue is drained off onto a filter paper in a funnel held below. It is then washed with alcohol and upon drying is ready for mounting and examination.

When fine sands or silts have to be examined, the normal flotation method described above is too slow. In such cases, it is necessary to use a centrifuge to hasten settling. Since the average heavy-liquid separation produces a larger quantity of heavy mineral grains than can be conveniently handled on a microscope slide, some method of selecting an average portion for examination must be employed. The simplest method is to pour the sample on to a flat surface in the form of a cone, flatten this and divide it into four quarters. Two opposite quarters are removed and rejected, and the remaining two quarters mixed together and the process repeated until a sample of the desired size remains.

Filtered Canada balsam is the medium usually used for mounting specimens, as mentioned on p. 188. It has a refractive index of 1·54, and about half the normal range of light minerals have a lower refractive index than this, whereas nearly all the heavy minerals have a higher index. Piperidine, a low melting-point crystalline alkaloid, is sometimes used instead of Canada balsam, since it has a refractive index of 1·68, which is about half-way in the heavy mineral range.

When the slides of the heavy residues are examined under the microscope, a systematic attempt can be made to identify the minerals present. The examination is primarily visual and records colour, crystal shape and morphology, and any special crystallographic features such as twinning, zoning, etching or striations. The shape of the grains, their angularity or roundness, with any evidence of cleavage, parting or fracture are noted, and optical properties such as refractive index, pleochroism, birefringence and interference figures determined if necessary[13].

Electromagnetic methods can also be used to identify minerals according to their magnetic properties, the sample being spread out between the poles of an electromagnet and gently agitated. The strongly magnetic particles are separated, and by increasing the current a "spectrum" of decreasingly magnetic minerals can often be recognised (Table 36).

A number of specialised chemical reagents have been developed for qualitative and quantitative micro-analysis of minerals, a few drops of which are sufficient to detect the presence of very small traces of such elements as magnesium, copper, nickel and lead. These reagents can be used in suitable circumstances as an auxiliary means of mineral identification.

Various stains can also be used to identify mineral grains, and in particular to differentiate between calcite, aragonite and dolomite, etc., as described earlier.

As each mineral of an assemblage is noted and identified in turn, it becomes necessary to assess the frequency of its occurrence, since the diagnostic assessment of the sediments concerned will depend upon the relative quantity of each heavy mineral present.

All methods of determining frequency depend on a visual examination of the mineral grains and upon counting the actual particles present. For statistical correctness, every grain on a slide should be counted. But as this would take so long and be so exhausting in practice, it has been shown that counting 300 grains gives sufficiently accurate results. Accuracy increases as the square root of the number of grains counted, but above 300 the increase of accuracy obtained is not large enough to warrant further expenditure of time.

The average grade-size of the total mineral assemblage and that of one or more prominent species can also be used as correlation evidence. For this, it is necessary to have a special eyepiece co-ordinate ruling in the microscope, previously calibrated. The number of squares of this ruling covering all the grains of one species is noted, and since the average grain can be considered as two-dimensional for this purpose, the square root of the total number of squares will give their average diameter.

TABLE 36

CLASSIFICATION OF MINERALS FOR HEAVY LIQUID AND MAGNETIC SEPARATION

Heavy Liquid	Highly Magnetic	Moderately Magnetic	Weakly Magnetic	Non-magnetic	
Bromoform — Acetylene Tetrabromide — Methylene Iodide — Thallous Formate — Thallous Formate-Mallonate		2·3 Glauconite		2·0-2·4 Zeolites 2·3 Gypsum 2·5 Leucite, Kaolin 2·5-2·6 Alkali Felspars 2·61-2·76 Soda-lime Felspars 2·6-2·75 Scapolite group 2·62 Chalcedony 2·63 Nepheline 2·64 Cordierite 2·65 Quartz 2·72 Calcite 2·85 Dolomite, Phlogopite	"LIGHT" MINERALS
		3·1 Iron-rich Biotite 3·2-3·6 Iron-rich Amphiboles and Pyroxenes 3·8 Siderite 4·1 Almandine 4·3 Melanite 4·4 Chromite 4·5 Xenotime 4·8 Ilmenite 5·1 Haematite	3·1 Biotite Amphiboles Pyroxenes 3·2 Tourmaline 3·4 Epidote 3·5 Olivine 3·6-3·7 Staurolite 3·8 Pleonaste 4·0 Garnet 4·2 Ferriferous Rutile 5·0 Monazite 5·3 Columbite	2·90 Muscovite 3·0 Tremolite 3·1 Enstatite 3·15 Apatite 3·18 Andalusite 3·2 Fluorite 3·23 Sillimanite 3·5 Sphene 3·52 Topaz 3·6-3·7 Spinel 3·6 Kyanite 3·9 Anatase 3·94 Brookite 4·0 Perovskite, Corundum 4·2 Rutile 4·5 Barite 4·7 Zircon 5·0 Pyrite	"HEAVY" MINERALS
	4·65 Titanoferrite, Pyrrhotite 5·17 Magnetite				

RESERVOIR PRESSURES AND FLUID LEVELS

In the process of delineating a petroleum reservoir, considerable importance must be attached to the measurement of the pressures and levels of the fluids encountered in the wells.

All the fluids in the pores of a reservoir rock are under a pressure which is best referred to as "reservoir pressure", rather than "rock pressure" or "formation pressure", two terms which can have other implications. The origins of reservoir pressures have been widely discussed, and it is recognised that there are various valid explanations, depending on the circumstances.

The term "pore pressure" is also in use, but it has rather wider connotations than reservoir pressure: it covers the pressures in pore fluids in non-reservoir rocks such as clays, as well as reservoir rocks. During recent years, considerable attention has been paid to "over-pressured" clays and shales, for such rocks present a hazard in drilling operations, if suitable precautions are not taken. Efforts have been made to predict the levels at which such over-pressured clays are likely to be present. The over-pressured clays have a porosity greater than that of the adjacent clays, and therefore they differ in density, electrical resistivity and sonic velocity from clays with porosities more nearly consistent with the normal condition for their depth of burial. It can be argued that this excess porosity implies a somewhat exceptional fluid pressure; hence the label "over-pressured clays"[3].

The expression "abnormal pressure" has been employed when the reservoir pressure significantly exceeds the pressure due to a column of salt water (0·465 psi/ft) extending from the surface of the field to the reservoir level.

When the fluids in a reservoir rock are stationary, the change in their pressures with elevation will depend only on the densities of the fluids and the heights of the gas and gas/oil solution columns. "Tall" gas columns will show some increase in density from top to bottom; likewise "tall" oil columns may show increases in density downwards. The very low compressibility of water means that for many purposes the change in density with elevation can be neglected. When flow is taking place, the flowing pressure gradients will be a function of several factors; the rate of flow, the viscosity of the fluid and the permeability of the rock. The resultant hydrodynamic pressures are superimposed on the hydrostatic pressures referred to above.

For some purposes, the reservoir pressures that are of interest are those which existed prior to any fluids having been removed from the reservoir through wells; these are the "initial reservoir pressures". However, even at this stage it is possible for the flow of water to be taking place through the reservoir rock.

Subsurface pressure measurements are most commonly made by means of a recording pressure gauge, of which there are several different forms. One has a helical bourdon tube which is sufficiently long to rotate an attached recording stylus over the full inside circumference of the cylindrical chart holder. A clock moves the chart longitudinally. Other types use a piston, acting through a stuffing box and being restrained in some manner by a helical spring. A stylus on the inner end of the piston records the pressure on a cylindrical chart holder, which is rotated by a clock. In each case, the pressure element and the entire recording mechanism are encased and sealed against external pressure, apart from an opening which allows the external pressure to be applied to the pressure element. The charts are usually of brass, copper or aluminium, because, unlike paper, they are not affected by humidity. Coated metal is preferred to bare metal, since this permits recording with less friction for the sharp-pointed stylus. A finer line is obtained with a black coating than with a white coating. The charts are read with a magnifier and finely graduated steel scale. Appropriate and frequent calibration of the gauges, using a dead-weight tester, is essential for reliable data.

The gauges have overall external diameters of between 1 in and $1\frac{5}{16}$ in, and their length is in the region of 4–6 ft.

The gauge is lowered into the well on a wire-line, and a record of its depth is kept against a surface clock, so that matching in terms of time is possible when the chart is eventually removed from the device for examination.

When equilibrium is attained between the fluid in the borehole and that in the rock, i.e. when no flow is taking place, the pressure recorded will be that which exists in the reservoir rock fluid at the particular level. From a knowledge of the density of the reservoir fluid under the prevailing reservoir conditions, it is then possible to calculate the pressure at any other level in the same fluid in the absence of flow. Provided that the level of the contact between two fluids is known, together with their individual densities, it is also practicable to calculate the pressure at a level which would be in a different fluid from that at the reservoir level at which the pressure measurement was made.

In certain studies, it is necessary to adjust the pressure measurements from a series of wells in one or more reservoirs to a given level ("pressure datum") and fluid; usually the selected fluid will be water. For a single continuous reservoir rock, such an adjustment will result in all the values being the same in the absence of flow; if there is flow, the values will change laterally, provided that the wells are not in a line at right-angles to the direction of flow. When the reservoir rock is discontinuous, so far as free communication through the given fluid is concerned, there can then be differences in pressure at the selected datum, even in the absence of flow. Should different reservoir rocks show the same pressure at the same pressure datum, this is suggestive of free fluid communication.

Clays and shales have very low permeabilities, and hence the pressure gradients in any fluids they contain can be considerable, even for small rates of flow. Two reservoir rocks separated by a clay may show different pressures at a given datum because flow is taking place between them through the clay. Alternatively, pressure measurements made in the two reservoir rocks at points one directly above the other, will show a difference which does not correspond to the pressure arising from a column of water equal in height to the difference in level between the two points of pressure measurement. (This assumes that there are no osmotic effects arising from differences in salinity of the waters in the two reservoir rocks, with the intervening clay functioning as a semi-permeable membrane).

In considering the explanation of reservoir pressures, attention has generally been paid to the pressure/depth ratio—the mean gradient between the land surface above the hydrocarbon accumulation (or in some cases sea-level) and the reservoir. Quite apart from the problem associated with selecting a mean surface elevation when the ground above the accumulation is not flat, it will be apparent that the gradient will differ according as the pressure measurement is made in the gas column, the oil column or the water column; and that it may vary with position in the gas and oil columns, and may also change with position in the water column (Fig. 61). Nevertheless, excluding the cases of tall hydrocarbon columns, there is a tendency for pressure/depth ratios to cluster round a value equivalent to a water column, with an upper limit of about 1psi/ft and a lower limit of a small fraction of a pound per square inch per foot of depth.

Commonly, the deeper the reservoir the higher is its pressure, but Fig. 64 shows one possible exception. When the reservoir outcrops, but is cut off by a fault or by a wedge-out at depth, then the pressure in the water in the reservoir rock will be determined by the highest level of the outcrop at which the rock is water-saturated. This will give a pressure/depth ratio for a water column, depths being measured from this outcrop level. However, if the elevation of the surface of the ground above the accumulation is not the same as the outcrop level, its use to give a pressure/depth ratio (for a water-pressure datum) will result in a value not the same as that of a water column (Fig. 62). Should a reservoir rock outcrop at different levels on the opposite sides of a region in which there are hydrocarbon-bearing traps, water flow from the higher outcrops towards the lower outcrops will take place through the reservoir rock. Hence, the pressure/depth ratios for water pressure measurements in the traps will have values lying between that obtained by using the level of the highest outcrop and that derived by using the level of the lowest outcrop (Fig. 63). (This assumes there are no limiting values imposed by faults which intersect the reservoir rock and allow flow to or from the surface where they occur.) The pressure pattern in the reservoir rock resulting from water flow will determine the local pressure values, and it has to be noted that water flow may not arise solely from artesian circulation.

Should a rock be undergoing compaction of any kind, as a consequence of vertical loading or lateral squeezing, fluid will be expelled by reduction in its porosity. Hence, for compaction associated with vertical loading the rate of pressure change with elevation will differ from that associated with a stationary water column; a pressure gradient arising from the flow must be added algebraically to the hydrostatic gradient. Consequently, fluid pressures in reservoir rocks above and below the compacting rock will be affected. Compaction resulting from lateral squeezing will have comparable effects on fluid pressures, through the fluid expelled from the compactible rock.

When reservoir rocks outcrop, this condition imposes a limit on the pressures which can exist in the fluids they contain, although the control may be somewhat more complex in detail than this simple statement may suggest. Nevertheless, an escape route such as is provided by an outcrop may be considered to have some of the properties of a safety valve.

Lenticular reservoirs are surrounded by very low-permeability rock. As a result, compaction proceeding in the enclosing clays or shales will raise the pressures in the reservoir rocks to values which may considerably exceed the pressure equivalent to a column of water extending vertically from the surface down to the lens. Indeed, the fluid pressure within the lens may approach a value equal to that exerted by the overlying rock load, which would

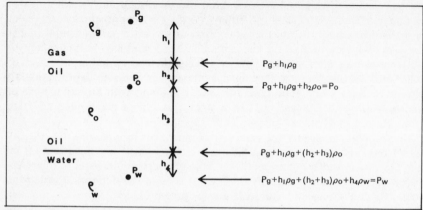

Fig. 61. Vertical pressure changes in a system in which gas in a reservoir rock overlies oil, which in turn rests on water.

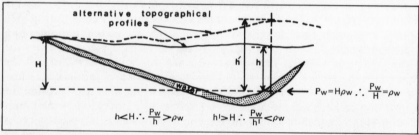

Fig. 62. Subsurface fluid pressure controlled by depth below outcrop, and not by depth below land surface directly above point at which the pressure is measured.

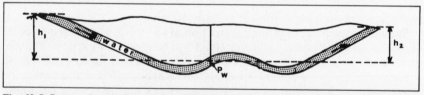

Fig. 63. Influence of water flow resulting from difference in level of outcrops of water-bearing rock. The water pressure at depth for an intermediate point lies between the two "hydrostatic" heads. $h_1\rho_w > P_w > h_2\rho_w$.

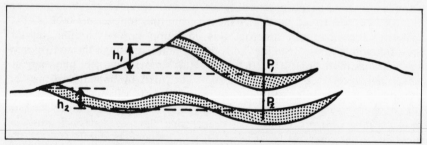

Fig. 64. Influence of outcrop level and stratigraphic position on subsurface pressures. $h_1 > h_2$. Hence, $P_1 = h_1\rho_w > h_2\rho_w = P_2$.

be the pressure on the fluid in the lens if it had been sealed by an impermeable skin (Figs. 65, 66).

In a sense, there are therefore two extremes: an "open" system in which the reservoir rock fluids have free communication with the surface, either by outcrop or via a suitable fault; and a "closed" system in which the reservoir rock is enclosed by very low permeability rock, so that fluid flow through the sealing rock is extremely difficult. In the latter case, when the distance from the surface to different parts of the lens varies considerably, the point of least cover will be most important in fixing the fluid pressure in the lens. Should the lens have negligible cover at some point, its fluid pressures might differ little from those of an "open" system.

It has been suggested that pressures can be "inherited" and this implies that the fluid was sealed in a reservoir lens when it was deeply buried. Subsequent erosion would necessarily have caused a drop in temperature, but the consequent contraction of the fluids and the reservoir rock has not been such as to maintain the same pressure/depth ratio as could be inferred to have existed at the time of sealing. As a result, the later pressure/depth ratio is anomalous with respect to the smaller depth of burial.

There is also the concept of "derived" pressures. This assumes the existence of two lenses at markedly different depths, each having a pressure in keeping with the respective rock load, or at least a pressure well above hydrostatic. The two lenses are then put into free communication by a fracture of, let us assume, negligible volume. They can then differ in pressure only by the equivalent of a vertical column of water extending from the level of the shallower lens to the level of the deeper lens. Hence, the pressure in the shallow lens will be raised and that in the deep lens will fall. The changes will depend on the relative volumes of storage space in the two lenses. The fracture will be filled by virtue of expansibility of the water in the two-lens system, and this requirement means that the rise in pressure for the shallow lens will be rather less and the drop in pressure for the deep lens will be rather more than would otherwise be the case. Indeed, for a relatively large volume fracture, both lenses could drop in pressure, the deeper one more than the shallow lens.

Various other factors have been proposed as sources of reservoir pressure: osmosis, deposition of cement, generation of hydrocarbons (especially gas), rise in temperature on deeper burial. However, it is at once apparent that the results of their action could only be felt in terms of pressure where a closed system is involved. Thus, there is evidence that clays can act as somewhat imperfect semi-permeable membranes. Hence, when a clay separates two reservoirs containing waters of different salinities, osmosis will cause water to flow from the reservoir with water of low salinity into that with high salinity water, unless there is a pressure difference which offsets the difference in osmotic pressures of the two waters. For an open system, flow could take place, but there would be no pressure build-up. When the more saline water is in a lens, osmotic flow would be possible until the pressure built up to such a value as to offset the difference in osmotic pressures of the two waters (as modified by the flow). Since in general mineral cements have low solubilities, extensive deposition of cement in a reservoir rock could result only from the flow of very large volumes of water through that rock. Such flow seems to be more likely for open than for closed systems, with the consequent probability that deposition of cement in itself is unlikely to give significant increases in pressure.

Rise in temperature will cause the liquids to expand, and to the extent that the increased depth of burial, by compressing the liquids, does not offset the tendency for them to expand on heating, there will be a rise in pressure in a closed system. Conversely, erosion will drop the temperature and lead to the opposite effect.

Breakdown of organic matter to give products of greater aggregate volume under the subsurface conditions than the parent materials could also cause some rise in pressure, but only in a closed system. Any effect is likely to be very small.

Applications of reservoir pressure data

Initial reservoir pressure data are of practical value, especially when available from more than one well in a single reservoir rock. By adjustment of the data to a given datum in the water, directions of water flow can be determined. From these, the direction of displacement of hydrocarbon accumulations under hydrodynamic conditions can be indicated. The likely directions of tilt of hydrocarbon/water contacts are also shown.

When nearby wells reveal the pressures in different fluids, it may be possible to deduce the level of the fluid contact between the two fluids. The subsurface densities of the two fluids must be known, and broadly the greater the density difference between the two fluids the more accurate will be the predicted level of the fluid contact. Thus, gas/water and gas/oil contact levels will be indicated more precisely than oil/water contacts. Uncertainty will be introduced if there is water flow, unless there are enough wells to enable suitable corrections to be made to the water pressure data.

Let P_g be the pressure in the gas, gm/cm^2
   P_w be the pressure in the water, gm/cm^2
   ρ_g be the density of the gas, gm/ml
   ρ_w be the density of the water, gm/ml
   H be the distance of the gas measurement above
      the water measurement, cm
   h be the distance of the gas/water contact above
      the water measurement, cm

$$Then, \quad P_w - h\rho_w = P_g + (H - h)\,\rho_g$$
$$P_w - P_g = h\,(\rho_w - \rho_g) + H\rho_g$$

$$Whence, \quad h = \frac{P_w - P_g - H\rho_g}{\rho_w - \rho_g}$$

If $P_w - P_g > H\rho_g$ h will be positive
If $P_w - P_g < H\rho_g$ h will be negative, so the fluid contact will be below the
      point at which the water pressure was measured.

If $P_w < P_g$ with H negative, h will be negative.

When one well gives a pressure in gas and a second well a pressure in water, in the absence of other information it will not be clear whether or not there is also an oil column. The prediction on the assumption of there being no oil will give a hypothetical gas/water contact level. This will be lower than the gas/oil contact level and higher than the oil/water contact level, if there is oil between the gas and the oil. By making assumptions about the subsurface density of any oil present, limiting levels for any gas/oil and oil/water contacts can be indicated. Predictions of the levels of fluid contacts will be of value in selecting sites for follow-up wells. Furthermore, in conjunction with provisional structural maps, the fluid contacts will enable preliminary estimates to be made of the amounts of hydrocarbons contained in the traps.

It is not necessary for the wells to be on the same structure for predictions to be possible, but there must be communication between the points of pressure measurement through the fluids. Under certain circumstances, it may also be reasonable to combine a pressure measurement in the hydrocarbon column with information on the elevation of the reservoir rock outcrop in order to make a prediction of the fluid contact level.

If the predicted hydrocarbon/water contact level is below the level of the pressure observation in the water, the well (or other observation point) providing the water pressure value cannot be on the same structure as the well from which the pressure measurement in the hydrocarbon column was obtained. Furthermore, in this case the predicted level of the hydrocarbon/water contact can be taken to fix an upper limit for the level of the spilling plane separating the two structures on which the wells (or the well and outcrop) are placed.

If the predicted hydrocarbon/water contact level is between the levels at which the pressure measurements were made in the hydrocarbon and in the water, the two wells *may* be on the same structure. Should the predicted hydrocarbon/water contact be higher than the level of the pressure measurement in the hydrocarbon column, it is evident that there is an error in the assumptions. In fact, there cannot be free fluid communication (not necessarily direct) between the points at which the two pressure measurements were made.

In the absence of flow, a difference in the pressures in a given fluid at the same datum level implies that the given fluid is not continuous between the two points in question. For measurements in gas or oil, this does not necessarily mean that the reservoir rock is not continuous between the two points; but the two points must be in separate traps. For measurements in water, a difference in pressure at the datum level normally indicates a break in continuity of the reservoir rock between the two points. However, water in an isolated depression in a continuous reservoir rock could have a pressure which differs from that in water at the same level elsewhere in the reservoir rock.

Fluid Contacts

In Fig. 67, G shows the pressure value and the depth of a pressure measurement in the gas, and the line AGC gives the pressure gradient in the gas column. The line DGB represents the pressure gradient for the formation water. Assuming that a gas/water contact exists, any pressure measurement

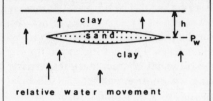

Fig. 65. Fluid pressure in sand enclosed by clay undergoing compaction, involving upward flow: $P_w > h\varrho_w$.

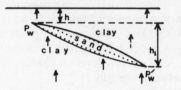

Fig. 66. Fluid pressure in inclined sand lens enclosed by clay which is undergoing compaction involving upward water flow: $P_w > h\varrho_w$, $P^1_w = P_w + h_1\varrho_w$ $> (h+h_1)\varrho_w$ (assuming negligible pressure drop for water flow in the sand). The presence of the sand leads to some water flow from the lower to the upper end of the sand lens.

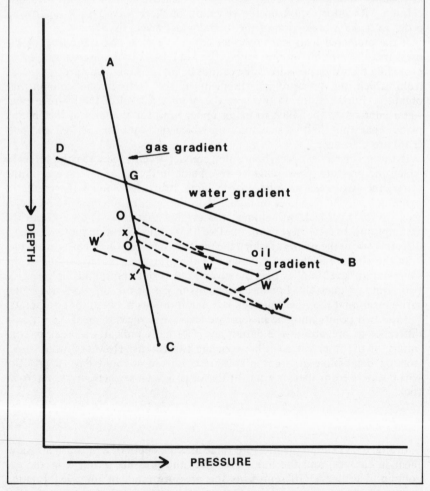

Fig. 67. Relationship between fluid pressures, fluid pressure gradients and fluid contact levels.

in water which falls in the sector BGC, e.g. W, could be on the same structure as the gas pressure observation, G. A pressure measurement represented by W' in the DGC sector could be in the same reservoir rock as G, but cannot be on the same structure. Pressure measurements in water corresponding to points above the line DGB cannot be in free fluid communication with the point corresponding to G, i.e. they are in different reservoirs, or at least isolated from G, if in the same reservoir rock. A line drawn through W with the slope of the water pressure gradient gives x as the gas/water contact for the G-W combination; similarly x' gives the gas/water contact for the G-W' combination. It is assumed throughout that there is no water flow.

If oil occurs between the gas and water and gives a gradient represented by ow (this gradient is clearly speculative when oil has not been detected in the trap), then o is a possible position for the gas/oil contact, with w the corresponding oil/water contact. o' and w' have comparable significance when the water pressure measurement is not on the same structure as G, i.e. for the G-W' combination.

For a particular system, it is evident that when there is oil, as well as gas and water, the gas/oil contact will be higher and the oil/water contact lower than the level predicted for a gas/water contact. It is also apparent that when oil is present in the G-W combination a line parallel to ow, and passing through W, intersects the gas gradient line at the highest possible level for a gas/oil contact. Hence, the maximum possible thickness for a speculative oil column can be predicted. In the case of the G-W' combination, a line drawn through G parallel to $o'w'$, the oil gradient line, intersects the water gradient line through W' at the lowest possible level for an oil/water contact, thus defining the height of the maximum possible oil column in the structure in which the gas pressure measurement, G, was made.

In the case of the G-W' combination, x' is the highest possible level for the saddle or other spill-under point separating the two structures on which the G and W' measurements apply.

Clearly, direct information on fluid contacts will be sought, but the inferences suggested above can be of value in preliminary assessments and in the selection of positions for obtaining critical well data at least cost.

Under certain circumstances, pressure measurements in single wells can provide other useful deductions concerning oil and gas reservoirs. These measurements call for the well to be shut-in after a period of constant rate of production, with the build-up of pressure against time after shutting-in being recorded; or for the declining pressure to be recorded against time during a phase of constant producing rate. When conditions are suitable, such observations can provide information on the permeability of the reservoir rock immediately around the well and in the more distant parts, on the presence or absence of an impermeable barrier such as a fault close to the well, or some indication of the size of the accumulation (a most valuable piece of information in the very early history following a discovery). There are, of course, on numerous occasions other pieces of information available at an early stage from which intelligent interpretation can derive some indication of the likely size of an accumulation. As is mentioned elsewhere, comparisons of sizes obtained by different and independent approaches can be helpful[16].

SUBSURFACE MAPS

Subsurface maps of various types are essential in the search for and in the development of oil- and gasfields. It has already been shown that the shape of the upper surface of a reservoir rock must be such that it provides a trap; and a knowledge of its actual or probable shape is important in attempts to site development wells correctly, while information on the thickness of the oil or gas zone is needed in making estimates of reserves by volumetric methods of calculation.

Although the ultimate goal is a map of the reservoir rock itself, during the exploration phase the structure maps available will only be tentative and may not represent the actual reservoir rock, although it is to be expected that they will be indicative of its general form.

Many types of subsurface maps are in use, some of them only for special purposes; but those which are indispensable are the *stratum or structure contour maps* and the *isopachyte maps*. Stratum contour maps show lines of equal elevation, positive or negative, relative to some horizontal datum, for the interface between two types of rock, say a shale and a sandstone; iso-pachyte maps show lines of equal thickness for a single stratum or group of strata. (It is commonest to interpret the thickness as that seen in a vertical well, not as the thickness measured normal to the bedding, although there have at times been suggestions that the term isopachyte should refer to the latter thickness—i.e. the so-called "true" thickness—rather than to the former[18].)

Structure contour maps were first used in the United States in the 1870's for delineating the anticlines of the Big Horn Basin in Wyoming, and form an essential preliminary to exploration drilling. They are based on observations of the value of the parameter under consideration at a series of discrete points, followed by interpolation between pairs of adjacent points or extrapolation via such points to give other points at a series of selected and uniformly-spaced values. Identical values are then connected by lines to give specific contours.

Elevations are expressed relative to sea-level, or in some circumstances relative to an arbitrary datum plane selected to avoid negative numbers, for example, 10,000ft or 20,000ft below sea level.

Contour intervals can vary with the purpose of the map and the amount of data available; in oil exploration operations, structure contour intervals of 100ft to 10ft are commonly used, with map scales of perhaps 1 : 10,000 or less. The contour interval employed must always be related to the quality and quantity of the observations, the range in values of the parameter contoured, and the fineness of detail that it seems justifiable to attempt to show.

Clearly, the greater the number of observation points in a given area the better will the contours be defined. It should be remembered, however, that subsurface contours are inevitably inferred, not traced, lines, so that poor quality or few data, or bad assumptions, may cause the contours to differ from reality over much or all of their paths. Good and plentiful data coupled with skilled utilization can yield maps which agree closely with new observations not included in the original interpretation. Naturally, all such maps are subjected to checking or revision when new information becomes available.

Surface geological mapping coupled with topographical surveying tech-ques can provide the elevations of a particular outcropping geological boundary or marker at known plan positions, and from those observations stratum contours can be derived to show the form of that boundary. A stratum contour is everywhere parallel to the strike of the surface contoured; hence the local dip is at right-angles to the contour line. Dip and strike information from adjacent beds can be used to aid in the construction of contours. Comparable constructions can also be made from aerial photographs.

Having drawn these contours, there may then be the assumption that the structure shown is some guide to the form of the deeper beds. However, outcropping beds are sometimes subject to disturbances of near-surface origin which may alter their form without affecting the deeper beds. Hence, in such areas caution must be applied in making subsurface projections from the observed form of the outcropping beds, and confirmation of the form or the supposed subsurface structure by a seismic survey is generally necessary, as mentioned below.

In areas for which contoured topographical maps exist, the intersections of a specific geological boundary with the topographical contours may be used to obtain the elevations of that boundary at known plan positions. Again, caution is necessary because the topographical contours themselves may have been interpolated, giving rise to possible quite modest errors which may, nevertheless, be significant when using the observations to define stratum contours in regions of gentle dip.

There are areas in which superficial deposits preclude the use of surface mapping to show the structure of the underlying rocks. It may then be decided to drill shallow holes and to log them to find the positions of one or more markers believed to be indicative of or helpful in predicting the structural form at depth. This provides plan positions and elevations from which stratum contours can be derived, and these can in suitable cases be associated with data obtained from surface mapping.

Digital computer procedures can be used to construct structure contour maps of stratigraphical surfaces if sufficient well-log data are available to feed into the system. The maps obtained will show structures of various scales as undulations—the larger undulations being represented by the lower spatial frequencies and the smaller undulations by the higher frequencies. Spatial filtering will then permit the preparation of subsidiary maps showing only those undulations which fall within a predetermined range of sizes.

The computation of such *spatially filtered maps* has been applied with success to the analysis of structure in areas of gentle dip such as the Interior Plains of southern and central Alberta. The maps obtained there define the presence of two orthogonal intermediate-scale structural trends, the most prominent of which runs northeast-southwest, while the other runs northwest-southeast. It is thought that many of the structures associated with these trends are tabular blocks of low relief bounded by basement-controlled faults, some of which have formed hydrocarbon traps.

In exploration operations, the aim is to obtain, before drilling deep and costly wells, the clearest possible picture of the structure at or near to potential reservoir horizons, since it is recognised that for various reasons surface or near-surface structures do not necessarily coincide with the position and form of deeper structures. Reflection seismic surveys provide a means whereby,

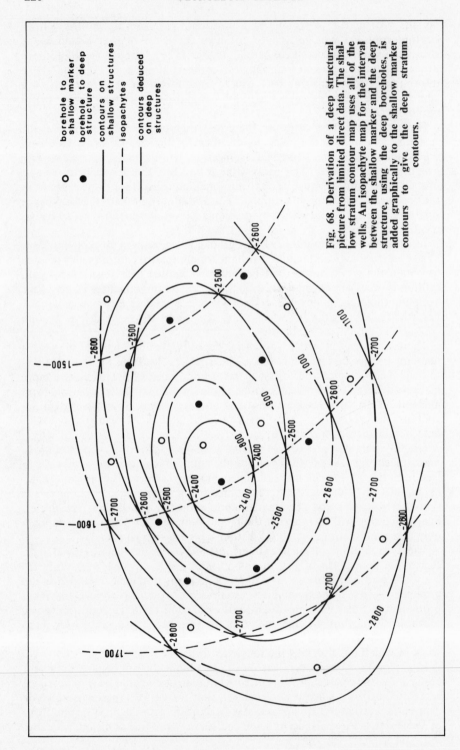

Fig. 68. Derivation of a deep structural picture from limited direct data. The shallow stratum contour map uses all of the wells. An isopachyte map for the interval between the shallow marker and the deep structure, using the deep boreholes, is added graphically to the shallow marker contours to give the deep stratum contours.

borehole to shallow marker

borehole to deep structure

contours on shallow structures

isopachytes

contours deduced on deep structures

under favourable conditions, the structure of the rocks at depth may be revealed Quite complex treatment of the basic reflection-time data may have to be applied in order to convert them to the depth data required by the geologist. It is obviously necessary to apply appropriate velocity information and, when the reflecting beds have strong dips, also to "migrate" the reflections, because the incident and reflected wave paths will then no longer be near vertical. Failure to undertake migration leads to depth and position errors.

Differences between surface and deeper structures may involve differences in position and/or shape, and a series of factors may be involved. Competent folds must change in intensity with depth, and in position also when they are asymmetrical. The effects are more marked the stronger are the folds. Incompetent folds can, in some cases, retain substantial similarity of form, and they may show less change in position at depth when they are asymmetrical.

The presence of angular unconformities between the shallow and deeper beds also leads to changes in structure with depth. Indeed, any lateral variations in the vertical thickness of a bed or group of beds, arising from whatever cause, must have an influence on the relationships between the structures of the beds above and below the zone of laterally changing thickness. Evidence of a broad nature on such changes in thickness may be available from careful measurements on outcropping beds, or from deep wells drilled earlier and penetrating the same beds. Such information can be employed for constructing isopachyte maps for the beds in question. Having constructed an isopachyte map in this way, it can then be superimposed on a structural map which depicts the form of the top of the group of beds in considerable detail, in order to carry out contour arithmetic to predict the form of the base of the groups of beds. Thus, where a stratum contour intersects an isopachyte contour, the numbers are added algebraically to give other numbers which are themselves the basis for deriving stratum contours for the base of the group.

Once again, it must be recalled that the shallow structure contours and the isopachyte contours are inferred lines, so the resultant third set of contours will be dependent on both sets of inferences. The values are certain to be correct only at the points where the deep wells meet the lower surface of the group. Elsewhere, for the reasons mentioned above, they must be viewed only as aiming at the best prediction possible from limited data. When the shallow stratum contours and the isopachytes are drawn with the same numerical intervals, the deduced deep stratum contours can intersect either of the other two sets of contours only where they themselves intersect (Fig. 68).

In any set of contours based on limited and unevenly spread data, there is justification for indicating different degrees of certainty in a general way by the use of different styles of line, e.g. full lines, broken lines, or broken lines with question marks. There is much to be said for retaining on the final contoured map the observation points, so as to indicate the amount and the distribution of the control on the contours.

Contour lines for fault surfaces are employed to show the position of non-vertical faults at depth, and thereby to permit the indication of the plan position of the line where a contoured stratum surface meets the fault surface.

The isopachyte map of a group of beds between two markers, forming the upper and lower boundaries of the group, can be interpreted as showing the approximate form of the lower marker at the time that the upper marker was laid down, provided that it is reasonable to assume that the upper marker was deposited on an essentially horizontal surface. Small thicknesses imply structurally high points, and vice versa. Likewise, the isopachyte maps of a series of superimposed groups of beds may be used as a guide to the structural evolution of the area during the total period represented by the groups.

By selecting a series of datum planes, stratigraphic isopachyte maps can be used in regional studies and can give valuable information about the geological history of sedimentary basins and the times of faulting and folding processes. Thus, a notable change of thickness of a formation at a fault shows that the fault was active while that formation was being deposited, its displacement during that time being equivalent to the difference in thickness of the bed on the upthrow and downthrow sides[10].

For dating folding movements also, the "isopachs" will show by the thinning of beds over the axis of a fold that uplift was occurring during the time they were being laid down—structures of this nature are termed "supratenuous". Correspondingly, the adjacent synclines will show a thickening of the same strata.

Other types of maps sometimes used in petroleum geology include *lithofacies* or *lithological ratio* and *clastic ratio* maps. The lithofacies map depicts by pictorial symbols, numbers, or contours ("isopleths"), the proportions of sandstone, limestone or clay in a particular bed at different points; it thus provides a convenient visual appreciation of how one rock type may grade laterally into another within the same formation or group of formations. The clastic ratio map shows by appropriate symbols the proportion of clastic to non-clastic sediments at various points; occasional variants are *sand-shale* ratio maps. A number of different sedimentary rock characteristics have been used for the construction of lithofacies maps, including variations of grain size, porosity, permeability, mineral content, etc; however, no single index can express the complete lithological nature of the bed in the same way as a structure contour expresses the level at a particular point.

The *biofacies map* attempts to depict variations in the biological as opposed to the lithological aspect of a stratigraphical unit—using details of the nature and numbers of the faunal assemblages present, both in terms of the proportions of the various groups found or as ratios of these groups to each other. Such biofacies maps can provide valuable additional information to equivalent lithofacies interpretations, indicating, for example, directions of variation of the environment of deposition.

In attempting to overcome the limitations of two-dimensional cartography, *cross-sections* are often used to show the nature and thickness of the strata existing at any point on a map relative to their nature and thickness at other selected points. They provide a method of adding the dimension of depth to a two-dimensional map. The lines of cross-sections selected depend on the availability of data. Thus, to define and illustrate an anticline, for example, the results of drilling wells fairly closely spaced along the major and minor axes would obviously be desirable, but approximate cross-sections can be made along these axes by projecting on to them (with suitable adjustments) data from wells drilled to one side or other of these lines, or combining data from outcrops. The vertical scale on such cross-sections is usually exaggerated in order to show a reasonable amount of detail in a small space, so that this practice inevitably gives a distorted impression of the degree of folding.

Various other methods of depicting three-dimensional information are also sometimes used; thus, *block diagrams* show perspective sections of three-dimensional blocks of sediments surrounding important wells, while the *isometric stratigraphic* (or *"fence"*) *diagram* consists of several lines of stratigraphic cross-sections arranged on an isometric base map with the correlation lines drawn in to give the impression of a network of fences viewed obliquely from above. Local or regional *palaeogeological maps* are sometimes constructed by combining well data with surface observations to depict the surface geology of a particular area as it existed in past time. Computer programmes can now be written which result in the output of printed maps and cross-sections incorporating selected geological parameters, thus providing a rapid and flexible source of information upon which development and exploration planning can be based[8].

REFERENCES

1. V. Apostolescu, "An attempt at zoning by ostracods in the Cretaceous of the Senegal Basin". *Rev. Inst. franc. Petrole,* **18**, 1675–1694 (1963).
2. J. Brooks, P. R. Grant, M. Muir, P. van Gijzel and G. Shaw (Eds.) "Sporopollenin". Academic Press, London 1971.
3. R. E. Chapman, "Clays with abnormal interstitial fluid pressures". *Bull. Amer. Assoc. Petrol. Geol.,* **56** (4), 790–795 (1972).
4. A. T. Cross (ed.) "Palynology in Oil Exploration". *Soc. Econ. Paleontologists Mineralogists Spec. Pub.,* **11**, 29–57, (1964).
5. J. H. Germeraad, C. A. Hopping and J. Muller, "Palynology of Tertiary sediments from Tertiary areas". *Rev. Palaeobotan. Palynol.* **6**, 189–348 (1968).
6. D. Gill, "Application of a statistical zonation method to reservoir evaluation and digitized log analysis". *Bull. Amer. Assoc. Petrol. Geol.,* **54** (5), 719– 729, (1970).
7. M. Glaessner, "Micropalaeontology". Melbourne Univ. Press (1945).
8. G. C. Grender, L. A. Rapoport and R. G. Segers, "Experiment in quantitative geologic modelling". *Bull. Amer. Assoc. Petrol. Geol.,* **58** (3), 488–498 (1974).
9. C. C. M. Gutjahr, "Carbonisation measurements of pollen grains and spores and their application". *Leidse Geol. Medl.,* **38**, 1, (1966).
10. W. E. Hintze, "Depiction of faults on stratigraphic isopach maps". *Bull. Amer. Assoc. Petrol. Geol.,* **55** (6), 871–890 (1971).
11. C. A. Hopping, "Palynology and the oil industry". *Rev. Palaeobotan. Palynol.,* **2**, 23–48 (1967).
12. A. S. Horowitz and P. E. Potter, "Introductory Petrography of Fossils". Springer-Verlag, Berlin (1971).
13. M. P. Jones and M. G. Fleming, "Identification of Mineral Grains". Elsevier Publishing Co. (1965).
14. W. C. Krumbein and L. L. Sloss, "Stratigraphy and Sedimentation", 2nd ed., W. H. Freeman, San Francisco (Chapters 3, 8, 13 and 14) (1963).
15. W. G. Kuehne, "Photoelectric separation of microfossils from gangue". *Proc. Geol. Soc. Lond.,* 1664, 221–222 (1970).
16. C. S. Matthews and D. G. Russell, "Pressure build-up and flow tests in wells". Soc. Petrol. Eng. AIME, New York (1967).
17. H. B. Milner, "Sedimentary Petrography" 4th Edn., Allen and Unwin Ltd., (1962).
18. P. E. Pennebaker, "Vertical net sandstone determination for isopach mapping of hydrocarbon reservoirs". *Bull. Amer. Assoc. Petrol. Geol.,* **56** (8), 1520–1529 (1972).
19. J. Rade, "Otway Basin, Australia: use of calcareous nannoplankton and palynology to determine depositional environment". *Bull. Amer. Assoc. Petrol. Geol.,* **54** (11), 2196–2199 (1970).
20. A. J. Rudman and R. W. Lankstone, "Stratigraphic correlation of well-logs by computer techniques". Bull. Amer. Assoc. Petrol. Geol., **57** (3) 577–588 (1973).
21. N. N. Subbotina, "Fossil Foraminifera of the USSR". Colletts (1953).
22. F. G. Tickell, "The Techniques of Sedimentary Mineralogy". Elsevier Publishing Co. (1965).
23. E. N. Tiratsoo, "Principles of Petroleum Geology". (Chapter 16), Methuen Ltd., London (1951).

CHAPTER 9

Petroleum in Space and Time

EARLIER chapters will have shown that for a petroleum accumulation to exist in the subsurface at the present day, four essential conditions must have been satisfied in the course of its geological history. First, a source rock of the requisite "maturity" must have generated a large volume of crude oil or closely-related "proto-petroleum"; secondly, a permeable reservoir rock must have been connected to the source rock via a channel suitable for fluid migration and must have been in the correct stratigraphic relationship with it; thirdly, an effective closed trap must have been formed in which the oil could accumulate; and fourthly, the filled trap must have been preserved intact for millions of years.

Furthermore, it is essential that each of these prerequisites should have existed in the correct relationship in *time* as well as in space. The factor of "timeliness" is therefore the fifth essential requirement for an oil accumulation to exist today; the fact that accumulations are relatively rare—and large accumulations rarer still—probably indicates that this requirement has only infrequently been satisfied.

The objective of petroleum geology is the discovery and development of commercial crude oil and gas accumulations, and it is therefore of obvious importance to consider where, in the light of the above conclusions, present-day oilfields and gasfields are most likely to occur.

THE FRAMEWORK OF THE EARTH

It is nowadays generally believed that the Earth is made up of two principal components: a *core* of about 3,500 km radius, the outer shell of which is made up of ultra-dense liquid rock surrounding a solid inner iron mass; and an overlying generally solid *mantle,* about 3,000 km thick, composed of very dense rocks—eclogites and peridotites—the upper layers of which (the "asthenosphere") can sometimes act as a highly viscous liquid.

Floating on the mantle, and distinguishable from it by a sharp change in seismic wave velocity,* is a thin *crust* whose composition and thickness vary

---

* *The Mohorovicic discontinuity.*

significantly depending upon whether it underlies the continents or the oceans. Thus, beneath the continents, which now make up about 30% of the Earth's surface area, the crust is 20–40 km thick, and is mainly composed of acid igneous ("sialic") rocks of densities 2·67–2·77 g/ml (much lighter than the high-density rocks of the upper mantle). Beneath the oceans, the crust is much thinner, and is of a nearly uniform thickness of about 8 km, being largely made up of basaltic ("simatic") rocks of average density 2·9 g/ml. The uppermost layers of both the continental and oceanic types of crust may be composed of irregular and varying thicknesses of sediments, which have been formed and distributed by geological processes; such layers may be up to 15 km thick on land, but only 1 km or less on the ocean floors.

The boundary zones between the continents and oceans form the *continental margins*. These have an area equivalent to about a quarter of the Earth's present dry land, and are made up of three components: the gently-dipping shelves, which vary in width from 5 to 250 miles and are the submerged edges of the continents; the steeper slopes, about 10 to 30 miles wide, the bases of which mark the transition zone from continental-type crust to oceanic-type crust; and the continental rises, which largely comprise "fans" of continental sediments. (The continental margins are discussed in more detail on pp. 257-9.)

The continents are made up of several geotectonic components, of which the most important are the "Stable Areas"—the Cratons and Shelves; and the contiguous "Mobile Areas".

I. The Stable Areas

(a) The "Shields" or "Cratons"

About a quarter of the present land area of the Earth consists of very ancient, highly-metamorphosed Pre-Cambrian rocks, which have seldom if ever been covered by the sea. These areas have been termed "Shields" or "Cratons" and form the stable nuclei of the continents.

Examples are the Canadian, Fenno-Scandian, Indian and Brazilian "Shields".

(b) The "Platforms" or "Shelves"

At various times, the "Shields" sustained gentle tilting and submersion of their edges, so that shallow seas encroached and deposited sediments around the central nuclei; the thickness of these sediments does not usually exceed 6 km. Much of the Earth's present land surface is made up of the resultant "Platform" or "Shelf" areas, which form gently-dipping, low-relief regions of sub-continental extent—e.g. Western Siberia and the Great Plains of North America.

As a result of deep-seated crustal processes and large-scale, vertical ("epeiro-genic") movements, the platform areas were often subjected to regional warping effects, and in some cases were broken up into smaller blocks.

Some of these blocks have remained stable since Pre-Cambrian times, while others have subsided and been covered by younger strata. The blocks are often bordered by belts of relatively gentle "paratectonic" folds, produced by the deflection of regional stresses around them. Examples of stable fault-blocks bordered by folds of this type are the Colorado Plateau of the United

States, the Moroccan Meseta, the high plateau of Algeria, and the Pannonian "block" of eastern Europe.

In some areas, the broad swells resulting from epeirogenic movements formed regional arches or "anteclises" (anticlinal structures often many miles long, as in Siberia) and sags—the intracratonic basins (or "synclises") mentioned below. More localised vertical ("taphrogenic") movements produced block-fault mountains and rift valleys. Taphrogenic movements also affected the underlying crystalline "basement" rocks in some parts of the world, producing deep-seated structures which are sometimes reflected in the overlying sediments.

"Shelf" areas are customarily divided into stable and less stable categories. The unstable shelves form the transition or "hinge" zones between the stable and mobile areas of the Earth's crust (described below), and are characterised by having been subjected to cyclical sedimentation.

II. The Mobile Areas[1-16]

Tectonic forces, the nature of which are still only partly understood, have acted at different times during the Earth's geological history so as to produce various types of *downwarps* in and around the stable areas. A characteristic of these depressions is that their floors continued to subside over lengthy periods of time, during the course of which the depressions were filled with great thicknesses of sediments.

(a) Geosynclines

One variety of these downwarps on the continental platforms took the form of a long, narrow and often sinuous "trench" or "trough", which could extend for hundreds of miles. Such troughs were not always continuous, but were sometimes made up of a series of separate linear depressions, which eventually became filled with great thicknesses of shallow-water sediments. At several times in the course of the Earth's history, the masses of sediments deposited in depressions of this kind were subjected to processes of "orogenesis"—violent folding, and uplift, accompanied by volcanic activity.

As a result, the uplifted and contorted sediments tended to spill outwards over the platform areas lying on the flanks of the troughs, and often subsequently became the roots of mountain chains, since, being composed of relatively light material derived from the weathering of the upper crust, they were isostatically uplifted whenever the original downward drag, believed to have caused the troughs, weakened.

The term "geosyncline" has been used to describe the long, narrow, subsiding depressions in which these thick piles of sediments accumulated and were subsequently affected by orogenesis. The stable flank which provided the bulk of the sediments filling the depression is usually termed the "hinterland", and the opposite flank the "borderland" of the geosyncline.

There is considerable variation in the nature of geosynclines; in particular, the processes of trough filling and subsequent orogeny were sometimes widely separated in time. Furthermore, the axis of a geosyncline could shift in the course of its history as a result of marginal deformation.

Attempts have been made to differentiate between varieties of geosyncline, principally on the basis of their contents of volcanic material. Thus, "eugeosynclines" have been defined as geosynclines, the sediments of which show evidence of intermittent volcanic activity during the filling process; whereas "miogeosynclines" were unaffected in this way. In the Pacific region, it has been observed that these two types of geosyncline often occur side by side in the form of a "couple" separated by a regional uplift—called in such cases a "geanticline".

The fundamental cause of the periodic outbursts of orogenic activity in the Earth's crust is not yet clearly understood, but must have been related to deep-seated geotectonic processes. There have in fact been three principal phases of this activity since the close of the Cambrian. As a consequence, geosynclinal troughs which were formed and filled with sediments in Silurian-Devonian, Carboniferous-Permian, and Tertiary times, were subjected to what are termed the Caledonian, Hercynian and Alpine phases of orogeny, respectively. The sediments were compressed into folds and isostatically uplifted into mountain belts, in the cores of which there was great igneous activity and magmatic penetration. Many of these mountain belts were later largely obliterated either by the foundering of their roots, by epeirogenesis, or by erosion.

Remains of Caledonian mountain belts occur in parts of Britain and Scandinavia, while in North America, the Appalachian mountains are a complex of two orogenic systems corresponding roughly to the Caledonian and Hercynian systems of Europe. A Hercynian belt still survives in Europe as the Ural Mountains, and remnants of a southern belt are found in southwest Ireland and the Harz and Bohemian Mountains of Central Europe.

Most of this southern Hercynian belt foundered to form the Tertiary "Tethys" geosyncline, which extended for many hundreds of miles between two ancient northern and southern landmasses (p. 241) in late Mesozoic times. Subsequently, a complex series of mountain-building movements began, producing the mountain chains which now extend from the Pyrenees via the Alps, Carpathians, Caucasus, Taurus and Himalaya ranges to the mountains of Indonesia and New Zealand. About the same time, another narrow geosynclinal belt was forming along the western coasts of the Americas and around the Pacific Coast of Asia. This in due course resulted in the mountain ranges which now encircle the Pacific, including the Rocky and Andes mountain systems.

The youngest of the mobile belts are predictably areas where earthquake shocks are common. In fact, a plot of the epicentres of major earthquakes shows close coincidence with the Mediterranean and circum-Pacific belts. Volcanic activity is also typical of the central, structurally weakest zones of these mountain chains.

Clearly, any petroleum accumulations formed in the sediments of a geosynclinal trough are likely to be destroyed during its later, violent tectonic history. Petroleum accumulations in orogenic areas are consequently relatively infrequent, and the importance of geosynclines from the point of view of petroleum geology is therefore indirect:

(a) the uplifted mountain chains have provided sources of sediments for deposition in younger basins;

(b) the orogenic forces which deformed the geosynclinal "pile" have often produced less strongly folded structural traps in other sedimentary rocks in the same region.

(b) Other Types of Sedimentary Basin

A geosynclinal trough may be considered to be a specialized type of sedimentary basin, the contents of which have been subjected to orogenesis. But there are many other types of sedimentary basin, which are distinguishable from geosynclinal troughs by their shapes (usually circular or oval in plan) and the fact that their sediments have not been directly affected by any orogeny.

Sedimentary basins are termed "active" or "dynamic" if they are downwarps in the Earth's crust, the floors of which continued to sink as the depressions were filled with sediments. The forces producing the downwarps are now thought to have been essentially gravitational, resulting from the loss of support consequent on the movement away of subcrustal material. The resultant basins are typically asymmetrical, since the sagging process would generally have been concentrated along one side of the basin, where a fault zone developed (Fig. 69).

In contradistinction to these structural types of basin, there are also "topographic" or "depositional" basins in which sediments accumulated long after the depression was formed. In some cases, two or more structural and depositional basins may be geographically superimposed to form a "vertically composite" basin, as is the case in the intermontane basins of the Rocky Mountains region.

Attempts have been made to classify sedimentary basins in various ways, on the basis of their shapes, histories or tectonic origins. The following is one suggested classification of the principal groups which are of importance in petroleum geology[8]:

I. Basins which are underlain by Continental-type Crust *

1. Cratonic Basins

These are primary dynamic basins which are circular or elliptical in plan, and which occur in the interior of cratonic regions—examples are the Michigan, Illinois or Williston Basins of North America. Basins of this type are usually filled with a single cycle of Palaeozoic sediments.

2. Intracratonic and Cratonic-Rift Basins

Secondary dynamic basins, resulting from the local breakdown of crust on the crests or flanks of major regional uplifts, are usually found towards the edges of the cratons. They may be filled with several cycles of sedimentation, some of which have been derived from local orogenic uplifts. Examples are the West Texas and Ural-Volga Basins.

This group also includes the grabens and half-grabens, ranging from narrow, Suez-type rifts to broad basins which are broken up into block-fault type structures, as in the Sirte Basin of Libya.

---

* In this context, it must be remembered that, perhaps as a result of "plate" movements (q.v.), some continental areas are apparently separating, while others are being drawn together. Hence, the present-day margins of the continents do not everywhere coincide with the limits of the oceanic and continental types of crust, so that cratonic basins may extend offshore (e.g. the North Sea basin), while intermediate-crust basins may occur in areas which are now cratonic (e.g. the South Caspian basin).

II. Basins near the Continental Margins which are underlain by Intermediate-type Crust

1. Embayments

These are downwarps which open out on to enclosed seas, gulfs or small ocean basins, e.g. Gulf of Mexico, the Caribbean Sea, Middle East Gulf, etc., and show multiple cycles of sedimentation developed in deep troughs along the margins of continents. Basins of this type are particularly rich in petroleum.

2. Marginal Basins[1,2]

An increasingly important group of basins from the viewpoint of petroleum geology are the "marginal" or "open" basins which occur on the margins of the continents—particularly on the flanks of the Atlantic, and to some extent the Indian Ocean. These basins may have resulted from the continental "rifting" processes described on p. 240; there are no "borderlands" discernible, while the "forelands" may be far inland. Basins of this type are usually filled with a wedge of sediments that is thickest along the present-day coastline but thins both inland and out to sea. Extensive lateral and step-faults are common on the seaward side of such basins, in which there are often salt domes. Several basins of this type have been discovered in recent years on the west coast of Africa and along both the east and west coasts of South America.

3. Basins parallel to geosynclinal axes

Many sedimentary basins are known which are roughly elliptical in plan, with their major axes parallel to the axes of nearby geosynclinal uplifts. The deformation of the sediments contained in such basins is clearly related to the folding movements originating in the geosyncline, and the structural trends that have resulted are therefore largely parallel to the main axes of orogeny. Predictably, most such basins are associated with lines of Tertiary uplift, since older basins will tend to have been destroyed by younger orogenies.

4. Upper Tertiary Deltas[17-19]

Thick masses of Upper Tertiary sediments have been laid down at points where some major rivers reach the continental shelves. These wedges of sediment can be considered as a special type of depositional basin, which is productive of petroleum off the mouths of the Niger and Mississippi rivers.

Simply because they are regions of thick sedimentation, sedimentary basins are likely to be areas which have been especially favourable for the generation of petroleum. Primary migration processes which develop as geostatic pressure increases may be expected to cause the oil and gas to move upwards and outwards from the deepest, central parts of the basins towards the edges, where the hydrocarbons would be trapped in structural folds resulting from gravity effects or nearby orogenic activity. Fluids passing out of the basin proper would probably be trapped in the hinge zone or on the adjoining continental shelf, where widespread gently-dipping sandstones and limestones would provide excellent reservoirs wherever local traps had been formed.

It is understandable why dynamic basins in particular should be so favourable for petroleum. They typically contain varied, shallow-water sediments which are similar in their nature to those which have gone to make up the neighbouring mountain chains, but which are unmetamorphosed and less deformed. The great thickness of clays and shales deposited in the deeper, central areas of the basins provide potential source rocks, while the sandstones and limestones lying on the shallower flanks make excellent reservoir rocks. Traps result from the many unconformities, or the folds produced by gravity slumping and lateral compression; periodic desiccation results in the deposition of evaporite beds which make excellent seals. All the basin requirements for the generation and accumulation of petroleum are therefore likely to be present, and dynamic basins in fact contain some of the world's most important oilfields.

Of course, there are great variations between the concentrations of oil that have been found in different sedimentary basins and in different parts of the same basin. In some basins where there are large traps, for example, relatively few of these may be filled with hydrocarbons, due presumably to one or other of the necessary conditions not being satisfied; in other basins, there may be many, but only relatively small, oil and gas accumulations; while in still others, there are only a few very large or "giant" fields.

Clearly, much will depend not only on the volume of the sediments laid down in a basin but also on their nature, the proportions of source and reservoir rocks, the frequency with which marine transgressions and regressions have produced or covered unconformities, and the general geothermal gradient of the region. A thick organic-rich sedimentary sequence and a suitable thermal gradient are obviously favourable factors as regards petroleum generation and subsequent primary migration. The periodic "silling" and desiccation of the basin to produce evaporites (for eventual reservoir seals) and anaerobic conditions are also favourable factors. Deformation of the basin sediments, either as a result of gravity factors or by the effect of neighbouring orogenic mountain-building processes will clearly favour the development of structural traps. "Stratigraphic" and reef-type traps are likely to be found in basins with frequent unconformities, overlaps and facies alternations.

All parts of a particular basin will not, in any case, be equally favourable for petroleum accumulation. Towards the central region, pressures and temperatures would tend to favour the occurrence of gas rather than oil, while any hydrocarbon accumulations that have been formed may lie too deep to be reached by drilling. There seems to be a tendency for older reservoirs to occur on the outer flanks, and younger oil to be found basinwards, presumably because these areas represented successive zones of optimum oil generation, i.e. of critical depth of burial. (Examples of this can be noted in the US Gulf Coast, where the oil is found ranging in age from Jurassic to Pleistocene in parallel belts, or in the Mesopotamian basin, where Jurassic oil on the "foreland" shelf is followed in turn by Cretaceous and Tertiary accumulations towards the deep basin.)

MODERN TECTONIC THEORIES[20–57]

In the preceding pages, "tectonic processes" have been mentioned as the source of the energy required to produce crustal deformations of various types, and there is still considerable controversy as to the nature of such processes. However, in recent years, there has been growing support for the theories of the "new global tectonics", which seek to explain the geotectonic development of the Earth in terms of the movement of a relatively few "lithospheric blocks".

It is thought appropriate, therefore, to give an account at this stage of the main conclusions that have been reached, while emphasising that acceptance of these theories is by no means universal, and many difficulties still remain unresolved[36, 39, 43].

The striking similarity of shape of the west coast of Africa and the east coast of South America has for long encouraged the belief that they were once united and have since moved apart. If fitted at the 500-fathom line the match between these continents is remarkably close, and supporting evidence for a former proximity includes the existence of similar geosynclinal features and granitic rocks in Brazil and Gabon, as well as some striking ecological and botanical similarities. Aeromagnetic surveys also show remarkable parallelism; thus, the NE-SW anomalies of Sierra Leone and the Ivory Coast, which are presumably related to deep-seated "basement" features, are repeated in Surinam and Guyana.

The evidence for the relative movement or "drift" of continents, which was first developed as a geological theory in 1912,[54, 55] has in recent years become increasingly convincing, due mainly to the accumulation of magnetic measurements that have been made on a wide variety of rocks since the early 1950's. These have been based on the observation that the ferromagnetic particles present in a cooling rock take up the direction of magnetism of the Earth's field at the moment the rock temperature drops past a characteristic temperature called the "Curie point".

Since the directions of palaeomagnetism observed in rock specimens obtained from different parts of the world varied considerably, it was initially thought that these differences could be explained by the fact that the Earth's magnetic pole had moved in the course of time with respect to its geographic or rotational pole, a process which might be termed "magnetic polar wandering".

Subsequently, as palaeomagnetic data accumulated from rocks of different ages and provenances, it became clear that the degree of magnetic polar wandering which would be needed to explain the widely varying results obtained would be impossible, and other explanations were therefore sought. Thus, it was suggested that the whole Earth might have moved in some way relative to its axis of rotation (this, of course, is fixed relative to the Sun, apart from the small variations due to nutation and precession). If such a process had taken place it would be true "polar wandering", and could have been responsible for the observed variations in the directions of rock palaeomagnetism. Another process suggested at about the same time was that the outer lithosphere of the Earth might have somehow "slipped" over the mantle, which itself had always remained fixed relative to both the magnetic and rotational axes.

Either of these theories, however, if correct, would imply that all rocks of the same ages, from wherever obtained, should indicate a common magnetic pole. In fact, however, rock specimens of the same ages but derived from different continents show widely varying polar directions, leading to the conclusion that the continents must have moved in relation to both the rotational and the magnetic poles, as well as to each other, over a period of at least the last 200 million years.

According to the theory of "plate tectonics" which has been developed since 1967, the surface of the Earth (both continents and oceans) is made up of a mosaic of rigid, aseismic lithospheric blocks which are slowly and intermittently in motion. About a dozen such major "plates" have been identified, as well as a number of minor and more complex smaller units. Most of the "plates" support at least one massive continental craton; but there are also oceanic plates which mainly underlie ocean basins.

The plates are separated by linear "active bands" in which vulcanicity and seismic activity seem to be concentrated. Plate boundaries are characterized by three general types of observable phenomena:

(a) When the plates are moving relatively apart (in the process known as "sea-floor spreading"), magmatic material from the underlying lithosphere wells up between them to form linear lava ridges. As the lava cools, it takes up the direction of magnetism of the prevailing magnetic field, and since (as explained below) this has reversed its polarity many times during the past, the result has been the production of parallel, linear "mid-ocean ridges" which are magnetized in opposite directions.

Sea-floor spreading implies that the continents have been split and spread apart as a result of the upwelling of new crustal material from the Earth's mantle and its incorporation into the ocean basins along both sides of the mid-ocean ridges which are now known to exist in the Atlantic, Pacific and Indian oceans. Measurements of the direction of magnetization of rock specimens have led to the inference that the rate of spreading may average several centimetres a year. Although the movement has been intermittent and irregular, it would be quite sufficient to produce the phenomena of continental drift in the course of geological time, probably best exemplified by the separation of Africa and South America, and the associated growth of the Atlantic Ocean.

(b) Plates can also apparently *slide past* one another along lines of lateral or "transform" faults without any resultant generation or destruction of plate material. Thus, the mid-ocean ridges in the Atlantic and Pacific Oceans, which run roughly north-south, are broken into a series of segments, each about 200 miles long and offset relative to each other. At the points of offset, other ridges branch off roughly at right angles to the main ridge. These have been termed "transform fault zones" and are notable for being the sites of "sea-mounts" which project above the deep ocean floors, as well as being earthquake and volcano zones.

(c) When plates converge, there are several possible results, depending on whether oceanic or continental-type plates are involved:

(i) When two oceanic plates collide, one plate is thrust under the other, and the lower plate descends towards the mantle, where it is partially destroyed by heat. Part of the slab continues to sink to a depth of as much as 700 km, where it seems to come to rest; part is converted to low-density magma, and this rises to merge with the upper plate (Fig. 70). In the Pacific region, the "subduction zone" of plate destruction is typically a "trench"—a deep, linear depression, usually many miles long, which may be partially filled with sediments. The Earth's crust is apparently being "digested" at rates of around 12 cm/p.a.

along the Aleutian, Japanese and other Pacific trenches—the rates of plate consumption in these regions and of plate production along the mid-ocean
ridges sum to approximately zero on an Earth of essentially constant radius[31].

The rising low-density magma derived from the partly-consumed descending slab, appears generally to form explosive-type volcanic islands which
are arranged in a curved line ("island arc") on the continental sides of the
trenches, the curvature of these arcs reflecting the curvature of the trenches.

(ii) When a continental-type plate is advancing towards and over-riding a
relatively stationary oceanic plate, a coastal mountain range will be formed along
the leading edge of the continental plate (e.g. the mountains of Chile), and
a marginal trench will be pushed ahead of the continent.

(iii) When an oceanic plate is advancing and passing beneath a relatively
stationary continental plate, island arcs and subduction zone trenches will be
formed (e.g. the Burma-Indonesia, Philippine and Taiwan-Luzon arcs).

(iv) Where two continental-type plates slowly converge, as is thought to
have occurred when the northward-moving plate carrying the Indian sub-
continent impinged on the main mass of Asia, major mountain ranges
(e.g. the Himalayas) will be formed as a result of the squeezing of the
sediments carried on the underthrust plate.

For how long have the processes of continental drift been active? One line
of evidence implies that some processes of crustal mobility were operating
3100 million years ago, although stable continental-type blocks may not
really have evolved until 500 million years later. Once having been formed,
these blocks were subsequently subjected to periodic phases of mobility. It is
obviously difficult to make meaningful hypotheses about "palaeo-plate-
tectonics", but it is possible that the Pre-Cambrian "Shield" of Canada has
been made up from at least five pre-existing continental components.

There is certainly persuasive evidence to support the theory that a major
process of continental break-up and movement started in Triassic times,
about 200 million years ago. At that time there were only two continents in
existence—the "super-continents" of "Gondwanaland" which lay between
the southern ice-cap and the tropics, and "Laurasia", which roughly straddled
the Equator.

They were separated by the Tethys Ocean and surrounded by long sedimentary troughs, which were the sites of considerable seismic and volcanic
activity. For some 40 million years, a general northward drift of the
continental masses continued intermittently; then, about 160 million years
ago, Gondwanaland began to break up into a number of separate components.
This period of break-up was accompanied by a general out-pouring of
basaltic material, the formation of downwarped basins which were invaded
by shallow seas, leading to the production of great thicknesses of evaporites[29],
and the spread of the Tethys Ocean southwards along the edges of the new
continents. The Indian Ocean began to open about 160 million years, and the
South Atlantic about 120 million years ago. South America rotated westwards
from Africa, and India rotated eastwards from Africa and moved northwards
until it collided with Asia. Australia and Antarctica first drifted away from
Africa and then finally separated about 50–60 million years ago. Africa itself
moved northwards to collide with Europe, and the resultant compression
of the Mediterranean Sea is still continuing.

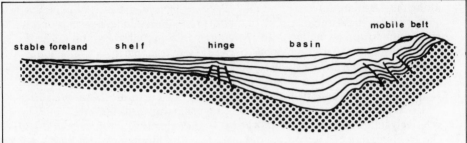

Fig. 69. Idealized cross-section of an asymmetric sedimentary basin lying between a stable "foreland" and a "mobile belt".

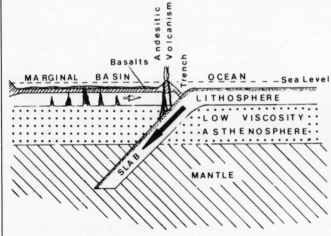

Fig. 70. Diagram illustrating subduction of an oceanic plate, with evolution of volcanic island arc, marginal basin and trench. (after Sleep and Toksoz, Nature, 547, Oct. 22, 1971).

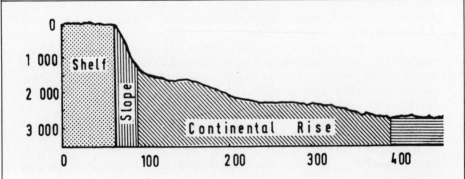

Fig. 71. Cross-section depicting the continental shelf, slope and rise off the U.S. Atlantic coast. (Horizontal scale in nautical miles, vertical scale in fathoms).

Activity in Laurasia was generally slower than in Gondwanaland. North America began to rift away from Africa about 200 million years ago but remained connected to Europe for a further 80 million years. Then, about 120 million years ago, the Atlantic opened to about a quarter of its present width, and after a period of relative quiescence, another active phase of opening started about 80 million years ago and is thought still to be continuing.

Mechanisms of Plate Motion[23, 26, 32, 46]

Considerable controversy surrounds the nature of the propelling forces which have caused the movements of lithospheric plates.

It was first believed that sea-floor spreading was the result of the growth of the Earth's core, which produced convection currents in the mantle material. These tended to break up and set in horizontal motion the relatively light, floating continental masses. Rising convection currents eventually reached the Earth's surface along the median ridges of the ocean basins, and the continental masses came to rest over areas where the currents descended back into the mantle. If, exceptionally, a rising magmatic current reached the Earth's surface beneath a continent, rather than as more usually along a mid-ocean ridge, a phase of continental break-up might then be initiated.

More recently, it has been proposed that plate movements are due to the presence of a number of thermal centres or "hot spots" fixed in the upper mantle, from which "plumes" of hot material rise intermittently to burn holes in the overlying crust. The continental plates are pushed away from these "hot spots", due to the creation of new ocean floor, and when the lithosphere above a "plume" is set in motion due to the process of sea-floor spreading, a "plume scar" is often left on the crust in the form of a line of volcanic cones. For example, it is suggested that Iceland now lies over one such "hot spot", and the volcanic islands of Scotland form a related "plume scar". Similar "hot spots" are believed to underlie the Hawaiian and other groups of Pacific islands. These are made up of lines of exposed volcanic cones, together with a number of submerged cones or "sea mounts", whose linear arrangements are thought to indicate the direction of local plate movement during a time-span measurable by the range of ages of the volcanic material which has been intermittently ejected from the cones. The ages of this material (obtained by various radiometric techniques) in fact generally increase in proportion to the distance of the volcanic island concerned from the conjectural "hot spot".

A number of other theories of the mechanism of continental drift have also been proposed. Thus, it is possible that the crustal plate propelling force is really only the outward "push" from the growing mid-ocean ridges which results from the "downhill" gravitational movement of the plates on each side of the ridges. Again, plate movement may be due at least in part to the downward drag of the outer plate edges as they sink into the oceanic trenches. However, some form of "convection cell" mechanism seems to be the most likely cause. It can in fact be theoretically shown that whenever a relatively thin sheet of material in which thermal conductivity is not symmetrical about a centre point floats on a heated viscous fluid, it will tend to be propelled by induced convective motion.

There is a considerable body of opinion which believes that besides explaining the evolution of the ocean basins, the theory of "plate tectonics" can also account for the growth and collapse of geosynclines and the consequent creation of mountain belts. Thus, a present-day, "live" geosyncline is thought to lie to the east of the United States, paralleling the coast for a distance of some 2,000 km. This takes the form of a plano-convex wedge of typical geosynclinal sediments (turbidites or "flysch") which may be as thick as 10 km at its maximum. This sedimentary wedge covers the continental rise to a width of 250 km, and may be equivalent to the eugeosynclinal member of an ancient geosynclinal couplet. The equivalent miogeosyncline is thought to be represented by a second sedimentary wedge which caps the continental slope to the west, attaining a thickness of 3–5 km under the coastal plain and continental shelf. The sediments of this wedge are very similar to those of the folded Appalachian Mountains of Pennsylvania.

It is suggested that similar growing geosynclines may be being formed along the edges of other continents, and that the orogenic cycle of mountain growth may indeed be explicable in terms of "plate tectonics": the continental margins form the areas where the plates which have been set in motion as a result of mid-ocean sea-floor spreading interact with the continental masses.

Initially, the geosynclinal wedges would be most likely to grow along the tectonically stable trailing edges of a moving continental plate; but subsequently, if the direction of plate motion were reversed, a trench or subduction zone would be formed along this continental margin and the geosynclinal wedge would eventually collide with the trench, and its sediments would be crumpled into a eugeosynclinal fold belt, which might then be sheared and thrust over the neighbouring miogeosynclinal belt. Low-density magmas would ascend from the descending continental plate and invade the rocks of the eugeosyncline. The collapsed, tightly folded and faulted geosyncline could be as thick as the continental plate and be underlain by oceanic rocks; whereas the neighbouring miogeosynclinal rocks, being nearer the stable continental mass, would be underlain by continental-type rocks and could be expected to be less intensely folded.

These theories lead to the belief that the continents grew in the course of time by processes of accretion of geosynclinal belts and island arcs; this seems to be especially applicable to Asia. In that continent, there is evidence for the accretion of at least nine separate blocks, and the orogenic belts between the major cratonic areas are probably the sutures along which the crustal fragments have collided and coalesced at various times between the Upper Carboniferous and the Early Mesozoic.

Plate Movements and Petroleum[27, 33, 51]

The intermittent movement of lithospheric plates must have had important consequences in the cycle of petroleum genesis, migration, accumulation and destruction, but this is a very complex story which still awaits clarification. It is evident that long-continued plate movements must inevitably have brought about far-reaching changes, especially in the development and distribution of many plant and animal species.

An important aspect of plate movements must be the resultant effects on local climates and environmental conditions[30, 35a]. In general, tropical and semi-

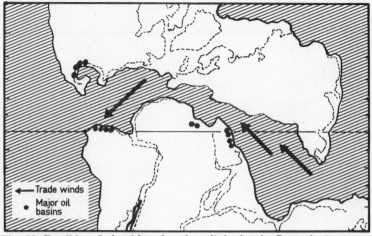

Fig. 72. Possible relationship of major oil basins in Laurasia (top) and Godwanaland (below) during the Cretaceous (after Irving *et al.* 33a).

tropical waters fringing the continents could be expected to contain the largest proportions of organic matter and thus to be the most likely zones of oil generation. But if the continents have moved in relation to each other and to the poles, it is clear that present-day oilfields may be found in geographical positions which are far removed from the latitudes in which their petroleum accumulations were originally formed.

The analysis of geomagnetic data collected from a number of major Tertiary (Oligocene to Pliocene) oilfields of Europe, Africa and Asia gives a mean palaeolatitude of 28·6°N compared with a present-day mean latitude of 35·6°N, indicating generally warmer waters at the time of origin than present-day latitudes would account for. Furthermore, the magnetic data available from a limited number of Mesozoic and Palaeozoic oilfields which lie today in intermediate latitudes seem to indicate that they originated in a belt of latitudes of 0–20° on either side of the Equator.

It has in fact been suggested [33a] that as much as 72% of all known oil may have been formed during late Mesozoic times. On this theory, a geologically short period of only some 80 million years (say from 140 to 60 million years ago) may have produced most of the petroleum which has accumulated in the present-day Middle East, Libyan, "circum-Atlantic" and West Siberian basins, as the consequence of a rare optimization of conditions favouring generation in the regions which are now the Middle East and Central North America, but which then, on "plate tectonic" theory lay astride the Equator. On this basis (Fig. 72), much of the oil now found in Tertiary sediments in these areas has migrated upwards from deeper and older source rocks.

While such theories help to explain the important oil accumulations of such apparently "cold" areas at the North Slope of Alaska, the North Sea basin or Siberia, it must be emphasised that sufficient data are not yet available to substantiate this palaeolatitude approach as generally valid in exploration; there are too many unknown and variable factors involved—not the least being the presence of enormously abundant organic phytoplankton in

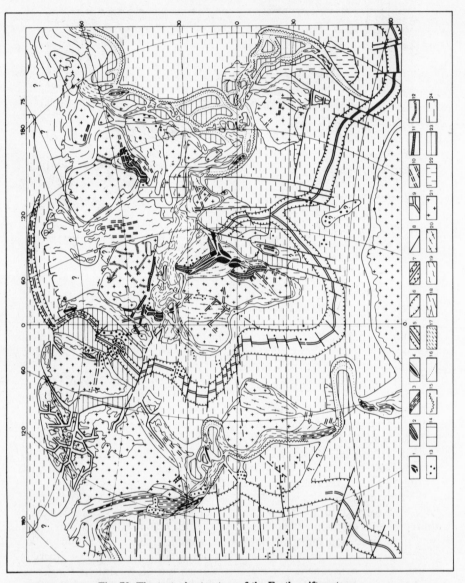

Fig. 73. The tectonic structure of the Earth—rift systems.
A. Recent rift belts and zones: (a) Continental rift belts: 1 = platform arch-volcanic rift zones;
2 = platform crevice-like rift zones; 3 = epeirogenic rift zones; 4 = intercontinental rift zones.
(b) Oceanic rift belts; 5 = mid-oceanic ridges with rift valleys; 6 = mid-oceanic ridges without
rift valleys; 7 = mid-oceanic ridges with volcanics; 8 = large-scale faults and wrench faults
active during the Cenozoic. B. Pre-late Cenozoic zones of extension crust fracture and graben
building: 9 = Late Mesozoic and Early Cenozoic; 10 = Early Mesozoic; 11 = Palaeozoic;
12 = Late Proterozoic; 13 = areas of Late Cenozoic volcanism; 14 = zones of Cenozoic
orogenesis; 15 = recent deep trenches; 16 = recent geosynclinal zones; 17 = Alpine and Lara-
mide fold zones; 18 = Mesozoic fold zones; 19 = Palaeozoic fold zones; 20 = late-Proterozoic
fold zones; 21 = Pre-late Proterozoic platforms; 22 = oceanic floor with continental-type crust;
23 = deep-sea depressions with oceanic crust; 24 = oceanic basins with oceanic-type crust.
(Source: E.E. Milanovsky, *Tectonophysics*, **15**, 1/2, 66 (1973),
by permission of Elsevier Publishing Co.).

present-day Antarctic waters, at least during the summer months. This seems to show that polar waters may also provide rich concentrations of organic material suitable for oil generation.

Perhaps a more easily arguable aspect of the effects of the theories of "new global tectonics" on petroleum geology may lie in a consideration of the effects of the geomagnetic polarity reversals that have been discovered to have occurred in past geological eras.

Polarity Reversals[58-61]

As mentioned above, analysis of the magnetic properties of rock samples derived from many parts of the world has revealed the remarkable fact that there have been a large number of reversals in the polarity of the Earth's magnetic field over a period of at least some 350 million years—i.e. extending back as far as the older Palaeozoic. If the present arrangement of the geomagnetic field is considered to be "normal", there is evidence that during the Palaeozoic it was generally "reversed", and in particular a "reversed" field prevailed for a very long period of about 50 million years during Late Carboniferous–Early Permian times. Subsequently, the geomagnetic field seems to have been "normal" for some three-quarters of total Mesozoic time, apart from one period in the Lower Triassic when there were a number of relatively rapid reversals.

There was another long period of predominantly "normal" polarity from the Upper Triassic to the Upper Jurassic (although the Triassic–Jurassic boundary is marked by a reversal). Another long "normal" interval extended for some 50 million years during the Cretaceous. The reversals that have taken place in the Cenozoic era have been relatively far more frequent—as many as 160 reversals having been detected over the period of 65 million years that has elapsed between the Palaeocene and the present, with a small preponderance of the "reversed" state. The present-day "normal" geomagnetic condition seems to have been initiated about 0·7 million years ago, and, since only 15% of "normal" intervals have been longer than this, a changeover may be relatively near.*

The measurements of magnetism which have been made on lava flows on land, correlate well over the last four million years with the reversal patterns found in the rocks of the sea-floor around the mid-ocean ridges. Further back in time, only sea-floor magnetic measurements are possible.

Little is known about what causes these periodic reversals of geomagnetic polarity. It is reasonable, however, to assume that the changes may be due to the Earth's orbital irregularities, which are also presumed to be the cause of major earthquakes, climatic changes and, in fact the whole process of sea-floor spreading.

It is possible, however, to speculate on the effects that geomagnetic reversals may have had in our particular field of petroleum geology. Thus, it has been suggested that the periodic "flipping-over" of the Earth's magnetic field may have been an indirect cause of the widespread destruction of surface life, because of the consequential removal during the reversal period of the ionized screens in the upper atmosphere which normally deflect some proportion of the

---

* *There have also been a number of relatively short-lived reversals in this period. If the Earth's magnetic field continues to weaken at the present rate it would become zero in about AD 4,000.*

incoming solar and galactic radiation. On the other hand, it is likely that even a total temporary dipole collapse would not mean that the Earth's surface received a radiation "dose" of more than about 10% above normal, since atmospheric absorption is the most important agent of cosmic ray protection. However, the warming effect of the incoming "solar wind", which is generally blocked by the Earth's magnetic field, might have important climatic consequences, and there is evidence to connect magnetic reversal effects with periods of upper-mantle geotectonic activity, vulcanism and the consequent release to the upper atmosphere of large amounts of dust. However produced, it is likely that such periods of relative faunal extinction may have been of importance from the point of view of petroleum geology, inasmuch as they would have brought about a reduction in the quantity of organic material deposited in the shallow seas around the continents—the potential biogenic source material for eventual petroleum generation.

PETROLEUM IN TIME[62–66]

It was once believed that petroleum was formed only under special circumstances, but nowadays it is generally accepted that oil generation is in fact a very common process, whereas its accumulation and preservation in large concentration are relatively rare and call for the concatenation of several favourable factors. The probability of the coexistence in the "right" relationship in space and time of "mature" source rocks, permeable reservoir rocks, suitable trapping mechanisms and adequate seals will be proportional to the product of the probabilities of their individual occurrence; hence, the existence of a major present-day accumulation of petroleum is understandably rare.

Commercial accumulations of petroleum have in fact been found in reservoir rocks of virtually every age, but the size and frequency of occurrence of the accumulations vary considerably. Any discussion about the age of petroleum accumulations is always complicated by the fact that while it is not difficult to identify and date a present-day petroleum reservoir rock, it is only occasionally that a source rock can be positively identified. Furthermore, there is no certainty as to when the processes of oil generation and migration began, and how long they were active before the accumulation was finally formed.

Efforts have been made to "date" crude oils by geochemical means—in particular by the examination of their trace metal content and the ratios of certain stable isotopes. Such studies have shown that at least in some cases, the crude was formed considerably earlier than the reservoir rock in which it is found today.

In spite of these uncertainties, it is nevertheless usually accepted that reservoir beds are, in most cases, of about the same ages as the oils they contain.

On this assumption, some interesting conclusions may be drawn. For example, a comparison of the past production plus the unproduced oil reserves of some of the larger US oilfields with the geological ages of their principal reservoirs indicated (Table 37) that in the USA the Ordovician, Cretaceous and Tertiary were the most favourable periods for oil accumulation; while relatively the least oil is found in the very old (Cambrian and Pre-Cambrian), the very young (Pleistocene), and the mainly terrestrial (Triassic) beds. However, the discovery of Triassic accumulations on the North Slope of Alaska shows that this formation is far from devoid of petroleum prospects.

In other parts of the world, the relationship of hydrocarbon distribution to rock age is different, although it is generally true to say that most hydrocarbons have been found in Tertiary and Mesozoic reservoirs.

The predominance of younger reservoirs is to be expected, since accumulations in these rocks will have been "at risk" from tectonic disturbances or erosion processes for the shortest times. Although at one time it seemed that the USSR was an exception to this general rule, with 52% of proved reserves in Palaeozoic rocks in 1959 and 61% in 1966, by 1970 the proportion had decreased to 37%—still greater, however, than for most of the rest of the world.

In the following Section a summary is given of the principal known occurrences of petroleum according to the age of the reservoir beds.

1. The Pre-Cambrian

Until recently, no traces of ancient organic life had been detected in the great thicknesses of Pre-Cambrian rocks known in many parts of the world, which represent a time interval of some 2,500 million years. The Pre-Cambrian was therefore generally considered to be of no interest from the point of view of petroleum geology. However, the discovery of "wet" gas in Central Australian Pre-Cambrian rocks and similar gas "shows" in the Siberian "Shield" region has caused some revision of thought, and it is now considered possible that hydrocarbons may be found in other sections of relatively unmetamorphosed Pre-Cambrian strata. These hydrocarbons would perhaps be abiogenic in origin; alternatively, in those distant times, the biogenic source organisms had not yet learned to use the calcium carbonate in the ancient seas to form hard skeletal structures, which would explain the absence of associated fossils. Pre-Cambrian hydrocarbon accumulations would presumably in any case be very rare.

2. The Palaeozoic

Palaeozoic oil accumulations have had to survive for so long (up to 500 million years) that it is remarkable that any of them still exist at the present time. Nevertheless, many important *Silurian* and *Ordovician* oilfields lie in a belt across North America through Kansas, Oklahoma, Texas and New Mexico, with accumulations occurring in the Arbuckle, Ellenburger and Simpson formations. The Ordovician Trenton Limestone of Ohio, Indiana, and Michigan has also produced large volumes of oil and gas. *Cambrian* oil is less common, but several billion barrels of oil and equivalent amounts of gas have been produced from Cambrian reservoirs in the United States, Eastern Canada, Siberia and North Africa.

Devonian oil has been produced in the Western Hemisphere from numerous oilfields in the Appalachian geosyncline, the Michigan Basin, and the West Texas Basin of the United States; from carbonate reef accumulations in Western Canada; and from sandstone reservoir rocks in Bolivia. In the Eastern Hemisphere, Devonian reservoirs produce oil in Algeria and Libya, but the most important Devonian oil-producing regions are those of the western USSR—notably the Pechora-Ukhta Basin, the Ural-Volga region,

TABLE 37

AGE DISTRIBUTION OF OIL ACCUMULATIONS

Geological System	Approximate age in millions of years	Oil in USA in billions of brl (past production +probable unproduced reserves)[a]	Oil in world's largest fields outside USSR (ultimate reserves in billions of brl)[b]	Oil found in USSR (proved reserves— percentages) 1959[c]　1966[d]　1970[d]		
Pleistocene	1	1	} 7·6			
Pliocene		5·2				
	14					
Miocene		7·3	} 62·3			
	31					
Oligocene		6·0				
	44					
Eocene		2·5	13·1			
	60					
CENOZOIC		22·0 (47%)	83·0 (38%)	15·8%	14·8%	13·0%
Cretaceous		10·1	} 114·3			
	125					
Jurassic		0·8				
	157					
Triassic		0·0				
	185					
MESOZOIC		10·9 (18%)	114·3 (53%)	32·1%	24·2%	50·0%
Permian		7·9	} 19·7	10·2%		
	223					
Carboniferous		13·5		17·5%		
	309					
Devonian		2·1		} 22·2%		
	354					
Silurian		0·5				
	381					
Ordovician		4·3				
	448					
Cambrian		0·1		2·2%		
	553					
PALAEOZOIC		28·4 (35%)	19·7 (9%)	52·1%	61·0%	37·0%

a. G. Hopkins, *J. Petrol. Technol.* **2**, 6, (1950).
b. G. Knebel and G. Rodriguez-Eraso, *Bull. Amer. Assoc. Petrol. Geol.*, **4**, 554 (1956).
c. Mirchink *et al., Neft. Khoz.* 4 (1959).
d. N. Eremenko *et al., Bull. Amer. Assoc. Petrol. Geol.*, **56** (9) 1717, (1972).
Note: Since the basis of each of these estimates is different, they can only be compared with caution.

and Kuibishev and Orenburg Provinces. Probably the largest Palaeozoic oilfield in the world is *Romaschkino* in the Tartar ASSR.

Carboniferous oil has been produced in Europe, in relatively small amounts in England and Holland, with gas in Germany; much larger volumes of oil have come from various parts of the USSR—most notably the Ural-Volga area. There is also Carboniferous oil in North Africa (Algeria and Egypt).

In the United States, large volumes of oil have been produced from *Mississippian* and *Pennsylvanian* (Carboniferous) age rocks in the Appalachian geosyncline region, in the East Interior Basin, in the Mid-Continent area and in the Rocky Mountains province.

Permian oil production is particularly important in two parts of the world—in the West Texas Basin, where no less than 57 of the US "giant" oilfields occur, producing mainly from Permian carbonate reservoirs; and in the Ural-Volga area of the USSR, most notably in Perm Province, where there are at least 40 major oilfields with Permian dolomite reservoirs. Smaller volumes of Permian oil have also been found in Argentina, while in Europe, Permian sandstones and dolomites form the reservoir beds for the North Sea and most of the West German gasfields.

3. The Mesozoic

The *Triassic* era is considered to have been a relatively unfavourable time for petroleum generation, due to the arid, terrestrial conditions which generally prevailed. However, small oilfields with Triassic reservoir rocks are known in South America (Argentina and Bolivia); in Europe (Austria and Sicily); in Indonesia (Ceram); and in Australia. There are also reported to be oil accumulations in China in Triassic aeolian sandstone reservoir rocks.

In North Africa, the huge *Hassi R'Mel* gasfield of Algeria has a mainly Triassic reservoir, and there are also several Triassic oilfields in the Central Basin of Algeria.

The discovery in recent years of the *Prudhoe Bay* oil accumulation on the North Slope of Alaska—the largest oilfield ever to have been found in North America—may considerably upgrade the Triassic from the point of view of petroleum geology, since some of the most important of the reservoirs of this field are reported to be Triassic sandstones and conglomerates.

The *Jurassic* was a favourable time for petroleum generation, and many large reservoirs of Jurassic age are known in every continent. In North America, Jurassic oilfields occur in the Rocky Mountains region, in Texas, Alabama and in Alaska. Considerable volumes of Jurassic oil have come from the Tampico region of Mexico, and smaller amounts from sandstone reservoirs in Western Canada. In Europe, Jurassic oil has been found in Germany, and to a small extent in England. There are relatively modest oilfields with Jurassic reservoir beds in Morocco and Australia. However, it is in the USSR and the Middle East that the most important Jurassic oil occurrences are known. Thus, in the USSR, large volumes of oil and gas are produced in West Siberia from Jurassic sandstone beds—notably along the Ural Mountains foothill zone, in the Middle Ob region and in Tyumen province. There is also Jurassic oil production in Byelorussia and the Ural-Emba salt dome province.

In the Middle East, the Jurassic "Arab Zone" limestones are the principal reservoir beds in Saudi Arabia and also in Qatar. These were laid down in four major cycles so that each of the thick Upper Jurassic carbonates is sealed by an overlying anhydrite bed. The resultant oil accumulations in the Ghawar region of Saudi Arabia are some of the largest in the world.

The *Cretaceous* was one of the most important eras for petroleum generation and entrapment in all parts of the world. It was in general a time when shallow seas covered large parts of the continents and oil-forming organisms abounded.

In North America, large volumes of petroleum have been produced from both Upper and Lower Cretaceous rocks in several parts of the United

States—most notably in Texas, Louisiana, Arkansas and Mississippi, and also in the Rocky Mountains Province. Oil and gas are also produced from Cretaceous sandstone reservoirs in Western Canada, and from Lower Cretaceous carbonates in Mexico.

In South America, considerable volumes of oil have come from Cretaceous reservoirs underlying the shallower Tertiary producing formations in Western Venezuela, and also from Argentina, Colombia and Peru.

In Europe, Lower Cretaceous rocks are productive in France, Holland and Spain; Upper Cretaceous formations make up the main reservoirs in Northwest Germany, and in the Danish and Norwegian offshore oilfields.

Cretaceous sandstones are productive in the West Ukraine, Astrakhan, Northeast Caucasus and Emba regions of the USSR. There is increasingly important oil output from Lower and Upper Cretaceous sandstones in the Middle Ob and Tyumen areas of West Siberia, and also, as yet to a smaller extent, in East Siberia.

North Africa has prolific Cretaceous reservoirs in some of the Libyan fields of the Sirte Basin; and in West Africa, the oilfields of Gabon and Angola have Lower Cretaceous sandstone and carbonate reservoirs, respectively. There is also production from Lower Cretaceous sandstones in Western Australia.

Perhaps the largest concentration of Cretaceous oil in the world lies in the Middle East. Many of the Saudi Arabian oilfields have Middle and Upper Cretaceous reservoir sandstones lying above the productive Jurassic carbonates. Similar Cretaceous reservoir rocks underlie the shallower productive Tertiary reservoirs in Northern Iraq and Iran; in Southern Iraq and in Oman, Cretaceous reservoirs are the most important producing zones.

Middle Cretaceous sandstones provide at least 1300ft of permeable oil reservoir rock in the *Burgan* field, as well as affording the most prolific reservoir formation in the other Kuwait oilfields. Lower and Middle Cretaceous sandstones and limestones are very important reservoir rocks in the offshore and onshore oilfields of the Neutral Zone.

4. The Tertiary

Most of the world's oil so far discovered has been found in Tertiary reservoir rocks—i.e. in beds roughly 1–60 million years old. It has been estimated for example that nearly half the ultimate output (past production plus probable unproduced reserves) of the USA lies in Tertiary rocks, and that the upper half of the Tertiary alone accounts for more than 30% of all the oil ever discovered in the world.

Let us consider the series in order of ascending age:

A number of important *Palaeocene* (as well as Eocene) oil reservoirs occur in Libya, and some of the newly-discovered North Sea oilfields have reservoirs of Palaeocene age.

Important volumes of oil have been recovered from *Eocene* reservoirs in Southeast Texas and Louisiana in the USA; from the North Caucasus and Ferghana regions of the USSR; from the "Main Limestone" reservoirs of Northern Iraq, which is partly Eocene and partly Oligocene in age, and from the Neutral Zone. Eocene rocks form the main non-associated gas reservoirs of Pakistan, and in the Far East produce oil in Sarawak, Kalimantan, and Southeast Australia.

Oligocene reservoir rocks (mainly sandstones), are known in Europe in France and Poland, in the USSR (West Ukraine, Caspian area and Central Asia), in Burma, in the US Gulf Coast region and in California.

Oligocene-Miocene and Oligocene-Eocene carbonate reservoirs are prolific producers in Northern Iraq and Iran. Oligocene-Eocene sandstones are important oil reservoirs in the Maracaibo fields and in East Venezuela.

The *Miocene* era was one of the most important there has ever been as regards the generation and accumulation of petroleum. Prolific Miocene-age reservoir rocks—generally sandstones—are known in every continent except Australia.

In the United States, there are thick and highly-productive Miocene sandstone reservoirs in California and in the onshore and offshore Gulf Coast areas of Texas and Louisiana. In South America, Venezuela is the leading oil-producing country, with many oilfields both in the Maracaibo and Maturin Basins which have Miocene or Oligocene-Miocene sandstone reservoirs. Trinidad is also a region in which much Miocene oil has been produced.

In Europe, there are Miocene oilfields in Austria, Sicily, Hungary and Romania. The Caucasus area of the USSR has many Miocene and Miocene-Pliocene sandstone reservoirs, both in the Middle Caspian and South Caspian Basins. Miocene sandstones are also productive in the Sakhalin Island oilfields in the Far East.

In North Africa, sandstones and limestones of Miocene age form the reservoir rocks in fields on both sides of the Suez Canal, and in Nigeria there are many prolific onshore and offshore oilfields with Miocene sandstone reservoir beds.

In the Middle East, Miocene-Oligocene carbonates form the most productive "Asmari" reservoirs of the Iranian oilfields; and in East Asia, Miocene sandstones are productive in Assam, Burma, Sumatra, Java and Borneo.

Beds of *Pliocene* age form particularly prolific oil-producing reservoirs in the Gulf Coast area of the United States (notably in the offshore oilfields of Louisiana and in California).

In Europe, there is considerable Pliocene gas in the Po valley fields of Italy, and Pliocene or Miocene-Pliocene sandstone oilfields and gasfields of modest dimensions in all the countries across which the Pannonian Basin extends (Romania, Austria, Czechoslovakia and Yugoslavia). In the USSR, there are many Pliocene sandstone reservoirs in the Baku region, and in the fields in the Nebit Dagh and Mangyshlak regions. In East Asia, many of the oilfields of Indonesia and Borneo have Pliocene sandstone reservoirs. There are also reservoirs of this age in Japan.

In South America, considerable reserves have been found in Pliocene sandstone beds in East Venezuela.

5. The Quaternary

Pleistocene oil accumulations are something of a rarity; there is some likelihood that the occurrences reported (in East Venezuela, California, offshore

US Gulf Coast, etc.) have in fact resulted from upward seepage out of older
accumulations. The youngest beds may not have been buried deep enough for
pressure and temperatures to have been reached sufficient to allow petroleum
to be generated and migrate within them. Possibly, also, the most recent phase
of the Alpine orogeny has destroyed accumulations which were not protected
by sufficient thicknesses of overlying sediments.

THE DISTRIBUTION OF PETROLEUM[62-66]

Accurate statistics about the distribution of petroleum are difficult to
assemble because of the imponderable parameter of "commercial value"
which obscures the picture. Thus, as was explained in Chapter 1, many oil
and gas accumulations have been discovered but not developed because of
technical, economic or political factors—unfavourable oil quality, distance
from market, depth, restricted permeability, etc. The best estimates agree that
about 30,000 "significant" oilfields have been discovered; but 26,000 (nearly
87%) of this number are in the United States, where the criteria for technical
and economic viability of small accumulations have always been very different
from those prevailing in other parts of the world.

The analysis of the geographical position of at least the larger of the world's
oilfields confirms that the great majority lie either in sedimentary basins of one
sort or another, or in the neighbouring platform and "hinge" areas.

One statistical analysis of the occurrences of major oil accumulations
outside the USSR has showed that the shelf areas contain more fields than the
hinge zones, but that the individual fields are less prolific. It is interesting to
note that in this analysis, hinge and shelf zones together— i.e. the "stable" as
opposed to the "mobile" side of the basins—accounted for 64% of the known
oil, about three times the total found in "mobile rim" accumulations.

The structural importance of folds produced by nearby orogenic uplifts also
becomes apparent in any geographical analysis of oilfield distribution:

Oilfields associated with the Caledonian and Hercynian orogenies are
understandably relatively infrequent, due to the time such accumulations have
been "at risk". The fields along the western front of the Ural Mountains in the
USSR provide some important examples, however, while in the United States,
a number of major Palaeozoic oil accumulations occur along the outer rim of
the Appalachian Mountains.

The structures produced by the Alpine orogeny, on the other hand,
provide many examples of petroleum accumulations in anticlinal traps. Thus,
in Europe, oil pools of varying sizes occur in the foothills of the northern
Alpine fold zone, from the Pyrenees to southern Germany, Austria, Transylvania and the Carpathian fields of Romania. Thereafter, the northern fold
line crosses the floor of the Black Sea, and emerges to form the structures of
the Caucasus oilfields, continuing into the Apsheron Peninsula. It then strikes
across the Caspian Sea into Central Asia by way of Cheleken and Turkmenia
to Ferghana, with a number of associated oilfields, and to the northern rim of
the Elburz-Hindukush ranges. The Kwen Lun and neighbouring ranges north
of Tibet continue across China to join the circum-Pacific girdle south of Japan.

The southern branch of the Alpine orogeny is associated with oil accumulations in the foothills of the Atlas Mountains, the Apennines of Italy, and the

Dinaric Alps of Yugoslavia; it then continues in the great arc of the Zagros Mountains, in the foothills of which are many prolific Middle East accumulations. From Arabia, the same trend may strike through Baluchistan, south of Bandar Abbas, or alternatively may link up with the Carlsbad ridge off Socotra.

In Pakistan and India, the Alpine fold line stretches from Karachi to the Salt Range and then along the southern Himalayas into Assam and Burma. Oilfields occur in the north Punjab, in Assam, and in the Arakan Yoma and Pegu Yoma hills of Burma. The same folding movement extends to the Barisan Mountains of Sumatra and then passes through Java, Madoera and the Soenda Islands. Another branch turns north through Borneo to Sakhalin, Kamchatka and Japan. Oil is found in varying quantities along both these belts.

On the other side of the Pacific Ocean, many important oilfields lie along the Rocky Mountains of North America, from Alberta through Montana and Wyoming into Colorado. A younger, late Tertiary belt lying farther west is associated with the oilfields of California and Mexico.

In South America, the trend of the Rockies is continued in the Andes, and this range with its subsidiary mountain belts provides the dominating structural feature. The Brazilian "Shield" has acted as a northeastern "foreland", and oil is found all the way along the Andean foothills, as far south as Tierra del Fuego, and in the foothill structures of the Cordillera ranges of Colombia, Venezuela and Trinidad.

Any statistical and geographical analysis of the occurrence of world oil and gas accumulations must emphasise the extraordinary importance of a relatively few "giant" accumulations, their unpredictable distribution and their rarity.

In 1973, 62 countries were producing commercial volumes of crude oil, as summarized in Table 8 (p. 22). However, a very large proportion of this output came from only a small number of countries (Table 9) and from a relatively few "giant" oilfields in those countries.

Table 38 shows that a mere eight of the "giants" among the 30,000 known "significant" oilfields have accounted for 47·2 B brl or nearly 17% of all the oil produced in the world between 1859 and the beginning of 1974 (estimated to be 280 B brl). Similar statistics show that 83·7 B brl or 29·9% of this overall total was produced by only 31 "giant" oilfields.*

TABLE 38

"GIANT" OILFIELDS WITH LARGEST CUMULATIVE PRODUCTIONS
(as at 1.1.1974)

	B brl
Ghawar (Saudi Arabia)	9·0
Lagunillas (Venezuela)	8·0
Kirkuk (Iraq)	6·0
Agha Jari (Iran)	5·5
Burgan (Kuwait)	5·4
Bachaquero (Venezuela)	4·6
Abqaiq (Saudi Arabia)	4·5
East Texas (USA)	4·2
TOTAL	47·2

---

* Using here the definition of "giants" as accumulations containing 500 MM brl of oil or equivalent amounts of gas.

It has been estimated that about 75% of the oil ever found in the world occurred in these "giant" accumulations. Even more remarkably, the 187 "giant" oilfields (amounting in number to only about 0·6% of the world's known "significant" oilfields) contained between them about 84% of the world's remaining recoverable reserves in 1970. More recent statistics show, in fact (Table 39), that only 19 of these "giants" together contained nearly half of the world's remaining known crude oil reserves at the beginning of 1974.

The factors controlling the occurrence of "giant" hydrocarbon accumulations are not yet understood. Obviously, very large traps, plentiful source material, efficient processes of oil migration, and unfailing preservation mechanisms must have been needed; but perhaps, in addition, there is some fundamental tectonic relationship which has not been properly defined. Thus, it has been proposed that the shear patterns produced by deep-seated wrench-faults (i.e. faults where the dominant movement of one block relative to the other is horizontal, and the fault-plane is nearly vertical) have influenced the linear distribution of petroleum traps in some parts of the world—for example, California, Alaska, Colombia–Venezuela and the Gulf of Guinea[67-70].

TABLE 39

RECOVERABLE RESERVES REMAINING IN WORLD'S LARGEST
"GIANT" OILFIELDS (ESTIMATED, 1974)

	B brl
Ghawar (Saudi Arabia)	66·5
Burgan (Kuwait)	56·1
Khafji (Neutral Zone)	17·3
Samotlor (USSR)	14·1
Safaniya (Saudi Arabia)	12·6
Rumaila (Iraq)	11·9
Kirkuk (Iraq)	9·7
Marun (Iran)	8·9
Qatif (Saudi Arabia)	8·6
Gach Saran (Iran)	8·1
Ahwaz (Iran)	8·1
Abqaiq (Saudi Arabia)	7·8
Sarir (Libya)	7·6
Bibi Hakimeh (Iran)	7·2
Romaschkino (USSR)	7·0
Abu Safah (Saudi Arabia)	6·4
Minas (Indonesia)	5·6
Berri (Saudi Arabia)	5·5
Fedorovskoye (USSR)	5·0
TOTAL (19 oilfields)	274·0 B brl

Certainly, the results of high-level photography from satellites seem increasingly to confirm the existence of many deep-seated "lineaments" on the Earth's surface—lines of faulting which may have existed since the most ancient times and which must therefore have considerably affected the overlying sedimentary structures. The unravelling of the complex history of the movement of lithospheric "plates", discussed above, may eventually provide an explanation of some of the factors involved.

A total of 261 "giant" oil- or gasfields are known (187 oilfields and 74 gasfields) of which 126 occur in cratonic basins, and 135 in intermediate crust basins. (However, by far the largest reserves of oil are attributable to the latter group, due to the much greater size of the individual accumulations.)

TABLE 40

SEDIMENTARY BASINS WITH LARGEST NUMBERS OF KNOWN
"GIANT" ACCUMULATIONS OF OIL AND GAS*[8]

Arabian Platform-Iranian Basin	56
West Siberian Basin, USSR	29
Sirte Basin, Libya	13
Volga-Ural Basin USSR	10
Gulf Coast, USA	10
West Texas Permian Basin, USA	9
Pre-Caucasus—Mangyshlak Basin, USSR	9
North Sea Basin	8[†]
San Joaqin Basin, California	7
Baku Basin, USSR	7
Maracaibo Basin, Venezuela	6
Bukhara Basin, USSR	6
Alberta Basin, Canada	5
Anadarko-Ardmore Basin, USA	5
Dnieper-Donets Basin, USSR	5
Niger Delta, Nigeria	5
	190
41 other known sedimentary basins containing 1–4 "giants" apiece	71
	261

* *500 MM brl of oil or equivalent of gas.*
† *Estimated.*

Table 40 lists the world's most important sedimentary basins from the point of view of their petroleum contents.

The most prolific basins yet discovered are those centred around the Middle East Gulf, the Gulf of Mexico and Lake Maracaibo in Venezuela; it should be noted however that some of the basins listed (e.g. West Siberia or the North Sea) are still only in the early stages of their development.

Are there other major sedimentary basins yet to be discovered on the Earth's surface? The opening-up of remote areas and the increasing pace of exploration for petroleum makes it seem unlikely that many large onshore basins still await discovery. The Arctic basin may be the last of the world's remaining major sedimentary basins to be found which is at least partly on land. Future discoveries will increasingly tend to be made in offshore areas—initially on the continental margins, and later in the ocean basins, about which knowledge is now gradually becoming available.

The continental slopes and rises lie at depths which are as yet beyond the limits of economic development, since they are covered by waters more than 1,000ft deep; but rapid technological progress will soon bring them within drilling range. Meanwhile, the Continental Shelf is being actively explored off the coasts of some 75 countries of the world. The Shelf covers an area of about 11 million sq miles out to its customarily-assumed limiting depth of 100-fathoms, which marks the approximate limit of wave and current action and is about as deep as daylight will penetrate[71-75] (Fig. 71).

The general worldwide average width of the Continental Shelf is about 42 miles, and the volume of sediments it contains is of the order of 50 million cubic miles. Within this huge volume of sedimentary material must be many source and reservoir beds similar to those known on the land, so that the Continental Shelf sediments are likely to form a very valuable addition to the

TABLE 41		TABLE 42	
OFFSHORE OIL PRODUCTION, 1973		OUTPUTS OF WORLD'S MOST IMPORTANT OFFSHORE OILFIELDS, 1973	
	MM brl		*MM brl*
Venezuela*	929·0	*Safaniya* (Saudi Arabia)	351
Saudi Arabia	705·0	*Lagunillas* †(Venezuela)	326
USA	391·4	*Berri* (Saudi Arabia)	227
Nigeria	204·2	*Bachaquero* †(Venezuela)	216
Iran	178·0	*Khafji* (Neutral Zone)	121
Abu Dhabi	166·0	*Tia Juana* (Venezeula)	120
Neutral Zone	143·0	*Zakum* (Abu Dhabi)	106
Australia	137·0	*Lama* (Venezuela)	105
Brunei-Malaysia	95·6	*Sassan* (Iran)	69
Qatar	88·0	*Wilmington* †(USA)	67
USSR	88·0	*Kingfish* (Australia)	65
Dubai	81·0	*Maydan-Mazan* (Qatar)	56
Egypt	58·4	*Halibut* (Australia)	55
Indonesia	52·6	*Umm Shaif* (Abu Dhabi)	55
Cabinda	51·0		
Trinidad	34·6		
Gabon	20·5		
10 Others	58·8		
Total	3,482·1		
*From Lake Maracaibo.		† Partly onshore.	

world's sedimentary areas in which petroleum has been generated and accumulated.

The sediments of the Continental Shelf appear to have had a complex history, which is only slowly being unravelled. Sedimentary embayments occur along the edges of the continents, in which the beds generally dip and thicken towards the oceans, before thinning further out to sea, thus forming a series of lenses, often cut by major faults with their downthrows seawards. The topmost beds are really "relict" strata, very shallow water sediments submerged by the Post-Pleistocene marine transgression caused by the melting of the ice. Off Europe in particular, the Shelf area is wide and includes a number of semi-enclosed sea bodies which have resulted from gentle subsidence caused by the growth of the ice mass, combined with subsequent over-filling of the shallow oceans due to its melting.

The underlying sediments, on the other hand, show evidence of a more complicated geological history. A number of older basins and swells seem to have been formed off the shores of the continents in the geological past, some of which run roughly parallel and others roughly at right angles to the present coasts, which they may sometimes overlap. Notable examples of such coastal basins are known around the coasts of Africa, where, particularly in the west, they may extend far inland. Most of these basins in Africa seem to have been initiated during Late Jurassic to Early Cretaceous time and to have been formed by downwarping of the basement, with, in the earlier stages of development at least, the marked deposition of evaporites.

A considerable body of evidence has been collected about the Shelf sediments lying off the eastern coast of the United States, but no entirely satisfactory theory has been evolved to account for their complexity. The continental margin in this area has been variously explained as a wave-built terrace growing seaward and upward by deposition; as the result of isostatic sub-sidence with resulting outbuilding and upbuilding by sedimentary deposition; or as the consequence of simple basement subsidence. It has also been suggested that the continental slope was structural in origin and is now lapped up against and mantled by sediments, or even that the Shelf is an erosional platform, the continental slope being the consequence of faulting. Continuous seismic reflection profiles now confirm that the continental margin of the US Atlantic coast grew seaward during Tertiary times, but the origin of the pre-existing slope is still not understood.

World offshore oil production totalled some 3,482 MM brl in 1973, or more than 17% of total world crude oil output (Tables 41, 42). In addition, about 6% of world natural gas output came from offshore gasfields. It has been estimated that both the absolute and relative volumes of offshore oil produced will continue to rise during the remaining years of this decade, so that perhaps 30–40% of total world crude oil output may be coming from offshore accumu-lations in the early 1980's.

REFERENCES

Geosynclines and Basins

1. Anon., "Marginal basins form by sea floor spreading". *Nature* **243,** 234 (June 1973).
2. Anon., "Origin of marginal seas". *Nature,* **246,** 447 (Dec. 1973).
3. J. Auboin, "Geosynclines". Elsevier Publishing Co. (1965).
4. J. Barrell, "Rhythms and the measurement of geologic time". *Bull. Geol. Soc. Am.,* **28,** 745–904 (1917).
5. R. E. Chapman, "Petroleum and Geology: a synthesis". *APEA J.,* **12** (1), 36–38, 1972.
6. K. Dallmus, "Mechanics of basin evolution". In: "Habitat of Oil", *Amer. Assoc. Petrol. Geol.,* Tulsa, 883–931 (1958).
7. L. de Sitter, "Structural Geology", 2nd Edn. McGraw-Hill, 394 (1964).
8. M. T. Halbouty, R. E. King, H. D. Klemme, R. H. Dott and A. A. Meyerhoff, "Factors affecting formation of giant oil and gasfields, and basin classification". In: "Geology of Giant Oil Accumulations", *Bull. Amer. Assoc. Petrol. Geol.,* Tulsa (1972).
9. M. Kamen-Kaye, "Basin Subsidence and hypersubsidence". *Bull. Amer. Assoc. Petrol. Geol.,* **51** (9), 1833–1842 (1967).
10. M. Kaye, "Geosynclinal nomenclature and the craton". *Bull. Amer. Assoc. Petrol. Geol.,* **31,** 1289–1293 (1947).
11. W. Schwarzacher, "Sedimentation in subsiding basins". *Nature,* **241,** 1349–1350 (June 1966).
12. J. W. Shelton, "Role of contemporaneous faulting during basin subsidence". *Bull. Amer. Assoc. Petrol. Geol.,* **52** (3), 399 (1960).
13. H. Stille, "Einfuhrung in den Bau Amerikas", (1940).
14. J. H. Umbgrove, "The Pulse of the Earth". M. Nijhoff, The Hague (1947).
15. B. H. Walthall and J. L. Walper, "Peripheral Gulf rifting in Northeast Texas". *Bull. Amer. Assoc. Petrol. Geol.,* **51** (1) 102–110 (1967).
16. L. G. Weeks, "Factors of sedimentary basin development that control oil occurrence". *Bull. Amer. Assoc. Petrol. Geol.,* 36 (10), 2071–2124 (1952).

Deltas

17. K. Burke, "Longshore drift, submarine canyons, and submarine fans in development of Niger delta". *Bull. Amer. Assoc. Petrol. Geol.,* **56** (10), 1975–1983 (1972).
18. R. H. Clark and J. T. Rouse, "A closed system for generation and entrapment of hydrocarbons in Cenozoic Deltas, Louisiana Gulf Coast". *Bull. Amer. Assoc. Petrol. Geol.,* **55** (8), 1170–1178 (1971).
19. G. T. Moore, "Interaction of rivers and oceans—Pleistocene petroleum potential". *Bull. Amer. Assoc. Petrol. Geol.,* **53** (12), 2421–2430 (1969).

Plate Tectonics

20. T. Atwater, "Implications of plate tectonics for the Cenozoic tectonic evolution of western North America". *Geol. Soc. America Bul.,* **81,** 3513–3535 (1970).
21. J. M. Bird and B. Isacks (Eds.), "Plate Tectonics"—Selected papers from *Journ. Geophys. Res.,* Amer. Geophys. Union, Washington, USA (1972).

22. J. C. Briden, "Ancient Secondary Magnetization in Rocks". *Journ. Geophys. Res.,* **70** (20), 5205–5220 (Oct. 1965).
23. J. C. Briden and I. G. Gass, "Plate movement and continental magmatism", *Nature,* **248,** 650–653 (April 1974).
24. E. C. Bullard, "Continental Drift". *Q. J. Geol. Soc.,* **120,** 1–34 (Feb. 1964).
25. C. A. Burk, "Global tectonics and world resources". *Bull. Amer. Assoc. Petrol. Geol.,* **56** (2), 196–202 (1972).
26. K. Burke, W. S. F. Kidd and J. Tuzo Wilson, "Relative and latitudinal motion of Atlantic hot spots". *Nature,* **245,** 133–137, (Sept. 1973).
27. E. R. Deutsch, "The Paleolatitude of Tertiary oilfields". *Journ. Geophys. Res.* **70** (20), 5193–5221 (1965).
28. J. F. Dewey and J. M. Bird, "Mountain Belts and the new Global Tectonics". *Journ. Geophys. Res.,* **75** (14), 2625–2647 (1970).
29. J. F. Dewey and B. Horsfield, "Plate tectonics, orogeny and continental growth". *Nature,* **225,** 521–525, (Feb. 1970).
30. W. R. Dickinson, "Global tectonics". *Science,* v.**168,** 1250–1259 (1970).
31. R. S. Dietz and J. C. Holden, "Reconstruction of Pangaea: breakup and dispersion of continents, Permian to present". *Journ. Geophys. Research,* **75,** 4939–4956 (1970).
32. R. A. Duncan, N. Petersen and R. B. Hargraves, "Mantle plumes, movement of the European Plate, and polar wandering", *Nature,* **239,** 82–86 (Sept. 1972).
33. E. Irving and T. F. Gaskell, "The paleogeographic latitude of oilfields". *Geophys. Journal,* **7** (1), 54–64 (1962).
33a. E. Irving, F. K. North and R. Couillard, "Oil, climate and tectonics". *Canad. J. Earth Sci.* **11** (1), 1–17 (1974).
34. B. Isacks and P. Molnar, "Mantle earthquake mechanisms and the sinking of the lithospheres". *Nature,* **223,** 1121–1124 (1969).
35. B. Isacks and J. Oliver, "Seismology and the new Global Tectonics". *Journ. Geophys. Res.,* **73** (18), 5855–5899 (1968).
35a. N. Jardine and D. McKenzie, "Continental drift and the dispersal and evolution of organisms". *Nature,* **235,** 20–24 (Jan. 1972).
36. H. Jeffreys, "Imperfections of elasticity and continental drift". *Nature,* **225,** 1007–1009 (March 1970).
37. X. le Pichon, "Sea-floor spreading and continental drift". *Journ. Geophys. Research* **73,** 3661–3697 (1968).
38. W. D. Lowry, "North American geosynclines—test of continental-drift theory". *Bull. Amer. Assoc. Petrol. Geol.,* **58** (4), 575–620 (1974).
39. A. J. Mantura, "Geophysical illusions of continental drift". *Bull. Amer. Assoc. Petrol. Geol.,* **56** (8), 1552–1556 (1972).
40. D. P. McKenzie, "Plate tectonics of the Mediterranean region". *Nature,* **226,** 239–243 (1970).
41. D. P. McKenzie and W. J. Morgan, "Evolution of triple junctions". *Nature,* **224,** 125–133 (1969).
42. D. P. McKenzie and R. L. Parker, "The North Pacific: an example of tectonics on a sphere". *Nature,* **216,** 1276–1280 (1967).
43. A. A. Meyerhoff and H. A. Meyerhoff, "The New Global Tectonics: major inconsistencies". *Bull. Amer. Assoc. Petrol. Geol.,* **56** (2), 269–336 (1972).
44. E. Moores, "Utramafics and orogeny, with models of the US Cordillera and the Tethys". *Nature,* **228,** 837–842 (Nov. 1970).
45. W. J. Morgan, "Rises, trenches, great faults and crustal blocks". *Journ. Geophys. Research,* **73,** 1959–1982 (1968).
46. W. J. Morgan, "Convection plumes in the lower mantle". *Nature,* **230,** 42–43 (1971).
47. F. K. North, "Characteristics of oil provinces". *Bull. Canad. Petrol. Geol.,* **19** (3), 601–658 (1971).
48. G. Pautot, J. M. Auzende and X. le Pichon, "Continuous deep-sea salt layer along North Atlantic margins related to early phase of rifting". *Nature,* **227,** 351–354 (July 1970).
49. R. A. Reyment, "Ammonite biostratigraphy, continental drift and oscillatory transgressions", *Nature,* **224,** 137–140 (Oct. 1969).
50. A. E. Ringwood, "The petrological evolution of island arc systems". *Jl. Geol. Soc. Lond.,* **130,** 183–204 (1974).
51. T. L. Thompson, "Hypothesis for petroleum generation at convergent plate boundaries". (Abs.) *Bull. Amer. Assoc. Petrol. Geol.,* **58** (7), 1959 (July 1974).
52. D. L. Turcotte and E. R. Oxburgh, "Mid-plate tectonics". *Nature,* **224,** 337–340 (Aug. 1973).
53. F. J. Vine and H. H. Hess, "Sea-floor spreading", in A. E. Maxwell, ed., The sea, 4, 587–662 (1970). Wiley-Intersci. New York,
54. A. Wegener, *Petermanns Geog. Mitt.,* **58,** 185, 253, 305, 1912.

55. A. Wegener, "Die Enstehung der Kontinente und Ozeane", Viewig, Brunswick (1941).
56. J. T. Wilson, "Evidence from islands on the spreading of ocean floors". *Nature,* **197,** 536–538 (1963).
57. J. T. Wilson, "A new class of faults and their bearing on continental drift". *Nature,* **207,** 343–347 (1965).

Polarity Reversals
58. P. J. Burek, "Magnetic reversals: their application to stratigraphic problems". *Bull. Amer. Assoc. Petrol. Geol.,* **54** (7), 1120–1139 (1970).
59. J. Gribbin, "Solar geogmagnetic disturbances and weather". *Nature,* **249,** 802 (June 1974).
60. C. G. A. Harrison, "Evolutionary processes and reversals of the Earth's magnetic field". *Nature,* **217,** 46–47, (Jan. 1968).
61. C. G. A. Harrison and J. M. Prospero, "Reversals of the Earth's magnetic field and climatic changes". *Nature, * **250,** 563–564 (Aug. 1974).

Age and Distribution of Petroleum
62. M. T. Halbouty, R. E. King, H. D. Klemme, R. H. Dott and A. A. Meyerhoff, "Factors affecting formation of giant oil and gasfields, and basin classification". In: "Geology of Giant Oil Accumulations", *Bull. Amer. Assoc. Petrol. Geol.,* Tulsa (1972).
63. M. Kamen-Kaye, "Geology and productivity of Persian Gulf Synclinorium". *Bull. Amer. Assoc. Petrol. Geol.,* **54** (12), 2371–2394 (1970).
64. G. Knebel and G. Rodriguez- Eraso, "Habitat of some oil". *Bull. Amer. Assoc. Petrol. Geol.,* **40** (4), 547–561 (1956).
65. K. K. Landes, "Eometamorphism, and oil and gas in time and space". *Bull. Amer. Assoc. Petrol. Geol.,* **51** (6), 828–841 (1967).
66. E. N. Tiratsoo, "Oilfields of the World", Chapters 3–11, Scientific Press Ltd., Beaconsfield (1973).

Wrench-faults and lineaments
67. T. P. Harding, "Newport-Inglewood Trend, California—an example of wrenching style of deformation". *Bull. Amer. Assoc. Petrol. Geol.,* **57** (1), 97–116 (1973).
68. T. P. Harding, "Petroleum Traps associated with wrench faults". *Bull. Amer. Assoc. Petrol. Geol.,* **58** (7), 1290–1304 (1974).
69. G. Parker, "New basement tectonics". *Bull. Amer. Assoc. Petrol. Geol.,* **57** (5), 956 (1973).
70. R. E. Wilcox, T. P. Harding and D. R. Seely, "Basic Wrench Tectonics". *Bull. Amer. Assoc. Petrol. Geol.,* **57** (1), 74–96 (1973).

Continental Margins
71. A.A.P.G. Reprint Series, No. 3 "Continental Shelves—origin and significance". *Amer. Assoc. Petrol. Geol.,* Tulsa (1972).
72. R. H. Beck and P. Lehner, "Oceans, new frontiers in exploration". *Bull. Amer. Assoc. Petrol. Geol.,* **58** (3), 376–395 (1974).
73. H. D. Hedberg, "Continental margins from viewpoint of the petroleum geologist". *Bull. Amer. Assoc. Petrol. Geol.,* **54** (1), 3–43 (1970).
74. L. G. Weeks, "Marine geology and petroleum resources". PD 2 (3), *8th World Petrol. Congr., Moscow,* (1971).
75. L. G. Weeks, "Offshore petroleum development and resources". *Journ. Petrol.. Technol.,* **21** (4), 377–385 (1969).

CHAPTER 10

Petroleum Production and Reserves

OIL AND GAS PRODUCING MECHANISMS

Primary Mechanisms[6]

SEVERAL major mechanisms can play a part in the oil and gas production process, either separately or jointly. Production is commonly dominated by one such mechanism; however, there can be changes in the emphasis during the life of a field, while the way in which a field is managed can alter the relative contributions from the various producing mechanisms for that field. These producing mechanisms depend extensively on the expansibility of the fluids in free communication in the reservoir rock, in association with gravitational effects and contraction of the pore space in the reservoir rock and in sealing rocks adjacent to the accumulation. The proportions of free gas, gas/oil solution (and its gas content) and water in the reservoir rock determine, together with their physical properties, the expansibility on pressure drop. Gas and gas/oil systems (below the bubble point) are much more expansible than a gas/oil system above the bubble point, and this is rather more expansible than the brine in a reservoir rock.

The oil-producing mechanisms dependent on expansibility are dissolved-gas drive (solution-gas drive), gas-cap drive, and water drive. Their individual effectiveness involves gravitational effects, and is also influenced by the producing rates and the degree of uniformity and size of the openings of the pore system in the reservoir rock.

If a system consisting of free hydrocarbon gas and crude oil is subjected to progressively increasing pressure, gas goes into solution and the remaining free gas is also compressed. As the gas dissolves, the volume of the liquid increases appreciably. This series of changes continues until the gas is all dissolved; the pressure at this stage is known as the *bubble-point* or *saturation pressure* of the system. Further increases in pressure will then cause the volume of the liquid (gas/oil solution) to decrease. If, at the starting point there is unit volume of crude oil, the ratio of the volume of liquid at any later pressure to this unit volume gives what is termed the *formation volume factor* of the crude oil. The rate of change of volume of the total system with change in pressure is far greater for pressures below the bubble-point than for pressures above that pressure. In the latter condition, the crude oil is under-saturated with gas; below the bubble-point pressure the crude oil is saturated with gas. Increasing the amount of gas in solution causes the viscosity of the liquid to fall, and the interfacial tension of the solution against brine to increase.

When a well is drilled into a reservoir and put on production, the pressure at the bottom of the well is dropped below the initial reservoir pressure and flow is thus directed towards the well. This reduction in pressure causes expansion of the gas/oil solution in the reservoir rock immediately adjacent to the well, leading to expulsion of gas/oil solution into the well bore. The expulsion allows the pressure drop to influence the fluid progressively further and further away from the well. Should the pressure at the bottom of the hole be above the bubble point, so that the solution in the reservoir rock is under-saturated, no gas will be released in the reservoir rock even though some fluid expansion, and therefore flow, will take place. However, if the bottom-hole pressure is below the bubble-point pressure of the gas/oil solution, gas will be released in the pore spaces of the reservoir rock (in the absence of super-saturation), and this, by increasing the volume of the gas/oil system, will also force some of the gas/oil solution into the well. Initially, only the gas/oil solution enters the well bore and this will have less than its original content of dissolved gas; but when the free gas saturation in the rock adjacent to the well reaches a value which depends on the characteristics of the rock, free gas as well as gas/oil solution will flow into the well, and the overall producing gas/oil ratio begins to rise. Because of increase in pressure with distance from the well, the free gas saturation in the rock will diminish away from the well. The presence of free gas reduces the capacity of the rock to allow flow of the gas/oil solution; and as the volume of free gas space increases the flow of free gas becomes increasingly easy once the free gas saturation exceeds a critical value, whereas the flow of oil becomes more difficult. Such are the main features of the mechanism of *dissolved-gas drive*.

Should an oil accumulation have a free "gas cap", then wells should not be completed in a way that would allow gas from the gas cap to enter them directly. The low pressure at the well first affects the adjacent gas/oil solution, but is soon felt by the gas-cap gas which responds by expanding and enhancing the tendency for the underlying gas oil solution to flow downwards. It will be evident that gas must also be released from the solution adjacent to the gas cap as a consequence of the lower pressure. The expanding gas-cap gas will eventually make contact with this released solution gas, increasing the mass of the gas cap. Also if, as a result of gas release, sufficiently tall continuous free gas masses develop in the gas/oil solution they will move upwards by buoyancy to join the gas-cap gas, involving internal interchange of position for gas and liquid arising from gravitational effects.

This mechanism is *gas-cap drive,* and it is clearly superimposed on dissolved-gas drive. Because of the greater volume of total free gas relative to gas/oil solution in the reservoir rock, the expansibility of the entire hydrocarbon system at a given pressure is greater than for the case of no initial free gas, i.e. no gas cap. There is more gas, free and in solution, available to help in expelling oil than for strict dissolved-gas drive. The greater amount of compressional energy may be expected to increase the oil recovery over that to be expected from dissolved-gas drive, other conditions being the same.

The expansibility of oilfield brines is decidedly less than even the expansibility of under-saturated gas/oil solution, and of course much less than for saturated gas/oil systems. Nevertheless, a reduction of pressure in the oil column, as a result of oil and gas entering wells, will lead to a drop of pressure in the adjacent water and to expansion of that water. The viscosity and compressibility of the liquid, and the porosity and permeability of the rock impose controls on the rate at which this influence can spread. When a very large volume of water is affected by the pressure drop, the increase in volume resulting from the expansion will be substantial. The expanding water will push oil towards the low-pressure points at the wells. Its viscosity is closer to the viscosity of the gas/oil solution than is the viscosity of compressed gas. Overall, the water is more effective in displacing oil from the rock pores than is free gas, which for various reasons will be more prone than water to finger through the gas/oil solution in the rock. Provided that the rate of production of oil is not such as to lead to a relatively rapid pressure drop, and thus encourage the action of dissolved-gas drive, advantage can be taken of the potential *water drive* which is decidedly more efficient in displacing oil.

Interfacial forces, acting in water-wet rocks containing both water and hydrocarbons, tend to cause invading water to displace the hydrocarbons preferentially from the finer-pored sectors at a given level in non-uniform rocks. Indeed, those forces acting alone would cause the water/hydrocarbon boundary to be higher in the fine-pored than in the coarse-pored sectors. This is the phenomenon of *imbibition*. However, when there is general advancing flow of the water, permeability considerations will tend to make its advance, and therefore the displacement of hydrocarbons, easier in the coarse-pored (more permeable) sectors than in the fine-pored sectors of the reservoir rock. Consequently, relatively rapid rates of water invasion can lead to the by-passing of oil in the fine-pored sectors of non-uniform reservoir rocks, with eventual lower overall oil recoveries than might have been attained by lower rates of water entry. Again, other physical factors can be involved, including the details of the structure of the reservoir rock and their relation to the wells and areas of water entry; and it has to be remembered that in economic terms the rate of production assumes importance, as well as the total amount of oil or gas finally produced. The rates of water advance which give the greatest overall displacement of the oil may be too low to provide economically satisfactory rates of oil production in some cases.

Drop in pressure on the fluids in the reservoir rock will allow a slight reduction in the volume of the pore space. It may also locally modify flow patterns of water being moved by compaction, with local acceleration of compaction. An adjacent accumulation in the same reservoir rock which is not being produced could feel the effects of the pressure drop, leading to expansion of its hydrocarbon system. This would cause its oil/water contact to be depressed, with corresponding influx of water into the trap where pressure was falling in the course of production.

Rock pore compressibility is small, so its effects will be significant only where production is being dominated by liquid expansion and there is no relatively large body of water in free communication with the grossly under-saturated gas/oil system. However, cases are known of fractured reservoirs in which reduction of pressure leads to narrowing of the fractures and a reduction of the bulk permeability of the rock.

Tall oil columns in rocks with high permeability across the bedding or good permeability with steep dips, and no effective water drive, may yield much of their oil by gravity drainage when the pressure has dropped considerably and dissolved-gas drive has fallen off in effectiveness.

There is evidence that true dissolved-gas drive reservoirs have oil recoveries which are essentially independent of both individual well rates and of total reservoir producing rates. The same may be true of very permeable, uniform reservoirs producing under very active water drives. However, there are reservoirs for which it is considered that there is a maximum efficient rate; above this rate there is a significant reduction in the ultimate oil recovery by the primary producing mechanism. These reservoirs are rate-sensitive, and the producing mechanisms include partial water drive and gravity drainage. A suitable practicable life can give a maximum oil recovery factor.

The various mechanisms described above are referred to as *primary producing mechanisms;* they use only original energy in the reservoir to drive oil into the wells. This energy may be sufficient to carry the oil to the surface at acceptable rates initially or even throughout the producing life of a well. However, if the rate of flow falls below acceptable values it becomes necessary to apply artificial lift, i.e. pumping, gas-lift, swabbing, or baling to lift fluid from the well bore, thereby reducing back-pressure on the fluids in the rock and so permitting continuing entry of an adequate volume of fluid into the well bore. Alternatively, or in addition, *formation stimulation* may be undertaken. In this kind of operation, the ability of the rock to allow the passage of fluids is increased by the creation of fractures, by enlarging existing openings in the rocks adjacent to the wells, or by removing deposits which have partially blocked the openings during earlier production. Thus, fractures may be created or widened by fluid injection (hydraulic fracturing) and kept open by introducing suitable "propping" materials. Explosives may also be used to create fractures. Acid may be injected to widen openings by solution of the rock, while organic solvents may be used to remove clogging waxy and asphaltic deposits, or to remove water which has invaded the rock as filtrate from the drilling mud. The reservoir rock permeability in the immediate neighbourhood of the well is especially important. Sometimes, this has been impaired during the course of drilling the well or during subsequent production processes. Fractures induced in this damaged zone will greatly improve flow without the need to restore it as a whole to its undamaged condition[10, 12].

Attempts to improve fluid flow have also been made by the use of subsurface heaters opposite the reservoir rock. These warm the rock and its contained fluids near the well bore. The oil is reduced in viscosity and therefore flows more rapidly into the well under a given pressure drop.

Acoustic energy in the ultrasonic range (frequency of 20 Kc/sec and intensity of 55 watts/sq cm) has been used to stimulate fluid flow through porous media. A combination of a "sonic pump" with a subsurface heat exchanger has been shown to boost production from shallow, unconsolidated sands containing heavy oils. The sonic vibrations keep the casing perforations clear, while superheated water at 250 °C is circulated through the heat exchanger.

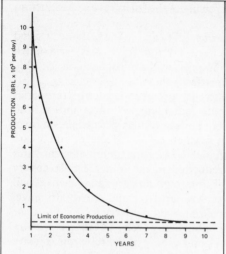

Fig. 74. A typical production decline curve for an oil-well on natural flow.

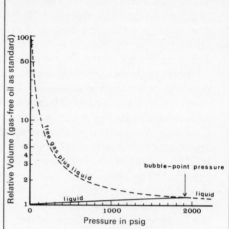

Fig. 75. Pressure-volume relationships for gas-oil system.

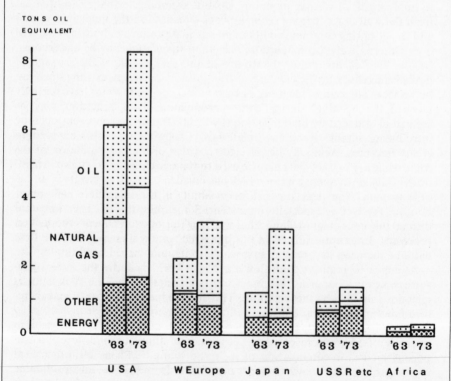

Fig. 76. The growth of world energy consumption in the decade 1963–1973 expressed in terms of tons oil equivalent per capita.

Heating of the rock and oil in these operations is dependent on conduction, in contrast with the injection of hot fluid into the reservoir rock which is discussed in the next section,

Secondary Recovery Techniques[20]

The fall in reservoir pressure causes gas release in the oil-bearing zone and an increase in oil viscosity, because there is less gas in solution, and also limits the possible pressure gradients towards the well. In combination, these three factors ultimately result in the rate of oil flow into the well bore becoming too low to be profitable, and so the well must then be abandoned unless means are available which can raise its output so as to more than offset the cost of applying them. This led to the application of techniques which involved the return of energy to substantial sectors of the producing reservoir, if not to the entire reservoir. Sometimes these techniques raised the reservoir's producing rate to levels as high as the peak attained by primary mechanisms; they became known as *secondary recovery* techniques, and were successful in raising the overall recovery to values beyond what the field could give by the primary recovery mechanisms. They also gave the best increases for fields controlled by the least efficient primary recovery mechanisms. Maintenance of the reservoir pressure at or near its original value naturally minimises the adverse effects of pressure drop on oil flow mentioned above. Consequently, some fields were the subject of pressure maintenance by injecting fluids, gas or water, at an early stage in their producing history.

It should also be noted that the various secondary recovery techniques always include an element of *pressure restoration* or at least a checking of the reservoir pressure decline, since they all call for the injection of fluids. This alone would offer some benefits, irrespective of the special features of a particular technique. Over the years, there has been a tendency to apply various secondary recovery techniques at progressively earlier stages in a field's producing life, for economic considerations indicate that in many cases it can be more profitable to obtain the oil early, rather than to spread it over a lengthy period. Thus, producing rates can be as important as high ultimate recovery.

These techniques call for injection operations to take place through one set of wells, while oil is produced from others. The injection wells are commonly former oil-producing wells, although in certain cases wells may be drilled especially for injection or production, so as to give more appropriate flow patterns than can be obtained by using the existing wells. Depending on the nature of the operation and the form and other characteristics of the reservoir, there may be "dispersed injection" or "line injection". In the first case, injection wells will be surrounded by producing wells, or vice versa, according to some simple pattern; in the second case, a line of injection wells will have a series of lines of producers on one side or the other of it. Typically, there could be a line of water injection wells low on the flank of a structure, or a line of gas injection wells high on the flank. Also, at a certain stage a line of injection wells may be abandoned, and their role transferred to an adjacent line of wells which hitherto have been oil-producing wells.

The extraction from a reservoir of any material that cannot be disposed of profitably, in a sense reduces the reservoir pressure unnecessarily. However, in order to recover the oil, production of gas is inevitable, and some fields also give increasing amounts of water with the oil as they grow older. Flaring of unwanted gas is a cheap way of disposal, but it is clearly an increasingly unacceptable waste of energy. Disposal of produced salt water can be an expensive operation. Subsurface disposal of water may be necessary, and if water or surplus gas can be returned to an oil reservoir, pressure will be maintained to some extent and there will be an increase in the ultimate oil recovery. Injection may have to be on a certain minimum scale to give worthwhile benefits from the point of view of increased oil recovery. Merely to return all the gas and water produced from a given oil reservoir clearly cannot maintain the original pressure, because oil has also been removed. Hence, when this technique is used, such gas and water must normally be supplemented by additional volumes of fluids. Gas may be available from other reservoirs or nearby fields, and the same may be true for water. On the other hand, it may be necessary to obtain this "make-up" water from other sources, either subsurface aquifers or surface supplies.

When gas is injected into a gas cap or water into the edge- or bottom-water zone at points remote from the fluid contacts, the producing behaviour of a reservoir will clearly resemble that of gas-cap drive or water drive, respectively. However, when injection is into the oil zone there will be some differences from the processes dependent on natural energy sources. Injection will be over a vertical section, and permeability layering will inevitably result in some sectors accepting the injected fluids at greater rates than others. There will also be a phase of pressure build-up wherein some disseminated gas goes into solution. The pressure and flow patterns created will differ markedly from those when injection was not taking place, irrespective of whether or not in the natural process there was a producing well at the site of the injection well.

It is evident that when secondary recovery or even pressure maintenance is to be undertaken, a great deal needs to be known about the detailed form and internal structure of the reservoir. Should there be discontinuities in the reservoir rock, then each structural unit requires separate consideration, as does a reservoir rock zone which is effectively a series of lenses. Otherwise, effort may be wasted in injecting fluid at sites which do not give the expected communication with adjacent producing wells. Nevertheless, the existence of an isolated reservoir unit may provide an opportunity for running a pilot field-scale test of the proposed technique.

Gas injection raises fewer problems than water injection. Water may require treatment before injection, since it must not cause clogging in the reservoir rock or its exposed surface in the borehole. Some reservoir rocks contain clays which swell or break down when in contact with water of salinity different from that in the rock at the time of first penetration by the drilling bit. When the water produced from the reservoir is injected together with the "make-up" water, the two waters clearly need to be compatible, i.e. not to form precipitates in the well or reservoir rock. Furthermore, the water should not carry down bacteria which may thrive and produce clogging slimes in the bottom of the hole where the water enters the reservoir rock.

When consideration is being given to the early use of secondary recovery techniques it is necessary to predict what would eventually be recovered by continued use of the primary mechanisms. This oil volume must be compared with the expectations from one or more secondary recovery techniques and note taken of the additional costs involved. Unless the additional recovery is likely to more than pay for applying the technique, there is obviously no advantage in applying secondary recovery. For late application, the amount remaining to be recovered by primary production is small, so the principal consideration in reaching a decision relates to prediction of the amount of oil which will be produced by the secondary recovery process.

For many years, gas injection and water injection were the techniques of secondary recovery. A number of new methods have been tried in recent years, and the terms *tertiary* and even *quarternary* recovery have been used for these. However, apart from being the third or the fourth method applied to a given field, such labels have little merit. Indeed, it has been recognised in some cases that their application immediately after the primary production stage would have provided a chance of better success. They are, in fact, really further secondary recovery techniques. (Examination of 172 projects of this kind showed that 107 were applied immediately after primary recovery operations and only 65 after some other type of secondary recovery technique[3].)

The recovery factors for primary producing mechanisms are inherently low for high-viscosity crude oils, and such oils are also dense. Hence, accumulations of these oils offer targets for considerable increases in the ultimate recovery factors, and the newer secondary recovery techniques tend to be directed particularly at high-viscosity crude oils. These methods have at least one of the following objectives: reduction in the viscosity of the oil by the use of heat, injected or generated in the reservoir; removal of the interfaces between the oil and the displacing fluid, for such interfaces contribute to the failure to obtain complete oil recovery by means of the primary producing mechanisms; making the viscosity of the displacing fluid closer to that of the oil being displaced, a procedure which cuts down fingering of the displacing fluid through the oil, thereby reducing the magnitude of a factor which is responsible for a low value for the overall oil displacing efficiency.

In *miscible displacement,* a slug of LPG (liquefied petroleum gas, mainly propane and butane) is injected, and its leading edge dissolves, and dissolves in crude oil, so that there is no interface between the two liquids. Then, natural gas is injected to force the LPG out towards the producing wells, and the reservoir temperature and pressure must be such as to make the gas freely miscible with the LPG, thereby eliminating any possibility of gas/LPG interfaces. This system has the capacity to remove oil very effectively from the pores contacted by the LPG, but because the LPG is less viscous than the crude oil and the gas less viscous than the LPG, the areal sweep efficiency is poor. A suitable quantity of LPG is 3% to 5% of the reservoir pore volume involved. LPG is not cheap, but when too small a quantity is used the follow-up gas breaks through to the crude oil, and then the objectionable gas/oil interfaces develop[21].

The same principle has been applied in several different ways, including high-pressure dry gas miscible displacement, which is practicable when the crude oil is rich in C_2 to C_6 components. These light components are vaporized and the enriched gas forms a zone equivalent to the LPG slug in its effects.

Unfortunately, this type of miscible drive has proved somewhat disappointing because of the poor sweep efficiency, although some improvement in this respect can be obtained when it is possible to drive the oil downwards, so deriving benefits from gravitational effects. One of the successes is at *Golden Spike,* a pinnacle-reef type reservoir in Alberta, where 95% of the oil is expected to be recoverable, using a dry-gas-driven bank of LPG derived from the crude oil.

An alternative system which has had very limited trial employs an alcohol slug which is miscible with the crude oil and with the water used to provide the main drive. This system has a better areal sweep than the LPG/gas system, and does not have the same pressure limitations.

Micellar slugs, basically formed of hydrocarbons, surfactants and water, with minor amounts of co-surfactant and electrolyte, have also been used for the miscible displacement of crude oils. Such slugs are followed by a mobility buffer, either a water-external emulsion or thickened water, then the main water which propels the slug and mobility buffer. Like LPG, the micellar slug completely removes oil from the zone contacted[7].

The *in situ* combustion technique is a means of reducing the viscosity of some of the oil by heat. Air is injected into the reservoir and ignition may be achieved by the use of a subsurface heater, or by chemicals dumped in the well. Alternatively, a mixture of hot combustion gas, air and steam may be injected until ignition occurs. In both cases, when combustion starts it is then necessary only to continue air injection and, indeed, cases are known in which injection of air alone, without any initial heating, leads to sufficiently rapid oxidation to create a burning front[21].

The ultimate drive is by the injection of air. Heat moves ahead of the advancing burning front in which the heavy components of the crude oil are consumed, as a result of conduction, moving hot gases, steam and the vaporized light components of the burnt crude oil. The crude oil ahead of the burning front is diluted by condensation of the light hydrocarbons and this, together with heat, reduces the viscosity of the bank of crude oil. Still further ahead is unaltered crude oil.

This is the "forward combustion process" in which the burning front finally breaks through to the producing wells after the main phase of oil displacement.

Hot water and/or steam have been injected to drive oil towards the producing wells, the increase in temperature reducing the viscosity of the oil in the zone affected by the heat ahead of the injected fluid. Very hot water was injected at *Schoonebeek* in Holland, starting in 1954. It is estimated to have improved the oil recovery quite considerably (perhaps by 50% more than if a cold water drive only had been used), at a modest cost. Cold water injected subsequently picks up residual heat from the warm reservoir rock and this aids in the continuing oil displacement process.

The "steam-drive" process has had considerable success in North America and Venezuela. However, the "steam-soak" technique (sometimes called the "huff-and-puff" technique) should really be viewed as a method of reservoir stimulation rather than a secondary recovery technique. It was discovered accidentally when a well in Venezuela, used for steam injection, had to be temporarily shut in. On opening-up, as a preliminary to resuming injection, the well produced oil for a time at a considerably increased rate. In this process, steam is injected into the reservoir for a period of about four weeks; then a period of the order of five days is allowed for soaking, after which the well is put on oil production for as much as twelve months. This cycle of events is then repeated, successive cycles broadly giving smaller maximum oil-producing rates and shorter periods of production than the earlier cycles at rates above the acceptable minimum. The peak rate may be as much as twenty times the rate immediately before the process was started[4, 17].

The steam injected improves the reservoir rock permeability by removing asphaltic and waxy deposits; it also reduces the viscosity of the oil. In thick or steeply dipping reservoirs, these two features make oil production by gravity drainage assume importance when the pressure is low. Otherwise, adequate reservoir pressure drives condensed steam and oil to the well bore. A significant fraction of California's oil production is currently obtained by "steam soak".

Colloidal substances such as partially hydrolysed polyacrylamides have been employed in small amounts to "thicken" water, and thereby to improve the areal sweep-out in water-drive operations by making the aqueous medium closer in its viscosity to the viscosity of the crude oil. In order to economise in the use of the chemical additive, a slug of the thickened water is followed by normal water, sometimes with a tailored transition zone between the two, the aim being to get the most favourable areal sweep-out possible.

There are variants of some of the above processes, one being the injection of carbon dioxide, which is miscible with oil at pressures over 700 psia, followed by water. In addition to complete oil sweep-out from areas swept by the carbon dioxide, it also causes viscosity reduction and swelling of the oil, two effects which help to increase oil recovery.

ESTIMATION OF OIL AND GAS RESERVES[1, 2, 11, 15, 22]

The term *reserves* is not and has not been used consistently. It may refer to the total amount of oil or gas in an accumulation or, more commonly, to the quantity known or expected to be recovered from an accumulation under stated or assumed technical and economic conditions. Since the proportion of the total crude oil originally in the reservoir rock that is ultimately obtained from an oil accumulation may range from a few per cent to as much as seventy per cent, it is obviously necessary when considering quoted figures to know which definition was used in arriving at the figures. The term "recoverable reserves" is unambiguous, but the qualifying terms "proved", "probable" and "possible" do not in themselves necessarily remove the ambiguity. (This is discussed more fully on pp. 278–9.) The recoverable reserves are related to the initial oil or gas in place through the "recovery factor".

The estimation of reserves is an important matter, for knowledge of the amount of oil or gas expected to be recovered from an accumulation is needed for various purposes, including business and technical decisions, and legal and fiscal matters. A reasonable estimate of the magnitude of the volume of oil or gas *in place* is especially important soon after the discovery of a field, for this will affect the plans for development and even the decision on whether or not to invest more money in the area. The two approaches in estimating reserves are as follows:-

1. Direct estimation of the amount of oil or gas in place, followed by multiplication by an appropriate recovery factor to give the recoverable reserves (i.e. the "volumetric methods").

2. Examination of the behaviour of the accumulation during production—in particular the relationship of the volume of oil (or gas) produced to some parameter—and extrapolation of this behaviour to the point at which it is no longer expected to be economic to continue production.

(The "material balance" method also depends on production data but gives a figure for the oil or gas in place, and therefore requires the application of a recovery factor in order to provide a value for the recoverable reserves[19].)

The volumetric methods (including the equivalent "comparative method") are the only means of estimating reserves capable of being applied at or before the start of production. Such methods can, however, be used throughout a field's producing life, with the general expectation of increased accuracy as the amount of relevant data increases. Those methods which depend in any way on production behaviour, and this includes the material balance approach, can be used with some expectation of being satisfactory, only after an appreciable fraction of the reserves has been produced. They, too, increase in reliability the greater the amount of appropriate information available. In both cases the real prediction obviously relates to a smaller amount of remaining reserves the greater the volume of oil or gas already produced.

Selection of recovery factors requires, among other things, consideration of the characteristics of the reservoir rock, the nature and proportions of the oil and gas present, reservoir pressure and the mechanisms likely to operate in driving oil or gas into the wells. The proportion of the oil in place which has been recovered from comparable abandoned fields, and in some cases reference to laboratory studies, provide data on recovery factors. Both approaches involve some measure of empiricism, even though in the case of oil the production-behaviour method may use quite complicated calculations or devices to match the production history, thereby providing values for the parameters needed to predict future behaviour. A major change in the mode operating an oil-well or oilfield will change the values of these parameters[18].

Volumetric methods

The first requirement is to ascertain the volume of storage space available to the hydrocarbons, gas or crude oil. This calls for information on the areal extent and thickness of the oil- or gas-charged rock, in fact for an isopachyte map of the hydrocarbon-bearing rock. It must be supplemented by information on the porosity and water saturation of the oil- or gas-charged rock, both of which parameters must be expressed as fractions, not as percentages. Then:

Volume of hydrocarbon-bearing space = Area × *Mean thickness* × *Mean porosity* × *(1-Mean water saturation)*

There are problems in dealing with the porosity and saturation data. Depending on the number and distribution of the observations, straight averages or some form of weighting may be employed.

For an oil accumulation, the volume of "stock-tank oil-in-place" is:

$$\frac{Volume\ of\ oil\text{-}occupied\ space}{Formation\ volume\ factor\ of\ crude\ oil}$$

and the reserves are:

$$\frac{Volume\ of\ oil\text{-}occupied\ space\ \times\ Recovery\ factor}{Formation\ volume\ factor\ of\ crude\ oil}$$

The recovery factor is expressed as a fraction. An equivalent expression can be used for gas, giving:

$$\frac{Volume\ of\ gas\text{-}bearing\ space\ \times\ Recovery\ factor}{Formation\ volume\ factor\ for\ gas}$$

Gas is commonly disposed of as a gas, and this involves long-term contracts to deliver it at a set rate of flow, otherwise a penalty is involved. When the pressure in the reservoir falls, it may no longer be practicable for the wells to produce gas at this minimum required rate. Then the field will have to be abandoned, or the contract modified, or the deficit made good from another reservoir. Hence, there is, in principle, a practical *abandonment pressure* for the particular circumstances, and this is what determines the recovery factor for a gas accumulation. It is possible to express the recovery factor in terms of the original and abandonment pressures, with note taken of whether or not water is expected to advance significantly during the producing life of the field.

Earlier discussion has indicated the considerations which will affect the value of the recovery factor in the case of oil.

Production-behaviour methods

Plots of mean reservoir pressure or of the ratio of that pressure and the corresponding compressibility factor against cumulative gas production may give a simple curve or even a straight line. If this curve or straight line is then extrapolated to the abandonment pressure, the position of the intersection gives the ultimate recovery, i.e. the reserves. The higher the proportion of the reserves that has been produced the better may the curve or line be expected to be defined, and consequently the more accurate should be the extrapolation. Also, the extrapolation is not so extensive because the volume of gas remaining to be produced is correspondingly smaller. Since it will not normally be practicable to shut in the entire reservoir to obtain pressure equilibrium, there will have to be some weighting or adjustment of the pressure measurements which themselves have an element of uncertainty. Hence, on the plots the production data will be more certain than the pressure values. These limitations must be recognised in making estimates of gas reserves.

Plots of production data for oilfields controlled by "natural" decline can also afford empirical methods for estimating reserves. These take various forms and their degree of applicability will depend on the extent to which certain controls have been applied to the reservoir, either by imposing varying restrictions on individual wells or changing the number of wells on production.

The plots include producing rate *versus* time, and log (producing rate) *versus* time. The area under the former curve is proportional to the cumulative production up to a given time. In the absence of certain manipulations of the wells the former plot may yield a curve with recognisable characteristics, while the second plot may approximate to a straight line. In both cases, if an abandonment producing rate is indicated, extrapolation is possible to give the time of abandonment and, by implication, the reserves. An alternative plot is producing rate *versus* cumulative production. Again, a simple form, even a straight line, may be defined, with the possibility of extrapolation to give the ultimate recovery.

Plots of mean reservoir pressure against cumulative production have also been used, as is done in the case of gas reservoirs. However, if a reservoir oil is grossly undersaturated with gas, the pressure will fall relatively rapidly at first as the amount of oil extracted increases; there will subsequently be a decidedly smaller rate of fall at higher values of cumulative production.

Other empirical plots which have been used include:

log (production rate) against cumulative oil production;

log (cumulative oil production) against water cut;

log (percentage of oil in total fluid produced) against cumulative oil production;

log (cumulative gas production) against log (cumulative oil production);

level of oil/water contact against cumulative oil production.

The suitability of the plot depends on the dominant producing mechanism for the particular field. The aim is eventually, at least, to get a "linear" plot and an important problem is the selection of a terminal value for oil rate, water or oil cut, cumulative gas, etc.

As in the case of gas accumulations, the definition of any usable relationship is better the greater the proportion of the reserves already produced, and the same is true when computational or instrumental means rather than graphical methods are used for matching the past history with a view to extrapolation into the future.

The concept of a *constant percentage decline law* has been applied both to single wells and to groups of wells or fields. However, even if individual wells accurately satisfied such a relationship, unless all of them had exactly the same decline rate, the aggregated output would not strictly satisfy a constant percentage decline law. This state of affairs can be considered to be one demonstration of the generally approximate nature of the relationship.

When purely graphical methods are used, and the plot does not give a straight line in simple extrapolation, it would be seem justifiable to give rather more weight to the recent past than to the distant past.

Material balance considerations[19]

The material balance concept considers the volume of the hydrocarbon-occupied space at a certain time, commonly the start of production, together with the pressure, and the contents of that same space after the production of known amounts of oil, gas and water in association with a known decline in reservoir pressure. One form is derived as follows:

N is initial number of volume units of stock-tank oil.

m is the ratio of gas-occupied and oil-occupied reservoir space at that stage.

$ß_o$, $α_o$ and S_o are, respectively, the oil formation volume factor, the gas formation volume factor and the gas solubility, at that stage.

$ß$, $α$ and S are the corresponding quantities after the production of n units of stock-tank oil.

R_n is the ratio of the gas produced and the oil produced.

w is the volume of water produced.

W is the total water entry into the reservoir space considered.

For the first state:

$Nß_o$ is the volume of the gas-oil solution. $mNß_o$ is the volume of the free-gas space.

For the later state:

$W-w$ is the effective water intrusion. $(N-n)ß$ is the volume of gas-oil solution. $mNß_o α/α_o$ is the new volume of the gas-cap gas present in the first state.

Under standard measuring conditions, i.e. at or near N.T.P:

The volume of gas produced is nR_n.

The volume of gas released from the oil produced is nS_o.

The volume of gas released from the remaining gas/oil solution is $(N-n)(S_o-S)$.

Hence, the total amount of gas released, but not produced, is $nS_o+(N-n) (S_o-S)-nR_n$, and this will have a subsurface volume of $α [nS_o+(N-n) (S_o-S)-nR_n]$.

Equating the volumes for the first and later states gives:

$$Nß_o+mNß_o = W-w+(N-n)ß+mNß_oα/α_o+α [nS_o+(N-n) (S_o-S)-nR_n]$$

Multiplying out and segregating the terms gives:

$$N=\frac{W-w+n (αS-ß-αR_n)}{ß_o-ß+mß_o(1-α/α_o)-α(S_o-S)}$$

Knowledge of the reservoir pressures and temperature for the two states, together with pressure-volume-temperature data for the given gas-oil system allows the selection of values of $ß_o$, $ß$, $α_o$, $α$, S_o and S. In the course of production n, R_n and w are measured: m is a constant, as is N the objective, but W is a quantity which will change as production continues, if there is water drive. Apart from the determination of m, the formula is in no way concerned with the details of the distribution of the free gas. It is obvious that if m were known exactly from volumetric measurements the of oil- and gas-occupied space, there would be no need to use the above relationship. However, if knowledge of porosities and water saturations is not available, but the gross volume of oil- and gas-occupied rock is known, a value for m could be obtained *assuming* that the porosities and water saturations are the same in the gas and the oil zones. If there is no water advance, or no gas cap at the first state, the relationship becomes simpler, and when both m and W are zero the formula is directly solvable.

The relationship can be written in a contracted form as:

$$N = \frac{W+A}{m\,B+C}$$

where A, B and C are obtained from the production and laboratory data at a particular pressure. It is apparent that certain pairings of values of m and W could yield the same figure for oil in place. In other words, independent data on these two factors are needed to give certainty to the value of N in general.

With continuous production records and pressure data for several succeeding states, a series of statements of the equation can be prepared, each using the same starting point. In the absence of water drive, values for m could be selected by trial and error which would lead to a series of statements giving the same value for N. However, when there is water advance as well as an initial gas cap, this trial-and-error approach would be more difficult and less likely to provide the right answer, for reasons noted in the preceding paragraph.

When there is no pressure drop on production, the relationship becomes indeterminate. Small pressure drops will lead to very small values for $(\alpha_o - \alpha)$, $(\beta_o - \beta)$ and $(1 - \alpha/\alpha_o)$. Coupled with limitations in the experimental data and and in the reservoir pressure survey data, for small pressure drops calculated values for N can be seriously in error, or erratic when more than one production period is considered. As with the empirical methods of estimating reserves, the smaller the fraction of the reserves already produced, the less accurate is the estimate likely to be.

Further practical problems arise when the oil column is tall. The pressures will—and the fluid properties may—then vary appreciably with level. A suggested solution for this difficulty is to consider several layers. Finally, the formation volume factors and solubilities differ according as the gas release from solution is a flash or a differential process. General considerations suggest that the process will differ in different parts of the total system involved in production, and that the relative incidence will change as production proceeds.

Although the use of a material balance relationship for the estimation of oil-in-place is not quite so straightforward as an equation might suggest, this should not be interpreted as meaning that the material balance relationship is useless. It certainly also has uses in other directions in oilfield studies. Moreover, in making estimates of reserves, the aim should be to use two or more *independent* methods whenever practicable. Reasonable agreement for the two methods would give confidence in the results; very different results would indicate errors in the assumptions or data used in one or in both methods. Sometimes, the further thinking engendered by the lack of agreement has led to worthwhile results beyond the derivation of a satisfactory figure for the oil reserves, the original objective.

Suppose that the assumption of no water drive has been made. When the calculations are made for progressively longer production periods, starting in each case from the same first state, the values of N should all be the same, provided that the assumption is correct. However, if the value of N increases as the length of the production period increases, this is taken to show that the assumption is wrong, and that there is in fact a water drive which is influencing production.

The terms in the general material balance relationship can be rearranged so as to segregate the produced fluids, with their volumes, except for water, expressed under subsurface conditions, while on the other side of the equation there are the contributions of the three producing mechanisms dependent on expansion:

$$n\beta + n\alpha \, (R_n - S) + w = \alpha N \, (S_o - S) - N(\beta_o - \beta) + mN\beta_o(\alpha/\alpha_o - 1) + W$$

Oil   Gas   Water

On the right-hand side, $mN\beta_o(\alpha/\alpha_o - 1)$ is the expansion of the original gas-cap gas; $\alpha N(S_o - S) - N(\beta_o - \beta)$ is the two-phase change of the original oil; W is the water intrusion.

The foregoing discussion on material balance has not referred explicitly to water compressibility or rock pore compressibility. However, both would be involved in the water entry term, W. When there is no free gas the total compressibility of the hydrocarbons is decidedly smaller than when a gas cap exists. Hence, when the accumulation is well above the saturation pressure it is appropriate to pay more specific attention to water and rock pore compressibilities.

Consider a volume of pore space of V, with an average water saturation of S, while the compressibilities of the water and gas-oil solution are, respectively, c_w and c_o vol/vol/unit pressure, and the rock compressibility is c_r pore vol/pore vol/unit pressure. For a pressure drop of p units, which does not take the gas/oil solution below its saturation pressure, the solution volume becomes $V(1-S) \, (1+pc_o)$, the water volume $VS \, (1+pc_w)$, and the pore volume $V(1-pc_r)$. This makes the new hydrocarbon storage space $V(1-pc_r)$ $- VS(1+pc_w)$, and at the new subsurface conditions the volume of the hydrocarbons exceeds this by: $V(1-S) \, (1+pc_o) - [V(1-pc_r) - VS(1+pc_w)]$, a volume of gas-oil solution which must be expelled from the reservoir rock through the wells. If the volume of stock-tank oil produced is n, its subsurface volume will be $n\beta$, where β is its formation volume factor. Therefore, $n\beta = V(1-S) \, (1+pc_o) - [V(1-pc_r) - VS(1+pc_w)]$

For initial stock-tank oil-in-place of N and a corresponding formation volume factor of β_o, $N\beta_o = V(1-S)$, whence $V = N\beta_o/(1-S)$.
Furthermore, $\beta_o(1+pc_o) = \beta$, and $\beta_o/\beta = 1/(1+pc_o)$
Thus:

$$n = \frac{V}{\beta}\left\{(1-S) \, (1+pc_o) - [(1-pc_r) - S(1+pc_w)]\right\}$$

$$= \frac{N\beta_o}{\beta(1-S)}\left\{(1-S) \, (1+pc_o) - [(1-pc_r) - S(1+pc_w)]\right\}$$

$$= \frac{N}{(1-S) \, (1+pc_o)}\left\{(1-S) \, (1+pc_o) - [(1-pc_r) - S(1+pc_w)]\right\}$$

Multiplying out and re-arranging the terms gives:

$$N = \frac{n(1-S) \, (1+pc_o)}{p[c_o + c_r + S(c_w - c_o)]}$$

On the assumption that c_o, c_w, and c_r are constant over the total pressure range considered, the quantity $(1-S)/[c_o+c_r+S(c_w-c_o)]$ will also be a constant, say K.

$$\text{Then: } N=\frac{nK(1+pc_o)}{p}$$

Since c_o is a small quantity, $(1+pc_o)$ will differ little from unity for moderate pressure drops. Consequently, there will be an approximately linear relationship between pressure drop and cumulative oil output.

Again, the formula tends to give a somewhat illusory impression that it is easy to get satisfactory answers. The value for S raises particular problems. Logs or cores taken during the drilling of wells could provide porosities and water saturations in the oil reservoir bed. Together with isopachytes of the oil zone, clearly such data could directly provide the volume of gas-oil solution and hence the amount of stock-tank oil. Edge or bottom water would contribute to the overall water saturation and influence the pressure behaviour of the accumulation.

Unless underlain by an areally more extensive hydrocarbon accumulation, there will be no incentive to drill wells which could define the size of the water mass in communication with the oil. Such water would affect the value of S, and if the water-bearing sector of the reservoir rock were large, the attainment of true pressure equilibrium could be a slow process. However, trial-and-error insertion of values for S could be a means of obtaining constant values for N, as is required.

Assuming the correctness of the various compressibility values and the pressure data, the problem is simpler than for the case of material balance below the bubble point (saturation pressure), since one unknown only, not two (one only being fixed), is involved.

The material balance concept leads ideally to an indication of the amount of oil-in-place and, like the simple volumetric method described earlier, requires the use of a recovery factor to give the reserves. It may, in its application, raise points which may suggest the presence or absence of water drive. On general grounds, the greater the amount of information which should exist when there has been more than a trivial amount of production, the more useful will be the deductions which should be possible.

PETROLEUM RESERVES

While the term "reserves" is conveniently used to denote the producible oil which is contained in a subsurface accumulation or within the limits of a particular geographical area, it has to be qualified by certain adjectives if it is not to be dangerously misleading.

The first group of qualifying adjectives are those which express the degree of confidence that can be attached to the presence of the reserves—i.e. whether they are in fact "proved" (less elegantly "proven"), "probable", "possible", or only "speculative". Unfortunately, these terms have been used rather loosely and are inevitably to some degree subjective and overlapping, although they express important differences.

It is normal practice for all operators to calculate the volume of reserves in any structure within their concession areas in which oil is discovered, and to improve and refine such estimates as appraisal and development of the field progresses. However, many of the factors which are required for such calculations can only be obtained as a result of considerable production history, so that the true extent of the oil reserves of a field may not be known until long after that field has been put into production. In general, initial estimates are progressively increased as time goes on, and the reserves which were originally only "probable" or "possible" are converted to the "proved" category as the result of appraisal wells drilled to define the extent of the productive rock volume and its parameters.

Some producing companies attempt to allow for the imponderables involved in development operations by using the concept of "discounted" reserves. These are assessed as a fraction—usually a half or a quarter—of the probable reserves of an incompletely evaluated oilfield or producing area.

In US practice, the area of an oil reservoir that is accepted as "proved" is defined (by the American Petroleum Institute) as: "(1) that portion delineated by drilling and defined by gas-oil or oil-water contacts, if any; and (2) the immediately adjoining portions not yet drilled but which can be reasonably judged as economically productive on the basis of available geological and engineering data. In the absence of information on fluid contacts, the lowest known structural occurrence of hydrocarbons controls the lower proved limit of the reservoir".

The second group of qualifying adjectives applied to the term "reserves" refers to their *degree of recoverability*. We have seen above that "primary" reserves can be defined as that portion of the original oil-in-place in a reservoir that can be expected to be recovered using the reservoir energy alone. Similarly, "supplementary" reserves denotes the additional oil that could be obtained by known secondary recovery proccesses. Obviously, the extent to which any such processes can be applied will be controlled by economic as well as technical factors; thus, the reserves which are recoverable "under current economic conditions" may be only a fraction of the reserves which might be physically recoverable without such restrictions. In fact, whether supplementary reserves become available at all must depend on the balance existing between the price to be obtained for the additional oil and the capital and operating costs of the installations required. In American Petroleum Institute practice, the reserves which can be "produced economically through application of improved recovery techniques (such as fluid injection) are included in the "proved" classification when successful testing by a pilot project, or the operation of an installed programme in the reservoir, provides support for the engineering analysis on which the project or programme was based".

When the attempt is made to extend estimates of reserves from single fields to large geographical areas, many difficulties intervene. For instance, only in North America is there any agreed basis for making such estimates[14, 18]. In the USA, numerous API committees regularly examine the reports submitted by all companies engaged in exploration, and after due consideration publish consolidated estimates of the proved reserves held at the end of every year in each state and in the United States as a whole. Of course, such

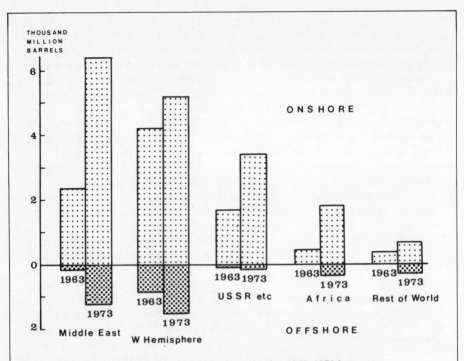

Fig. 77. Onshore and offshore oil production 1963–1973 by areas.
(*Source of data: BP Statistical Review, 1973.*).

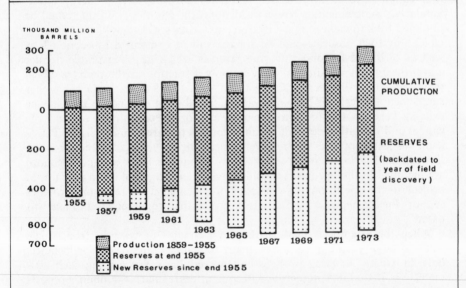

Fig. 78. World cumulative crude oil production and reserves (backdated to year of discovery)
(*Source of data: BP Statistical Review, 1973.*)

figures, however carefully compiled, provide nothing more than a kind of working inventory of the oil considered to be recoverable at a given time under the economics and operating conditions prevailing at that time. Such estimates refer to the primary oil plus the supplementary oil available as a result of fluid injection operations which are "known to be both actually in operation and economic".

An additional class of "indicated reserves" is also used in the United States to categorize reserves not "fully proved", and those also likely to be obtained from known reservoirs by the *future* application of supplementary recovery techniques. (Indicated additional reserves amounted to about 17% of the US proved reserves at end–1971.)

In the USSR, the systems of classification used are somewhat different. There, the "industrial reserves" of an economic mineral such as oil or gas are classified as the sum of three totals A_1, A_2 and B, where these classes are defined as: A_1—"reserves being currently exploited"; A_2—"reserves tested in detail by boreholes"; and B—"reserves established quantitatively, sufficiently accurately by prospecting. The form of the body or the distribution of the natural type of useful mineral or the technology of processing are insufficiently known."

In addition, the USSR classification includes groups C_1 and C_2, which are rather vaguely defined respectively as reserves adjacent to explored structures of a higher category, and reserves "determined by geological premises"; and D_1 and D_2, which are sub-groups of "predicted" reserves, believed to exist in relatively unprospected territory.

The 1968 OPEC definition divides "proved reserves" into two categories, and adds classes of "semi-proved" and "possible" reserves:

1.A. *Developed Proved Reserves:* Proved reserves currently subjected to exploitation through existing wells, reservoir outlets and production facilities.

1.B. *Undeveloped Proved Reserves:* Proved reserves that are not currently subjected to exploitation due to lack of wells, or lack of reservoir outlets to already existing wells, or lack of production facilities.

2. *Semi-Proved Reserves:*

The volume of hydrocarbons which geological and engineering information indicates, with a fair degree of certainty, to be recoverable from a tested reservoir, or those volumes of hydrocarbons technically recoverable, to a high degree of certainty, but the exploitation of which is deemed uneconomic.

3. *Possible Reserves:*

The volume of hydrocarbons expected to be recovered from untested reservoirs that have been penetrated by wells, or from the application of known supplementary recovery methods.

"National" Oil Reserves

Since all estimates of oil reserves must decline in reliability in proportion to the size of the area concerned, and because there are few equivalent official bodies issuing the same sort of carefully considered data as does the American Petroleum Institute, it follows that compilations of national oil reserves must be viewed with great caution. Nevertheless, attempts are regularly made to produce such data, since information about a country's oil reserves is of

obvious interest from many points of view. The information used inevitably
varies in its reliability from country to country and from year to year; and in
some cases it is simply only an arbitrary multiple—say 20 times—the last
year's crude oil production figure. Furthermore, for various commercial,
political or military reasons, published reserves estimates, where they are
available, may differ considerably from the information held in confidential
files by companies and governmental bodies.

A further difficulty is essentially one of terminology: it is seldom clear
whether the reserves quoted are in fact "proved", or whether they contain a
proportion of "probable" or even "possible" oil—added in for prestige
purposes or simply for lack of an agreed basis of definition. Similarly, there is
always some confusion as to the degree of recoverability that is implied in the
data from different sources.

In general, the compilations of reserves which are published refer to
"recoverable primary reserves"; but sometimes they also include an estimate
of the "supplementary reserves" that could be recovered by known secondary
techniques. Sometimes also, reserves of natural gas liquids are included in
the crude oil statistics.

Table 44 lists one estimate of the "published proved reserves" of the oil-
producing countries of the world, based on statistics which can only be as
reliable as is permitted by the limitations described above. Compilations such
as this provide at least a reasonable estimate of crude oil availability at any
time, and are of particular interest when compared with compilations made in
earlier years. Thus, it is remarkable to note that the published reserves have in
general increased with time—multiplying by a factor of more than ten over the
period since 1945 (Table 45). It seems, therefore, that in the past, crude oil
reserves have tended to be developed in relation to the commercial demand
for oil, rather than with regard to considerations of absolute availability.
However, this process may no longer be continuing, as is shown by examination of the reserves/production ratios summarized in Table 43.

The remarkable changes that have come about over the last quarter-
century in the proportions of world proved oil reserves held in different
producing regions are shown in Table 45. These are mainly due to the high
rates of consumption in areas such as the United States, Western Europe and
Japan, combined with the large new discoveries made and developed during
this period in the Middle East and Africa. Most remarkably, the United
States' proportion of world proved reserves declined from more than 35%
at the end of 1945 to only 5·9% at the end of 1973, while that of the Middle
East almost doubled over the same period and now exceeds 62%—most of
which lies in only six countries (Kuwait, Iran, Iraq, Saudi Arabia, Neutral
Zone and Abu Dhabi).

TABLE 43

RATIOS BETWEEN WORLD "PUBLISHED PROVED" RESERVES
AND ANNUAL PRODUCTION

1930	18
1940	16
1950	21
1960	39
1970	36
1972	31
1973	30

TABLE 44

WORLD "PUBLISHED PROVED" CRUDE OIL RESERVES

	MM brl		MM brl
USA	35,299·8	Algeria	9,930·0
Canada	9,628·9	Egypt	2,061·0
Mexico	2,846·8	Libya	23,208·0
		Morocco	1·2
North America	47,415·5	Tunisia	416·4
Trinidad	1,500·0	Africa, Northern	35,616·6
Others	1·2		
		Angola & Cabinda	1,310·0
West Indies	1,501·2	Congo (Brazzaville)	750·0
		Gabon	1,250·0
Argentina	2,312·4	Nigeria	18,250·0
Bolivia	222·7		
Brazil	774·0	Africa, Central & Southern	21,560·0
Chile	219·8		
Colombia	1,566·2	Abu Dhabi	18,700·0
Ecuador	5,912·8	Bahrein	276·0
Peru	544·2	Dubai	1,995·0
Venezuela	13,811·7	Iran	68,000·0
		Iraq	36,000·0
South America	25,363·8	Israel	1·5
		Kuwait	72,968·7
Austria	173·7	Neutral Zone	10,325·0
Denmark	250·0*	Oman	3,321·0
France	83·9	Qatar	5,624·0
Germany, West	542·0	Saudi Arabia	142,000·0
Italy	240·9	Syria	1,250·0
Netherlands	268·0	Turkey	143·3
Norway	4,500·0*	Others	500·0
Spain	75·0		
United Kingdom	6,300·0*	Asia, Middle East	361,104·5
Western Europe	12,439·5	Brunei-Malaysia	4,200·0
		Burma	43·0
Albania	76·0	China	14,800·0
Bulgaria	19·0	India	891·0
Czechoslovakia	18·5	Indonesia	11,500·0
East Germany	14·0	Japan	29·1
Hungary	205·0	Pakistan	32·4
Poland	45·0	Taiwan	17·6
Romania	1,405·0		
USSR	60,000·0	Asia, Far East	31,513·1
Yugoslavia	330·4		
		Australia	1,593·9
Eastern Europe	62,112·9	New Zealand	130·3
		Australasia	1,724·2
		TOTAL WORLD	600,351·3

* *Including North Sea.*

Note: Figures relate to end-1973. Data are based on "World Oil" statistics, August, 1974, with authors' amendments for some areas.

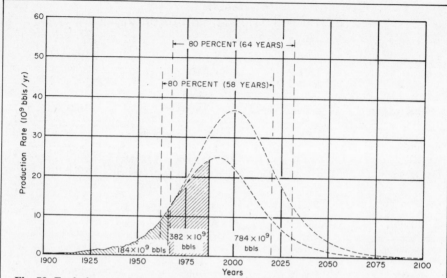

Fig. 79. Evolutionary cycles of world crude oil production for two projections of ultimate world output: upper curve—2100 B brl; lower curve—1350 B brl (after Hubbert[13]).

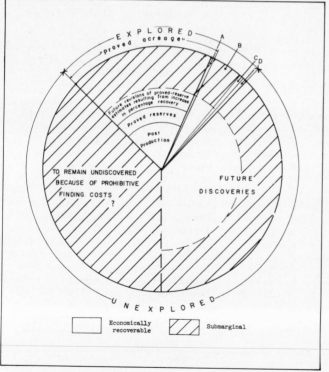

A. Future revisions resulting
   from net increase in
   estimates of oil in
   place in drilled acreage.

B. Future extensions to
   known pools, through
   development drilling.

C. In pools to be discovered
   in the course of
   development drilling of
   known pools.

D. In originally sub-
   commercial deposits and in
   drilled but undetected
   deposits that will be
   brought into production
   in the future.

Fig. 80. Diagram illustrating conceptual components of total oil in the ground. Not drawn to
scale. (Source: Hendricks[9]).

TABLE 45

WORLD "PUBLISHED PROVED" CRUDE OIL RESERVES 1945–1973

	end-1945[1]		end-1973[2]	
	B brl	% of total	B brl	% of total
North America	21·9	37·1	47·4	7·9
South America and West Indies	8·3	14·2	26·9	4·5
Western Europe (inc. North Sea)	0·8	1·4	12·4	2·1
USSR and Eastern Europe	8·0	13·6	62·1	10·3
Africa	0·1	0·1	57·2	9·5
Middle East	18·5	31·4	361·1	60·2
East Asia	1·3	2·2	31·5	5·2
Australasia	—	—	1·7	0·3
	58·9	100·0	600·3	100·0

1. 1945 production 2·8 B brl.
2. 1973 production 19·97 B brl.

Ultimate Resources and Energy Requirements[8, 9, 13, 23, 24]

The term "resource base" is used to describe the total volume of petroleum that lies beneath the surface of a particular geographical area. The proportion of this amount that will ultimately be producible under a specified set of economic and technical conditions comprises the "ultimate resources" of the area. This is equivalent to the "ultimate production" calculated at a given date (i.e. the past production plus proved reserves as at that date) together with the additional volume of oil which it is thought may be found and recovered in the future under forecast conditions of demand.

In spite of the obvious uncertainties involved, a number of estimates have been made of the "ultimate resources" of the world's major petroleum provinces, and attempts have also been made to extend these estimates to the whole world. The volume of suitable sediments in undeveloped basins has been estimated and their potential oil contents calculated by comparison with developed areas.

The hazards and guesswork involved in such operations are obvious, but for want of a better technique they have been thought worth trying. Unfortunately, since so much of the world's oil has been found in a few "giant" fields whose occurrence is statistically unpredictable, any "basin richness" compilations must be regarded with extreme caution.

In general, modern calculations of this type point towards ultimate world oil resources of around 2,000 B brl, whereas in the 1950's the equivalent totals were only 500–600 B brl. These estimates have obviously been expanded in proportion to the growth of proved reserves over this period, perhaps with little more than theoretical or mathematical justifications. Certainly, much of the world's undiscovered remaining oil may lie in very remote areas, deep waters, or be held in only small individual accumulations; there can be little more than guesswork to support calculations of what might be the economic costs and limits of its recovery in the distant future.

As the demand for energy increases[5, 16], it will obviously become more and more important that the recovery factor from oilfields still under production should be improved. At the present time, even in the United States where oilfield technology has reached the most advanced level, two barrels of oil are still left unproduced in the ground for each barrel that is produced—i.e. the overall recovery factor is only a little more than 32%. In other countries, the factor is generally considerably lower, depending on the sophistication of the secondary recovery systems that have been installed and the local demand for supplementary oil. It is possible that recovery factors may reach 40% or even 45% in the United States in the foreseeable future, but very improbable that anything like this figure will be obtained in other parts of the world. However, the importance of achieving improvements in recovery rates is shown by the calculation that only a 1% increase in average recovery from the world's currently known oilfields would be equivalent to adding a year's production—say 20 billion barrels—to the ultimate recovery likely to be obtained from them—equivalent perhaps to all the oil that could be recovered with great effort and expense from a new, major, but as yet undiscovered, petroleum province.

Another point that should be remembered is that compilations of crude oil reserves necessarily pay no attention to questions of crude oil quality. Some oils are clearly chemically more desirable than others from the refining point of view, while changing demands for different groups of refined products mean that oils which were not commercially attractive previously may subsequently become worth producing. Certainly, there are many known undeveloped deposits of crude oils in different parts of the world not currently considered to be of commercial quality, which will probably be added to the inventory of the world's crude oil reserves at some future time as the result of changes in technical and economic factors. There is thus likely to be an increasing incentive to develop the large accumulations of so-called "non-commercial" heavy, viscous oils which have been discovered in many areas.

The remarkable changes that occurred in the international oil industry in the last months of 1973, coupled with the inexorable growth of world primary energy consumption, make it clear that other than "conventional" sources of crude oil will be needed in the foreseeable future; the extraction of oil from the huge "tar sand" deposits of Alberta has already commenced and is likely to accelerate.

The large volumes of natural gas which are inevitably produced as a co-product with crude oils—volumes of the order of several hundred to several thousand cubic feet with every barrel or oil that reaches the surface—are already providing an important additional source of petroleum-based energy and chemical raw material. In earlier times, this associated gas, which is usually released from the oil during its passage through surface separators, was largely flared and wasted. Nowadays, however, increasing efforts are being made to utilise it, either immediately for reservoir pressure maintenance after compression and reinjection, or for industrial and commercial consumption.

A number of large non-associated gasfields have also been discovered in different parts of the world. They are of special note inasmuch as their gas outputs are usually available at much higher pressures than is the case with associated gases after separation, so that gas can be piped directly into long-distance distribution systems without the need for compression. Furthermore, since most non-associated gases are largely composed of methane, they can be liquefied by pressure combined with a refrigeration cycle to form a liquid only 1/600th of the volume of the original gas, and may then be transported as LNG by insulated tanker to suitable markets.

Natural gas provided about one-fifth of the world's primary energy requirements in 1973, when consumption was about 42 trillion cu ft; by 1985, it is estimated that the proportion will have risen to 25% and that world consumption then will be at least 90 trillion cu ft p.a.

One other source of future liquid fuel should also be mentioned—the shale oil which has been produced in relatively small quantities for several centuries, and which may now have a very important future.

The deposits of oil shale which occur in many parts of the world have no relationship to petroleum and thus do not really fall within the field of study covered by the present work. They are sedimentary deposits, formed by the deposition of multitudinous unicellular algae and resistant plant spores in very shallow seas, lakes and swamps, which can yield "oily" liquids (similar to, but by no means identical with crude oils) upon distillation in closed vessels at temperatures of 300°–400°C.

Shale oil has been commercially obtained from the distillation of oil shales in a number of countries—notably Scotland, Esthonia, China and Brazil; but so long as crude oil has been plentiful and cheap, all shale oil extraction processes have been relatively uneconomic except under special circumstances, since they involve the movement and treatment of large amounts of solid materials, with concomitant problems of waste disposal. However, the extraordinary increases that have taken place in the price of crude oil since 1973 may very well alter this situation in the future. In particular, the large deposits of Eocene oil shales in the Green River area of the western United States are likely to be developed within the next few years.

In this context, it must be borne in mind that all figures for "reserves" of shale oil are in quite a different category from any type of crude oil reserves; thus, *no oil at all* is recovered from the shale until it is retorted, and only if completely retorted is all its oil released. Hence, "reserves" only refer to the oil content of a particular shale deposit which *could* be released if there were sufficient economic incentive for all the shale to be suitably treated.

On this basis of definition, it has been estimated that there is the equivalent of some 2,500 B tons of shale oil potentially available in known, accessible oil-shale deposits capable of yielding 10–40% of their weight of oil by distillation. In addition, there are much larger amounts—perhaps as much as 47,500 B tons of oil—tied up in shales which would yield less than 10% by weight of oil but which perhaps could be used directly as pulverised furnace fuel for energy production[6a].

REFERENCES

1. A. A. Arps, "Estimation of primary oil reserves", *Trans. AIME, Petroleum Transactions,* **207,** 182–191 (1956).
2. J. J. Arps, "Estimation of primary oil and gas reserves", in: "Petroleum Production Handbook" (Ed. T. C. Frick), Vol. II, pp. 37–1 to 37–56 (1962).
3. W. B. Bleakley, "Journal survey shows recovery projects up," *Oil & Gas Journ.,* **72** (12), 69–78 (1974).

4. J. Burns, "Review of steam-soak operations in California," *J. Petrol. Tech.*, **21** (1), 25–34 (1969).
5. G. Chandler, "Energy: the changed and changing scene," *Petrol. Rev.*, 265–270 (July, 1973).
6. B. C. Craft and M. F. Hawkins, "Applied Petroleum Reservoir Engineering", Constable & Co., Ltd., London (1959).
6a. D. C. Duncan and V. E. Swanson, "Organic-rich shale of the United States and World Land Areas". *U.S. Geol. Surv. Circular 523*, Washington (1965).
7. W. B. Gogarty and W. C. Tosch, "Miscible-type waterflooding oil recovery with micellar solutions", *J. Petrol. Technol.*, **20** (12), 1407–1414 (1968).
8. G. D. Grant, A. L. Flood, L. F. Keating and W. A. Loucks, "A brief by the Alberta Society of Petroleum Geologists on Science and Technology in the supply and utilization of energy in Canada", *Bull. Canad. Petrol. Geol.*, **21** (2), 260–284 (1973).
9. T. A. Hendricks, "Resources of oil, gas, and natural gas liquids in the United States and the World", *U.S. Geol. Surv. Circ.*, 522 (1965).
10. G. C. Howard and C. R. Fast, "Hydraulic Fracturing", Soc. Petrol. Eng., Texas.
11. R. V. Hughes, "Oil property Valuation", John Wiley & Sons Inc., New York (1967).
12. M. J. Jeffries-Harris and C. P. Coppel. "Solvent stimulation of low-gravity oil reservoirs". *J. Petrol. Technol.*, **21** (2), 167–175 (1969).
13. M. King Hubbert, "Energy resources", in: "Resources and Man", W. H. Freeman, San Francisco (1969).
14. W. F. Lovejoy and P. T. Homan, "Methods of estimating reserves of crude oil, natural gas and natural gas liquids", Johns Hopkins Press, Baltimore (1965).
15. R. E. Megill, "An Introduction to Exploration Economics", Petroleum Publishing Co., Tulsa (1971).
16. J. D. Moody, "Petroleum demands of future decades," *Bull. Amer. Assoc. Petrol. Geol.*, **54**, 2239–2245 (1970).
17. C. B. Pollock and T. S. Buxton. "Performance of forward steam drive project—Nugget reservoir, Winkleman Dome field, Wyoming", *J. Petrol. Technol.*, **21** (1), 35–40 (1969).
18. J. T. Ryan, "An analysis of crude-oil discovery rate in Alberta", *Bull. Canad. Petrol. Geol.*, **21** (2), 219–235 (1973).
18a. J. T. Ryan "An estimate of the conventional crude-oil potential in Alberta", *Bull. Canad. Petrol. Geol.*, **21** (2), 236–246 (1973).
19. R. J. Schilthuis, "Active oil and reservoir energy", *Trans. AIME*, **118**, 33–(1936).
20. R. A. Smith, "Mechanics of Secondary Oil Recovery", Reinhold Publishing Corp., New York (1966).
21. Soc. of Petrol. Eng., Texas, *Reprint Series:* "Miscible Processes", "Oil and Gas Property Evaluation and Reserve Estimates", "Thermal Recovery Techniques", "Waterflooding".
22. W. F. Stevens and G. Thodos, "New Method for estimating primary oil reserves", *World Oil,* 113–115 (Dec. 1961).
23. H. R. Warman, "Future problems in petroleum exploration", *Petrol. Rev.,* 96–101 (March 1971).
24. A. D. Zapp, "Future petroleum producing capacity in the United States", *U.S. Geol. Surv. Bull.,* 1142 H (1962).

TABLE 46 TIME-STRATIGRAPHICAL UNITS—EUROPE   289

Era	System	Series	Epoch	Stage	Substage
CENOZOIC	QUATERNARY		Holocene		
			Pleistocene	Calabrian	
	TERTIARY	Neogene	Pliocene	Piacenzian	Levantinian
					Dacian
				Pannonian	Pontian
					Meotian
			Miocene	Sarmatian	
				Vindobonian	Tortonian
					Helvetian
				Burdigalian	
				Aquitanian³	
		Paleogene	Oligocene	Chattian	
				Rupelian	
				Lattorfian (Sannoisian, Tongrian)	
			Eocene	Priabonian	Bartonian Ledian (Auversian)
				Lutetian	Bruxellian
				Ypresian	
				Landenian	Sparnacian Thanetian
			Paleocene	Montian	
				Danian³	
MESOZOIC	CRETACEOUS²	Upper	Senonian	Maestrichtian (Dordonian)	
				Campanian	
				Santonian	
				Coniacian	
			Turonian	Angoumian	
				Ligerian	
			Cenomanian	Vraconian	
		Lower	Albian		
			Aptian	Gargasian	
				Bedoulian	
			Barremian⁴		
			Hauterivian		
			Valanginian	Berriasian	
	JURASSIC	Upper (Malm)	Portlandian⁵ (Tithonian)	Purbeckian	
				Bononian	
			Kimmeridgian	Virgulian	
				Pterocerian	
			Lusitanian (Corallian)	Sequanian (Astartian) Argovian (Rauracian)	
			Oxfordian		
		Middle (Dogger)	Callovian		
			Bathonian		
			Bajocian		
			Aalenian		
		Lower (Lias)	Toarcian		
			Charmouthian	Domerian
				Pliensbachian
			Sinemurian		
			Hettangian		
	TRIASSIC	Upper	Rhaetian	Keuper	
			Norian		
			Carnian		
		Middle	Ladinian	Muschelkalk	
			Anisian		
		Lower	Skythian	Bunter	

PALAEOZOIC	**PERMIAN**	Upper	Kazanian	Thuringian (Zechstein)	
		Middle	Kungurian	Saxonian (U. Rotliegendes)	
		Lower	Artinskian	Autunian (L. Rotliegendes)	
			Sakmarian		
	CARBONIFEROUS²	Upper	Stephanian (Uralian)		
			Westphalian (Moscovian)		
			Namurian		
		Lower	Dinantian	Visean	
			(Avonian)	Tournaisian	
	DEVONIAN	Upper	Famennian		
			Frasnian		
		Middle	Givetian		
			Couvinian (Eifelian)		
		Lower	Emsian		
			Siegenian		
			Gedinnian		
	SILURIAN (Gothlandian)		Ludlovian		
			Wenlockian		
			Llandoverian		
	ORDOVICIAN		Ashgillian		
			Caradocian		
			Llandillian		
			Llanvirnian		
			Arenigian		
			Tremadocian		
	CAMBRIAN	Upper			
		Middle			
		Lower			

Notes

1. Since the stratigraphy of Europe was the first to be worked out in detail, the names given here are generally used to describe strata of equivalent ages in other parts of the world. However, local names are often applied to the smaller stratigraphic sub-divisions.

2. In the United States, the Cretaceous is divided into Upper (Gulfian) and Lower (Comanchean). The Carboniferous is divided into Upper (Pennsylvanian) and Lower (Comanchean).

3. Names given in brackets are synonyms. The Table represents a reasonable working synthesis, but there are a number of controversial questions unresolved—e.g. some stratigraphers place the Aquitanian in the Oligocene, the Danian in the Upper Cretaceous, etc.

4. "Neocomian" is generally taken to include the Barremian-Berriasian sequence of the Lower Cretaceous, but is imprecise.

5. "Wealden" is a rock Unit of Lower Cretaceous to Upper Jurassic age.

GENERAL INDEX

(Names of oilfields in italics)
Bold figures indicate first page of major sections dealing with subjects.

294   INDEX